The Genome Editing Frontier

The rapid evolution of biotechnology has ushered an era in which the genome is no longer viewed as a static blueprint, but as a dynamic and editable system. *The Genome Editing Frontier: Emerging Methods and Applications* offers a comprehensive and forward-looking exploration of this transformation, tracing the field's progression from foundational molecular tools to today's high-precision, AI-integrated genome engineering platforms. This book presents an integrated perspective on how genome editing is reshaping medicine, agriculture, environmental science, and biotechnology. By bridging historical context with emerging innovations, it uniquely connects conceptual foundations, technological advances, and real-world applications within a single cohesive framework. Spanning the evolution from pre-CRISPR systems to next-generation digital and single-cell approaches, this volume highlights both the technical sophistication and the societal responsibilities of modern genome editing. Its interdisciplinary structure distinguishes it from conventional texts, combining functional genomics, computational biology, precision breeding, and ethical governance into a unified narrative of innovation.

Key Features and New to This Edition:

- Comprehensive coverage from ZFNs and TALENs to advanced CRISPR systems, multiplex editing, and epigenome engineering.
- Dedicated sections on AI-driven genome design, predictive repair modeling, and computational optimization of editing efficiency.
- Expanded coverage of single-cell genome editing and high-resolution functional genomics.
- New chapters on AI-enabled CRISPR immunotherapies and deep learning applications in genome engineering.
- Integrated perspectives on ethical, regulatory, and biosafety frameworks guiding responsible innovation.

Bringing together theory, methodology, and application, this volume serves as both a foundational reference and a forward-looking guide. It is ideally suited for graduate students, academic researchers, biotechnology professionals, and policy-engaged scientists seeking a rigorous and accessible roadmap to the future of genome editing.

The Genome Editing Frontier

Emerging Methods and Applications

Edited by

Neelanjana Choudhury, Jyoti Prakash Sahoo,
Dilip Kumar Dash, and Satyabrata Nanda

CRC Press
Taylor & Francis Group
Boca Raton London New York

CRC Press is an imprint of the
Taylor & Francis Group, an **informa** business

Designed cover image: Shutterstock

First edition published 2027
by CRC Press
2385 NW Executive Center Drive, Suite 320, Boca Raton FL 33431

and by CRC Press
4 Park Square, Milton Park, Abingdon, Oxon, OX14 4RN

CRC Press is an imprint of Taylor & Francis Group, LLC

ISBN: 9781041230403 (hbk)
ISBN: 9781041230472 (pbk)
ISBN: 9781003735113 (ebk)

DOI: 10.1201/9781003735113

Typeset in Times
by codeMantra

Contents

Preface

The rapid evolution of biotechnology has paved the way in an era where the fundamental code of life is no longer a fixed script, but a dynamic canvas for innovation. *The Genome Editing Frontier: Emerging Methods and Applications* serves as a comprehensive exploration of this transformation, capturing the transition from early molecular interventions to the high-precision, AI-driven era of genetic engineering. This book is designed to provide researchers, students, and practitioners with a holistic understanding of how these tools are reshaping medicine, agriculture, and environmental science.

The journey begins by establishing a strong conceptual and historical framework. Chapter 1 introduces the broader genome editing revolution, outlining the core tools and future horizons that define the field. To understand the current state of the field, the architectural milestones of the "Pre-CRISPR" era must be revisited; thus, the development of ZFNs and TALENs, the early pioneers of site-specific DNA cleavage, is detailed in Chapter 2. This sets the stage for the dramatic entry of CRISPR technology. Chapter 3 expands on the current CRISPR toolbox, while Chapter 4 transitions into functional genomics, showcasing how CRISPR-based screening methodologies are identifying gene functions at an unprecedented scale. As the field matures, the focus shifts from single modifications to complex system-wide interventions. Chapter 5 explores the strategies of multiplex genome editing, a critical leap for polygenic trait modification. The application of these high-level edits is then examined in the context of human health and specialized biology. Chapter 6 discusses the transformative potential of editing in stem cells and regenerative medicine, while Chapter 7 applies these same principles to Precision Breeding 2.0, focusing on the acceleration of agricultural crop improvement. The integration of computational intelligence marks the next frontier of the discipline. Chapter 8 illustrates how AI is being harnessed for resilient crop disease control, and Chapter 9 extends these next-generation tools to environmental and industrial biotechnology. Technical refinement remains a priority, as seen in Chapter 10, which addresses the precision design and delivery of guide RNAs. The scope of editing then moves beyond the DNA sequence itself, and Chapter 11 explores the emerging world of epigenome editing for gene regulation and therapeutics. The final segments of this book address the most granular and technologically advanced aspects of the frontier. Chapter 12 looks at single-cell genome editing, providing a high-resolution view of genetic variability. Chapters 13–15 form a dedicated suite on the role of digital transformation, covering computational design, AI-driven CRISPR immunotherapies, and deep learning models for predicting repair pathways and editing efficiency. These digital insights are grounded in biological reality in Chapters 16 and 17, which examine functional genomics in zebrafish and the remodeling of root system architecture for drought resilience in agricultural crops, respectively. Finally, Chapter 18 serves as a vital anchor, addressing the ethical, regulatory, and biosafety considerations that must guide these powerful technologies to ensure they are deployed responsibly for the benefit of humanity.

Ultimately, this volume is presented as a comprehensive synthesis of the current state of genomic science, reflecting both its technical brilliance and its profound responsibilities. It is hoped that the insights shared within these chapters will serve as a catalyst for further innovation, providing a foundational resource for those who will continue to push the boundaries of the genome editing frontier. By bridging the gap between theoretical frameworks and practical applications, this book seeks to empower the scientific community to navigate the future of genetic engineering with both precision and purpose.

– Editors

About the Editors

Dr. Neelanjana Choudhury, PhD, is currently serving as an Assistant Professor in the Department of Biotechnology at C. V. Raman Global University (CGU), Bhubaneswar, India, with nearly 10 years of combined teaching and research experience. Her research expertise covers molecular biology, genomics, proteomics, gene expression studies, bioinformatics, eco-toxicology, aquatic toxicology, and applied biotechnology. She is an active contributor to intellectual property development, with one granted utility patent and three published patents (utility and design) in areas including novel drug development, bioactive compound extraction, green energy, biopolymers, and sustainable biotechnological systems. Her granted patent on anthocyanin isolation represents a significant contribution toward novel formulations for cancer and diabetes management. Dr. Choudhury has made notable contributions to molecular research through more than 18 international GenBank accessions, strengthening global biological databases and enhancing India's representation in international repositories. She has published over 30 peer-reviewed articles including Scopus- and SCI-indexed journals, along with two international books. She has successfully guided 31 postgraduate students and is currently supervising 4 PhD research scholars. An internationally recognized academician, she has served as an invited, plenary, and keynote speaker at global conferences held in Canada, Japan, and the USA. Her achievements have been recognized through prestigious awards, including the Young Biotechnologist Award (2023), Eminent Scientist Award (2023), and Research Excellance Award (2026). She has also been a recipient of the University Research Fellowship and Best Poster Award in National Symposium on Plant Health Management for food security and safety (2016) by the Indian Phytopathology Society, and has secured the first rank in the Biotech Talent Search Examination conducted by BCIL, India. She also actively serves as a reviewer and editor for multiple international journals, contributing significantly to global scientific collaboration.

Dr. Jyoti Prakash Sahoo, PhD, is a dynamic researcher and academician currently serving as an Assistant Professor in the Department of Biotechnology, Faculty of Agriculture and Allied Sciences at C. V. Raman Global University, Bhubaneswar, India. Dr. Sahoo completed his MSc (2016) and PhD (2023) in Agricultural Biotechnology from Assam Agricultural University, Jorhat, India and Odisha University of Agriculture and Technology (OUAT), Bhubaneswar, India, respectively. His notable works include dissecting genetic architecture for *Cercospora* leaf spot, yellow mosaic virus, charcoal rot disease resistance in mung bean, molecular breeding strategies for stress resilience in crops, sustainable management of viral pathogens in okra, drought tolerance genetics in rice using microsatellite markers, BADH2 gene-specific sequence variation and phylogenetic relationship in aromatic rice, and morpho-biochemical and molecular characterization of beneficial soil microbes. His scholarly contributions have received over 1,100 citations, reflecting strong national and international research impact. He has published over 50 peer-reviewed research and review articles indexed by Scopus and SCI, while also authoring 10 edited books and over 35 important chapters for widely referenced agricultural biotechnology books. He is an active contributor to intellectual property development with 3 granted design patents. Dr. Sahoo has made notable contributions to molecular research through more than nine NCBI GenBank accessions, strengthening global biological databases and enhancing India's representation in international repositories. He has successfully guided 13 MSc (Biotechnology) students and is currently supervising two PhD (Biotechnology)

students. He actively contributes as an editorial board member for *Nature-Scientific Reports* and *BMC Genomics*, and serves as a potential peer reviewer for more than 80 prestigious international journals. He also holds life memberships in professional societies including the Indian Society of Genetics and Plant Breeding, Society of Biotechnology and Bioinformatics, India and Indian Society of Plant Breeders. His outstanding contributions to plant molecular biology and biotechnology have been recognized with prestigious honors, including the Young Scientist Award (2021), the University Best PhD Thesis (Academic Year: 2022–2023) Award of OUAT (2024), and multiple best oral presentation awards (2023) at national and international conferences.

Prof. (Dr.) Dilip Kumar Dash, PhD, is currently serving as Dean, Faculty of Agriculture and Allied Sciences at C. V. Raman Global University, Bhubaneswar, India. With an illustrious academic career spanning over 42 years, he has made outstanding contributions to teaching, research, academic leadership, and policy advisory roles in the field of horticulture. He obtained his doctoral degree from the Odisha University of Agriculture and Technology, Bhubaneswar, Odisha, India. Throughout his career, Prof. Dash has held several prestigious academic and administrative positions, including Dean, College of Horticulture, and College of Agriculture, Chiplima, Odisha University of Agriculture and Technology (OUAT), and Professor and Head, Department of Fruit Science and Horticulture Technology, College of Agriculture, Bhubaneswar, OUAT. Prof. Dash's research expertise is centered on crop improvement in major fruit crops, particularly coconut, papaya, and mango. His work spans diverse areas such as combining ability analysis, genetic divergence, molecular and morphological characterization, gamma radiation sensitivity studies, characterization of indigenous germplasm, plant tissue culture (callus induction and indirect regeneration), and evaluation of organic amendments for enhanced tree growth and yield. His research also addresses tree rejuvenation, canopy management, drip fertigation, and growth dynamics in perennial horticultural systems, contributing to sustainable and climate-resilient horticulture. A highly prolific researcher, Prof. Dash has published more than 100 research papers in Scopus-, SCI-, and NAAS-indexed national and international journals. He is a co-author of 4 books, has contributed 2 book chapters, and has edited 51 undergraduate practical manuals across horticulture and agriculture disciplines. Beyond academia, Prof. Dash has rendered significant service to the government and national agricultural bodies in India. He has served as a Member, Technical Committee, Directorate of Horticulture, Government of Odisha, India; Member, Technical Advisory Committee of the Coconut Development Board (CDB) Farm, India; Member, State-Level Monitoring Committee for implementation of CDB schemes, India; and Member, Seed Nut Procurement Committee of CBD, India. He has also contributed as a Resource Person for horticultural training programs; Co-Scientist (Horticulture) in the DBT-supported Rural Bio-Resource Complex, Puri District, Odisha, India; Member of the Monitoring Committee under the Technology Mission; and Team Member for the identification of new areas for oil palm cultivation in Odisha, India. Through his visionary academic leadership, impactful research, and sustained engagement with policy and extension activities, Prof. Dash continues to make enduring contributions to horticultural science, agricultural education, and sustainable rural development.

Dr. Satyabrata Nanda, PhD, is a progressive researcher and academician in Biotechnology, currently serving as Associate Dean at the School of Biotechnology, Centurion University of Technology and Management, Odisha, India. With over a decade of sustained teaching and research experience, he has established himself as a recognized expert in molecular plant–pest interactions, functional genomics, RNA interference (RNAi), and precision pest management. Dr. Nanda has authored more than 75 peer-reviewed research articles in reputed SCI- and Scopus-indexed journals, covering insect functional genomics, RNAi-based biopesticides, plant stress biology, gene family evolution, and sustainable crop protection. His scholarly contributions have received over 1,500 citations, with a Scopus h-index of 21, reflecting strong national and international research impact. Dr. Nanda has successfully secured competitive extramural research funding from agencies such as SERB, Government of India, and international postdoctoral grants from China. He has led multidisciplinary projects on brown planthopper–rice interactions, microRNA-mediated regulation, and RNAi-based pest control strategies. His research excellence has been recognized through multiple prestigious awards, including the Young Scientist Award, Distinguished Achiever Award, Abel Wolman Memorial Award, Eminent Achiever's Award, and an Overseas Postdoctoral Fellowship from the Chinese Academy of Agricultural Sciences. Demonstrating strong translational impact, he holds four patents, including one utility patent and multiple design patents focused on IoT (Internet of Things)-enabled pest monitoring systems and sustainable agri-technologies. He serves on the editorial boards of *BMC Plant Biology* and *BMC Genomics*; acts as Review Editor for *Frontiers in Plant Science*; and mentors doctoral, postgraduate, and undergraduate students while providing effective academic leadership as Coordinator of the Centre for Genetics and Genomics.

Contributors

Aqleem Abbas
College of Agricultural Engineering and Food Science
Shandong University of Technology
Zibo, China

Selorm Adablanu
Faculty of Science Education
Department of Information and Communication Technology
University of Education
Winneba, Accra, Ghana
and
Faculty of Computer Technology
Assam down town University
Guwahati, India

Sujaan Aehsas
Department of Biotechnology
Faculty of Agriculture and Allied Sciences
C. V. Raman Global University
Bhubaneswar, India

Afreen Anis
Department of Biotechnology
Faculty of Agriculture and Allied Sciences
C. V. Raman Global University
Bhubaneswar, India

Laxmipreeya Behera
Department of Agricultural Biotechnology
College of Agriculture
Odisha University of Agriculture and Technology
Bhubaneswar, India

Prabhudatta Behera
Department of Biochemistry
Kalinga University
Nava Raipur, India

Rakhi Biswal
Department of Agricultural Biotechnology
College of Agriculture
Odisha University of Agriculture and Technology
Bhubaneswar, India

Devendra Kumar Choudhary
Department of Plant Pathology
College of Agriculture
Indira Gandhi Krishi Vishwavidyalaya
Raipur, India

Rutupurna Das
Department of Zoology
School of Applied Sciences
Centurion University of Technology and Management
Odisha, India

Dhritismita Deka
Department of Science
Assam down town University
Guwahati, India

Prasanti Devi
Department of Botany
Gandhi Institute of Engineering and Technology University (GIET University)
Gunupur, India

Tuward J. Dweh
Department of Science
Assam down town University
Guwahati, India

Yanyin Guo
College of Agricultural Engineering and Food Science
Shandong University of Technology
Zibo, China

Shanshan Hao
College of Agricultural Engineering and Food Science
Shandong University of Technology
Zibo, China

Nana Ji
College of Agricultural Engineering and Food Science
Shandong University of Technology
Zibo, China

K. Keerthana
Life Sciences Department
Atria University
Bengaluru, India

Appanna V. Koppad
Department of Life Science
Atria University
Bengaluru, India

Sandeep Kumar
Department of Plant Pathology
College of Agriculture
Odisha University of Agriculture and Technology
Bhubaneswar, India

Nisha Kumari
Department of Environmental Science and Engineering
Indian Institute of Technology (Indian School of Mines)
Dhanbad, India

Upasana Mahopatra
Department of Life Science
Atria University
Bengaluru, India

Sanjoy Majumder
Department of Biotechnology
School of Biotechnology
Centurion University of Technology and Management
Odisha, India

Abinash Mishra
Department of Plant Breeding and Genetics
College of Agriculture
Odisha University of Agriculture and Technology
Bhubaneswar, India

Ashok Mishra
Department of Agricultural Biotechnology
College of Agriculture
Odisha University of Agriculture and Technology
Bhubaneswar, India

Deviprasad Mishra
Department of Agricultural Biotechnology
College of Agriculture
Odisha University of Agriculture and Technology
Bhubaneswar, India

Neeti Sanan Mishra
Plant RNAi Biology Group
International Centre for Genetic Engineering and Biotechnology
New Delhi, India

Ratikanta Mohapatra
College of Agriculture
Odisha University of Agriculture and Technology
Bhubaneswar, India

Radhakanta Nag
Department of Microbiology
Odisha University of Agriculture and Technology
Bhubaneswar, India

Yogesh Dashrath Naik
Agricultural Science Center
New Mexico State University
Clovis, New Mexico

Asha Velayudhan Nair
Life Sciences Department
Atria University
Bengaluru, India

Alok Ranjan Nayak
Department of Biotechnology
Faculty of Agriculture and Allied Sciences
C. V. Raman Global University
Bhubaneswar, India

Gagan Kumar Panigrahi
Department of Zoology
School of Applied Sciences
Centurion University of Technology and Management
Odisha, India

Gyanendra Kumar Patra
Department of Agricultural Biotechnology
College of Agriculture
Odisha University of Agriculture and Technology
Bhubaneswar, India

Rakesh Kumar Patra
Department of Agricultural Biotechnology
College of Agriculture
Odisha University of Agriculture and Technology
Bhubaneswar, India

Subhashree Pattnaik
Faculty of Agriculture and Allied Sciences
C. V. Raman Global University
Bhubaneswar, India

Adeeb Rahman
Plant RNAi Biology Group
International Centre for Genetic Engineering and Biotechnology
New Delhi, India

Ramyashree Devi G.S.
ICAR-Indian Grassland and Fodder Research Institute
Regional Research Station
Dharwad, India

Abhijit Ratha
Department of Biotechnology
Faculty of Agriculture and Allied Sciences
C. V. Raman Global University
Bhubaneswar, India

Kailash Chandra Samal
Department of Agricultural Biotechnology
College of Agriculture
Odisha University of Agriculture and Technology
Bhubaneswar, India

Anshuman Sarangi
School of Biotech Sciences
Trident Academy of Creative Technology
Bhubaneswar, India

Sitanshu Sekhar Satapathy
Faculty of Agriculture and Allied Sciences
C. V. Raman Global University
Bhubaneswar, India

Sougat Satapathy
Department of Plant Biotechnology
Centre for Plant Molecular Biology and Biotechnology
Tamil Nadu Agricultural University
Coimbatore, India

Gowsiya Shaik
Department of Life Science
Atria University
Bengaluru, India

Harvinder Kumar Singh
Department of Plant Pathology
College of Agriculture
Indira Gandhi Krishi Vishwavidyalaya
Raipur, India

Sunil Kumar Sunani
Department of Plant Pathology
College of Agriculture
Odisha University of Agriculture and Technology
Bhubaneswar, India

Marylin Taye
Department of Science
Assam down town University
Guwahati, India

Vishal
Department of Plant and Soil Science
Institute of Genomics for Crop Abiotic Stress Tolerance
Texas Tech University
Lubbock, Texas

Huihui Xu
College of Agricultural Engineering and Food Science
Shandong University of Technology
Zibo, China

Abbreviations

5mC	5-Methylcytosine
AAV	Adeno-associated virus
ABA	Abscisic acid
ABE	Adenine base editor
AdV	Adenoviral vector
AI	Artificial intelligence
ARF	Auxin response factor
BE	Base editor
bZIP	Basic leucine zipper
CAR-T	Chimeric antigen receptor T-cell
Cas	CRISPR-associated (protein/nuclease)
CBE	Cytosine base editor
ChIP-seq	Chromatin immunoprecipitation sequencing
CITE-seq	Cellular indexing of transcriptomes and epitopes
CNN	Convolutional neural network
CoDA	Context-dependent assembly
CpG	Cytosine-phosphate-guanine (dinucleotide site)
CRISPR	Clustered regularly interspaced short palindromic repeats
crRNA	CRISPR RNA
CRS	Csy4 recognition site
dCas	Catalytically inactive ("dead") Cas
dCas9	Dead Cas9
DL	Deep learning
DMD	Duchenne muscular disorder/Duchenne muscular dystrophy
DNMT	DNA methyltransferase (e.g., DNMT3A)
DREB	Dehydration-responsive element binding
DSB	Double-strand break
EPA	Environmental Protection Agency
ESC	Embryonic stem cell
FDA	Food and Drug Administration
GABA	γ-Aminobutyric acid
GMO	Genetically modified organism
gRNA	Guide RNA
GWAS	Genome-wide association study
HAT	Histone acetyltransferase
HDR	Homology-directed repair
HSC	Hematopoietic stem cell
HSPCs	Hematopoietic stem and progenitor cells
IDLV	Integrase-defective lentiviral vector
Indel	Insertion or deletion
iPSC	Induced pluripotent stem cell
ITR	Inverted terminal repeat
KRAB	Krüppel-associated box (repressor domain)
LEA	Late embryogenesis abundant (proteins)
LNP	Lipid nanoparticle
MAGE	Multiplex automated genome engineering
MAS	Marker-assisted selection

miRNA	microRNA
ML	Machine learning
MMEJ	Microhomology-mediated end joining
MOI	Multiplicity of infection
MSC	Mesenchymal stem cell
nCas9	Nickase Cas9
NCED	9-cis-Epoxycarotenoid dioxygenase
NGS	Next-generation sequencing
NHEJ	Non-homologous end joining
NLP	Natural language processing
NSC	Neural stem cell
OPEN	Oligomerized Pool ENgineering
PAM	Protospacer adjacent motif
PE	Prime editing
PEG	Polyethylene glycol
pegRNA	Prime editing guide RNA
PTG	Polycistronic tRNA-gRNA
PTM	Post-translational modification
QTL	Quantitative trait loci
RNN	Recurrent neural network
RNP	Ribonucleoprotein
ROS	Reactive oxygen species
RSA	Root system architecture
RT	Reverse transcriptase
RVD	Repeat variable diresidue
SAM	Synergistic activator mediator
scATAC-seq	Single-cell assay for transposase-accessible chromatin
ScGE	Single-cell genome editing
scRNA-seq	Single-cell RNA sequencing
SDN	Site-directed nucleases
sgRNA	Single-guide RNA
SNP	Single nucleotide polymorphism
TALE	Transcription activator-like effector
TALENs	Transcription activator-like effector nucleases
TET1	Ten-eleven translocation methylcytosine dioxygenase 1
TF	Transcription factor
tracrRNA	Trans-activating CRISPR RNA
UAV	Unmanned aerial vehicle (drone)
USDA	United States Department of Agriculture
VLP	Virus-like particle
WGS	Whole genome sequencing
WHO	World Health Organization
WUE	Water use efficiency
XAI	Explainable artificial intelligence
ZFA	Zinc finger array
ZFIN	Zebrafish information network
ZFN	Zinc finger nuclease
ZFP	Zinc finger protein

1 The Genome-Editing Revolution

Concept, Tools, and Future Horizon

Deviprasad Mishra, Laxmipreeya Behera, Rakesh Kumar Patra, and Ashok Mishra

1.1 INTRODUCTION

Genome editing has emerged as one of the most transformative innovations in modern biological sciences, fundamentally reshaping the way researchers interrogate, engineer, and manipulate genetic information. By enabling targeted, predictable, and highly efficient modifications at specific genomic loci, genome-editing technologies have accelerated progress across diverse fields including functional genomics, plant and animal breeding, regenerative medicine, and synthetic biology. The capacity to directly rewrite genetic sequences provides a level of precision and control, unmatched by earlier genetic engineering techniques, thereby offering unprecedented opportunities to understand gene function, correct pathogenic mutations, and design organisms with tailored traits (Jinek et al. 2012).

Historically, genetic manipulation relied on classical mutagenesis, which induced random mutations through chemical or physical agents, or on transgenic approaches that introduced exogenous DNA without regard to genomic context. While these methods contributed significantly to foundational genetic knowledge, they suffered from low specificity, variable outcomes, and unintended genomic disturbances. The emergence of programmable nucleases, initially Zinc-Finger Nucleases (ZFNs), followed by Transcription Activator-Like Effector Nucleases (TALENs), and ultimately CRISPR-Cas systems, marked a paradigm shift in genome engineering. These platforms enabled researchers to direct nuclease activity to predefined DNA sequences, triggering endogenous DNA repair pathways that facilitate either gene disruption or precise sequence correction (Ran et al. 2013).

Over the past decade, the field has rapidly advanced from conceptual demonstrations to real world applications. In medicine, genome editing supports the development of personalized therapies for monogenic disorders, cancer immunotherapies, and antiviral interventions. In agriculture, it provides avenues for generating high-yielding, stress-tolerant, and nutrient-enhanced crops, contributing to food and nutritional security. In industry, genome-edited microbial systems enhance biosynthesis of enzymes, bioplastics, and pharmaceutical precursors, driving sustainable biomanufacturing. These developments underscore genome editing's role as a cornerstone technology of the 21st century, driving innovation across biological, environmental, and clinical domains (Chang et al. 2017).

As genome editing continues to mature, the scientific community faces both extraordinary opportunities and complex challenges. Advances in tool fidelity, delivery systems, and computational design promise to further enhance the precision and scope of genomic interventions. Simultaneously, ethical, regulatory, and societal considerations regarding its application, particularly in human germline modification and ecological engineering, highlight the need for robust governance and global dialogue. Thus, while genome editing represents a technological revolution, its responsible development and deployment remain essential to ensuring sustainable benefits for science and society (Mao et al. 2008).

DOI: 10.1201/9781003735113-1

1.2 CONCEPT OF GENOME EDITING

Accurate target recognition is the first and most critical step in genome editing. Programmable nucleases identify specific genomic loci through either protein–DNA or RNA–DNA interactions. ZFNs use engineered C2H2 zinc-finger domains, each recognizing a 3-bp DNA motif, to assemble arrays capable of binding extended sequences (Urnov et al. 2010). TALENs employ TALE repeat domains, each corresponding to a single nucleotide, enabling more predictable and modular DNA targeting (Bogdanove & Voytas 2011). CRISPR-Cas systems, particularly CRISPR-Cas9, rely on guide RNA (gRNA) sequences that base-pair with complementary genomic DNA, offering a simpler and more scalable targeting mechanism (Jinek et al. 2012; Nayak et al. 2026).

Once bound to the target locus, the nuclease component introduces a DNA lesion. Double-strand breaks (DSBs) are the most common and potent form of nuclease-induced damage, historically used in ZFNs, TALENs, and conventional CRISPR-Cas9 systems (Doudna & Charpentier 2014). Single-strand nicks, generated by Cas9 nickase or Cas12 variants, present a more controlled alternative that can reduce off-target effects while still initiating high-fidelity repair routes (Ran et al. 2013). These site-specific breaks signal the cell to activate intrinsic DNA repair pathways, ultimately shaping the type and fidelity of the genomic edit. Following nuclease-induced DNA damage, eukaryotic cells recruit endogenous repair systems that define the final editing outcome. Two major pathways dominate. Non-Homologous End Joining (NHEJ) is the primary DSB repair mechanism in most somatic cells. It directly ligates DNA ends without requiring a homologous template, making it rapid but error-prone. The repair process frequently results in small insertions or deletions (indels), which disrupt coding sequences or regulatory elements, an outcome exploited for gene knockout applications (Chang et al. 2017). Homology-Directed Repair (HDR) is a high-fidelity pathway that repairs DSBs using a homologous DNA template. Investigators supply donor templates to direct precise sequence insertion, replacement, or correction (Mao et al. 2008). Although highly accurate, HDR is restricted to certain cell-cycle phases (S/G2), limiting its efficiency compared to NHEJ.

1.3 TOOLS OF GENOME EDITING

1.3.1 Zinc-Finger Nucleases

ZFNs represent one of the earliest programmable nuclease platforms and consist of modular Cys2-His2 zinc-finger motifs fused to the FokI endonuclease domain. Each zinc finger recognizes approximately three nucleotides, enabling the design of multi-finger arrays that specify unique genomic targets (Kim et al. 1996; Urnov et al. 2010). Upon dimerization, the FokI nuclease generates a site-specific DSB, which is subsequently repaired by endogenous DNA repair pathways. Although ZFNs established the conceptual foundation for targeted genome engineering, their application is limited by challenges in protein design, context-dependent binding effects, cytotoxicity arising from off-target cleavage, and the requirement for complex protein engineering expertise (Carroll 2011). Consequently, ZFNs have largely been superseded by more user-friendly platforms such as TALENs and CRISPR-Cas systems.

1.3.2 Transcription Activator-Like Effector Nucleases

TALENs are engineered nucleases composed of DNA-binding domains derived from *Xanthomonas* TALE proteins fused to the FokI nuclease. TALE repeats exhibit a simple and predictable one-repeat–one-nucleotide recognition code determined by variable di-residue (RVD: Repeat Variable Diresidue) motifs, enabling flexible and precise targeting across diverse genomic loci (Moscou & Bogdanove 2009). Compared to ZFNs, TALENs display higher target-site adaptability and reduced design constraints, resulting in robust gene-editing efficiencies in plants, animals, and human cells (Joung & Sander 2013). However, TALEN construction remains labor-intensive due to the highly repetitive nature of TALE arrays and the need for careful optimization to reduce recombination during assembly.

1.3.3 CRISPR-Cas Systems

CRISPR-Cas systems, adapted from microbial adaptive immune mechanisms, have revolutionized genome editing due to their simplicity, scalability, and affordability. The most widely used nuclease, SpCas9, is guided by a single-guide RNA that base-pairs with the target sequence adjacent to a protospacer adjacent motif (PAM) to induce a DSB (Jinek et al. 2012). Variants such as Cas12a (Cpf1) offer alternative PAM specificities, staggered DNA cuts, and reduced off-target effects, broadening the CRISPR toolbox (Zetsche et al. 2015). CRISPR's programmable RNA-guided mechanism eliminates the need for complex protein engineering, making it accessible for high-throughput and multiplexed editing applications. The results of the comparison among different genome-editing technologies are indexed in Table 1.1, and major CRISPR-Cas variants, features, and applications are indexed in Table 1.2.

1.3.4 Emerging Tools

1.3.4.1 Base Editors

Base editors (BEs) are next-generation precision genome-editing tools that enable direct, irreversible single-nucleotide conversions without inducing DSBs. Cytosine base editors (CBEs) convert C→T transitions, while adenine base editors (ABEs) mediate A→G transitions through the fusion of catalytically impaired Cas proteins with deaminases (Komor et al. 2016; Gaudelli et al. 2017). By avoiding DSBs and donor templates, BEs significantly reduce insertion–deletion (indel) formation and minimize cytotoxicity, making them ideal for therapeutically relevant point mutation correction.

1.3.4.2 Prime Editing

Prime editing is an advanced, versatile genome-editing system that allows targeted insertions, deletions, and all types of base substitutions without generating DSBs. It employs a Cas9 nickase fused to an engineered reverse transcriptase and a prime editing gRNA that encodes both target specificity and the desired genetic modification (Anzalone et al. 2019). Prime editing offers improved precision and fewer by-products than HDR-based methods, enabling high-fidelity correction of pathogenic alleles and complex genomic alterations previously inaccessible with earlier editing technologies.

1.3.4.3 CRISPR-Cas Variants

The continuous discovery and engineering of CRISPR-Cas systems have significantly broadened the scope, precision, and applicability of genome editing. While CRISPR-Cas9 remains the most widely used nuclease, numerous natural and engineered Cas variants, such as Cas12b, CasΦ, and Cas13, have emerged with altered PAM requirements, enhanced specificity, reduced molecular size, and expanded editing modalities. These innovations address key limitations of early Cas9 systems, including PAM dependency, off-target cleavage, and challenges associated with delivery in therapeutic contexts. Cas12b (formerly C2c1), a type V-B nuclease, is valued for its compact size and robust activity across a wide temperature range. Unlike Cas9, Cas12b generates staggered cuts and recognizes AT-rich PAMs, making it advantageous for plant genome editing where such PAM environments are common (Strecker et al. 2019). Engineered variants such as BhCas12b and AaCas12b have been optimized for mammalian editing, offering high fidelity with reduced off-target effects (Strecker et al. 2019). CasΦ (CasPhi), derived from jumbo bacteriophages, represents one of the smallest DNA-targeting nucleases identified to date (~70 kDa), which enhances its compatibility with adeno-associated virus (AAV) delivery systems (Pausch et al. 2020). Despite its small size, CasΦ retains efficient dsDNA (double-stranded DNA) cleavage activity and recognizes flexible PAM sequences, enabling broader genomic accessibility. Cas13 proteins, belonging to the type VI CRISPR family, are unique in their RNA-targeting capacity. Cas13 variants such as Cas13a, Cas13b, and Cas13d exhibit programmable RNA cleavage with minimal sequence constraints, supporting transcriptome engineering, RNA knockdown, and diagnostic applications including SHERLOCK-based pathogen detection (Anzalone et al. 2019).

TABLE 1.1
Comparison among Different Genome-Editing Technologies

Properties	ZFN	TALENs	MegaTAL	CRISPR-Cas9	Prime Editing	Base Editing
Mechanism	Engineered proteins with zinc-finger domains recognize specific sequences and induce double-strand breaks (DSBs).	Custom-designed proteins target specific DNA sequences.	Combination of TALENs and meganucleases for higher precision.	RNA-guided nuclease induces DSBs.	Uses a reverse transcriptase fused to Cas9 nickase to insert a new genetic material.	Directly converts one base to another without DSBs.
Targeting Precision	High precision, but difficult to design for complex targets.	Very precise, minimal off-target effects.	Extremely precise due to hybrid nature.	High, but off-target effects may occur.	Highly precise for small edits.	Precise but limited to specific base changes.
Ease of Design	Difficult due to complex protein design.	Complex, requires protein engineering.	Requires advanced engineering.	Easy; relies on guide RNA (gRNA).	Moderate; design involves RNA and template.	Simple for common base conversions.
Efficiency	Moderate to high.	High but lower than CRISPR in some cases.	High, though specialized use cases.	High, varies based on cell type.	Moderate to high, but dependent on cell conditions.	High for specific mutations.
Delivery Methods	Similar to TALENs (often electroporation).	Viral vectors or electroporation.	Viral delivery or plasmid-based.	Viral vectors, electroporation, liposomes.	Viral vectors or direct plasmid transfection.	Same as prime editing.
Off-target Effects	Low but possible.	Low off-target risk.	Very low off-target effects.	Occasional off-target activity; improving with new variants.	Minimal due to lack of DSBs.	Minimal; base-specific mutations only.
Edit Types	Knockouts, insertions, deletions.	Mostly knockouts and insertions.	Custom insertions and deletions.	Knockouts, insertions, deletions.	Insertions, corrections, small edits.	Base substitutions (e.g., C→T, A→G).
Limitations	Complex design, costly.	Time-consuming design and synthesis.	Requires expertise for design.	Potential immune response, off-targets.	Limited to small insertions/edits.	Limited to point mutations.
Application	Rare disease studies, gene therapy.	Functional studies, gene therapy.	Niche areas requiring high specificity.	Gene knockouts, disease modeling, gene therapy.	Disease correction, gene therapy.	Treating SNP-based diseases.

TABLE 1.2
Major CRISPR-Cas Variants, Features, and Applications

CRISPR-Cas Variant	Cas Type	Key Features	PAM Requirement	Advantages	Typical Applications
SpCas9 (*Streptococcus pyogenes* Cas9)	Type II	Most widely used nuclease; high efficiency	NGG	Well-characterized, robust activity	Gene knockouts, HDR-based editing, therapeutic research
SaCas9 (*Staphylococcus aureus* Cas9)	Type II	Smaller size (~3.2 kb)	NNGRRT	AAV delivery-compatible, high specificity	In vivo gene therapy, compact vector systems
Cpf1/Cas12a (e.g., AsCas12a, LbCas12a)	Type V	Generates staggered cuts; requires only crRNA	TTTV	High precision, low off-targets, easier multiplexing	Gene editing, multiplex engineering, plant genome editing
Cas12b	Type V	Thermostable, compact	Varies (e.g., TTN)	High fidelity, suitable for diverse organisms	Therapeutic editing, bacterial systems, diagnostics
Cas12f / Cas14	Type V-F	Ultrasmall (400–700 AA); ssDNA targeting	Variable	Can fit into small delivery vectors	Nanopore editing, compact delivery systems
Cas13a, Cas13b, Cas13d	Type VI	RNA-targeting nuclease; no DNA editing	No PAM (uses PFS)	Specific RNA knockdown, transient editing	Antiviral therapy, RNA diagnostics (SHERLOCK), transcriptome engineering
Cas9 Nickase (nCas9)	Type II	Mutated Cas9 cutting only one DNA strand	NGG	Reduced off-target effects; used in base editing	Base editing, high-fidelity editing, prime editing scaffold
dCas9 (Dead Cas9)	Type II	Catalytically inactive DNA-binding protein	NGG	Gene regulation (CRISPRi/a), epigenetic editing	Transcriptional modulation, chromatin studies
Base Editors (CBE, ABE)	Cas9 fused to deaminases	Single-nucleotide changes without DSBs	NGG (varies by Cas)	Precise point mutation editing	Disease correction, functional genomics

1.4 APPLICATIONS ACROSS SECTORS

1.4.1 Medicine

Genome editing has reshaped the landscape of modern medicine by enabling precise and durable correction of genetic abnormalities, engineering of therapeutic cells, and development of novel strategies against infectious diseases. The most established clinical applications involve correction of monogenic disorders, where pathogenic mutations can be directly repaired or functionally compensated. CRISPR-based therapies for β-thalassemia and sickle cell disease represent landmark achievements. Ex vivo editing of hematopoietic stem cells to disrupt the BCL11A erythroid enhancer reactivates fetal hemoglobin production, mitigating disease symptoms. These therapies exemplified by the CRISPR treatment Casgevy have demonstrated durable clinical benefits and recently received regulatory approval, marking the transition of genome editing from experimental to mainstream clinical practice (Frangoul et al. 2021; FDA 2023). In cancer immunotherapy, genome editing facilitates the engineering of chimeric antigen receptor T (CAR-T) cells with enhanced specificity, persistence, and reduced risk of graft-versus-host disease. Multiplex editing using CRISPR enables the removal of endogenous T-cell receptors, PD-1, or immune checkpoint regulators to generate universal "off-the-shelf" CAR-T therapies that circumvent donor–recipient matching (Stadtmauer et al. 2020).

Genome editing is also advancing antiviral therapeutics, particularly against persistent viral infections. CRISPR-Cas9 and CRISPR-Cas13 systems have shown promise in targeting latent HIV proviral DNA, hepatitis B viral cccDNA (covalently closed circular DNA), and emerging pathogens by cleaving viral genomes or degrading viral RNA transcripts. Further, significant progress has been made in in vivo genome editing, where gene correction occurs directly within patient tissues. Lipid nanoparticle (LNP) and AAV-delivered CRISPR systems have demonstrated efficacy in treating metabolic liver disorders such as transthyretin amyloidosis and hypercholesterolemia through direct hepatic gene disruption (Gillmore et al. 2021). Collectively, these developments illustrate the transformative potential of genome editing in medicine, offering durable, mechanism-based therapies for previously intractable diseases.

1.4.2 Agriculture

Genome editing has emerged as a powerful tool in agricultural biotechnology, enabling rapid and precise genetic improvement of crops and livestock to meet the challenges of food security, climate change, and sustainable production. One of the most transformative applications is the development of disease-resistant and pest-resistant crops. CRISPR-Cas-mediated disruption of susceptibility (S) genes, such as *MLO* in wheat and barley or *OsSWEET13* in rice, has generated resistance to fungal and bacterial pathogens without introducing foreign DNA, offering durable and non-transgenic solutions for crop protection (Wang et al. 2014). Similarly, editing insect-recognition receptors and defense regulators has enhanced pest resistance in crops such as maize and tomato. A second major application is the engineering of climate-resilient crop varieties. Targeted modifications of genes involved in drought response (*DREB*, *NHX1*), salinity tolerance, and heat shock pathways have produced plants capable of maintaining productivity under abiotic stress conditions.

Genome editing accelerates the development of such traits, particularly in orphan crops and minor cereals where conventional breeding is slow. Genome editing also supports biofortification efforts aimed at improving nutritional quality. CRISPR-mediated editing has been used to enhance provitamin A, folate, γ-aminobutyric acid, and mineral content in staples such as rice, maize, and tomato. For example, knockout of negative regulatory genes (*SlALMT9*, *OsSPL16*) and targeted manipulation of biosynthetic pathways have led to nutritionally enriched varieties with significant health benefits. Beyond crops, genome editing is increasingly applied to livestock improvement, where precise modifications offer opportunities for enhanced disease resistance, growth efficiency, reproductive performance, and animal welfare. Notable examples include PRRS (Porcine Reproductive and Respiratory Syndrome)-resistant pigs engineered by editing the *CD163* receptor and cattle with heat-tolerant slick-coat variants introduced via HDR-mediated editing (Whitworth et al. 2016). Together, these advancements highlight the transformative potential of genome editing in agriculture, enabling sustainable and resilient food systems while reducing reliance on chemical inputs, lengthy breeding cycles, and transgenic technologies.

1.4.3 Synthetic Biology and Industry

Genome editing has significantly accelerated innovations in synthetic biology and industrial biotechnology by enabling rapid, precise, and programmable modifications to microbial and cellular systems. One of the most influential applications is the engineering of microbial biofactories for sustainable production of high-value compounds such as enzymes, biofuels, bioplastics, nutraceuticals, and pharmaceuticals. Techniques such as CRISPR-Cas9, CRISPR-Cas12a, and multiplex genome editing allow simultaneous targeting of multiple genomic loci, facilitating large-scale rewiring of metabolic networks in microbes like *Escherichia coli*, *Saccharomyces cerevisiae*, and *Streptomyces* spp.

These engineered strains have been used to optimize industrial production of precursors, including polyketides, terpenoids, and biodegradable plastics such as polyhydroxyalkanoates. Genome editing also enhances metabolic pathway optimization, where precise genomic modifications fine-tune flux through biosynthetic pathways, eliminate metabolic bottlenecks, and increase precursor availability. CRISPR-mediated promoter editing, gene knock-ins, and dynamic regulation circuits enable high-throughput screening of pathway variants, accelerating the development of industrial strains with improved yield, stability, and production efficiency (Jakočiūnas et al. 2015). The integration of CRISPR tools with computational design and metabolic modeling further strengthens targeted pathway engineering.

Moreover, genome editing supports the design of novel biomolecules and synthetic pathways, enabling synthetic assembly of artificial operons, programmable gene networks, and orthogonal biosynthetic modules. Cas-based tools have facilitated the construction of synthetic chromosomes (e.g., Sc2.0 project in yeast) and enabled de novo biosynthesis of complex molecules such as aromatic compounds, alkaloids, and therapeutic peptides (Mitchell et al. 2017). Recently, RNA-targeting CRISPR systems like Cas13 have expanded the toolkit for post-transcriptional regulation, supporting dynamic control of gene expression in synthetic circuits. Collectively, genome editing empowers synthetic biology to deliver environmentally sustainable biomanufacturing solutions, reduce dependence on petrochemical processes, and accelerate the creation of programmable biological systems for diverse industrial applications.

1.5 ETHICAL, REGULATORY, AND SOCIETAL CONSIDERATIONS

The rapid advancement of genome editing has raised complex ethical, regulatory, and societal questions that demand careful deliberation to ensure responsible scientific progress. A central ethical concern involves human germline editing, where modifications to embryos, gametes, or early embryonic stages can produce heritable genetic changes. While somatic editing is widely accepted for its therapeutic benefits, germline interventions raise profound questions about consent across generations, unintended off-target effects, and the potential for nontherapeutic "enhancement" applications. The international response to the unauthorized CRISPR-edited babies in 2018 sparked global debate and led to renewed calls for moratoria and strengthened oversight (Greely 2019). Issues of equitable access to genome-editing therapies also pose significant societal challenges. Many CRISPR-based treatments remain prohibitively expensive, raising concerns that technological advances may exacerbate existing health inequities between and within nations (Jasanoff & Hurlbut 2018).

In ecological contexts, the deployment of gene drives, engineered systems designed to spread specific genetic traits rapidly through populations, has potential applications in controlling vector-borne diseases and invasive species. However, gene drives pose ecological and biosafety concerns, including the risk of uncontrolled spread, disruption of ecosystems, and irreversible biodiversity impacts. As a result, ecological risk assessment and containment strategies are essential (Esvelt & Gemmell 2017). The integration of genome editing with personalized genomics and big data platforms introduces additional concerns regarding genomic privacy, data security, and informed consent. Globally, regulatory frameworks for genome editing vary widely, from permissive environments encouraging innovation to restrictive regimes prioritizing precaution (Esvelt & Gemmell 2017). International bodies such as the WHO, UNESCO, and the National Academies have issued recommendations emphasizing risk–benefit assessment, public engagement, and inclusive decision-making. Ultimately, transparent governance, interdisciplinary ethical review, and ongoing public dialogue are essential for maintaining public trust and guiding the responsible development of genome-editing technologies (Esvelt & Gemmell 2017).

1.6 FUTURE HORIZONS

The future of genome editing is poised to be shaped by advances in molecular engineering, computational biology, ecological biotechnology, and translational medicine. A major focus of ongoing research is the refinement of editing tools, where next-generation platforms aim to enhance precision, minimize off-target cleavage, and improve spatiotemporal control of editing activity. The development of high-fidelity Cas9 variants, compact Cas enzymes suitable for in vivo delivery, and programmable epigenetic editors reflects a shift toward safer and more predictable genomic modifications. Parallel advancements in viral vectors, LNPs, and tissue-specific delivery systems are expected to expand clinical applicability while addressing current delivery bottlenecks (Anzalone et al. 2020). Another transformative direction is personalized medicine, where patient-specific genome edits could enable individualized treatments for cancer, rare genetic disorders, and complex multifactorial diseases. Integration of genomic sequencing with editing technologies promises precision therapeutics that correct pathogenic variants at their source, offering long-term or even curative outcomes. CRISPR-based diagnostics and RNA-targeting systems further extend the personalized medicine toolkit by enabling sensitive detection and dynamic regulation of disease pathways (Baig & Chien 2025).

In the agricultural domain, climate-resilient crop engineering will play a critical role in sustaining global food systems under increasing environmental pressures. Genome editing enables rapid enhancement of drought tolerance, heat resilience, salinity tolerance, and nutrient-use efficiency, supporting adaptation to extreme weather and degraded soils. The development of climate-smart crops will be central to global strategies for agricultural sustainability. The potential use of gene drives represents another frontier, with applications in controlling vector-borne diseases such as malaria, reducing agricultural pests, and managing invasive species.

However, their ecological risks including irreversible population changes and unintended ecosystem effects necessitate robust risk assessment frameworks, molecular safeguards, and international governance mechanisms (Esvelt & Gemmell 2017). Finally, the integration of artificial intelligence (AI) and nanotechnology is anticipated to revolutionize genome editing. AI-driven models are accelerating gRNA optimization, off-target prediction, Cas protein engineering, and design of synthetic circuits. Concurrently, nanomaterials, including polymeric nanoparticles, DNA nanostructures, and lipid-based carriers, promise precise, nonviral delivery of genome-editing components, improving safety and tissue targeting. Together, these interdisciplinary advances will define the next era of genome engineering, expanding its reach across medicine, agriculture, and biotechnology.

1.7 CONCLUSION

Genome editing has fundamentally transformed modern biological research, offering unprecedented precision and efficiency in modifying genetic information across diverse organisms. Its applications now span medicine, agriculture, environmental biotechnology, and industrial biosciences, driving innovations such as personalized gene therapies, climate-resilient crops, and sustainable bioengineering solutions. Yet, the pace of advancement justify the importance of responsible innovation, where ethical deliberation, transparent regulatory frameworks, and societal engagement guide technological deployment. As next-generation tools become increasingly accurate, multiplexed, and integrated with complementary fields such as AI, synthetic biology, and nanotechnology, genome editing is poised to serve as a foundational pillar of future scientific progress. Ultimately, the long-term impact of this revolution will depend not only on technical breakthroughs but also on humanity's ability to steward these powerful capabilities with equity, caution, and collective benefit at the forefront.

REFERENCES

Anzalone, A. V., Randolph, P. B., Davis, J. R., Sousa, A. A., Koblan, L. W., Levy, J. M., Chen, P. J., Wilson, C., Newby, G. A., Raguram, A., & Liu, D. R. (2019). Search-and-replace Genome Editing without double-strand Breaks or Donor DNA. *Nature*, 576(7785), 149–157.

Anzalone, A. V., Koblan, L. W., & Liu, D. R. (2020). Genome editing with CRISPR–Cas nucleases, base editors, transposases and prime editors. *Nature Biotechnology*, 38(7), 824–844.

Baig, M. M. F. A., & Chien, W. T. (2025). Nanotechnological approaches for the targeted delivery of CRISPR-Cas systems for genomic modifications, biomolecular sensing, and precision medicine. *Biomaterials Science*, 13(17), 4597–4638.

Bogdanove, A. J., & Voytas, D. F. (2011). TAL effectors: Customizable proteins for DNA targeting. *Science*, *333*(6051), 1843–1846.

Carroll, D. (2011). Genome engineering with zinc-finger nucleases. *Genetics*, 188(4), 773–782.

Chang, H. H., Pannunzio, N. R., Adachi, N., & Lieber, M. R. (2017). Non-homologous DNA end joining and alternative pathways to double-strand break repair. *Nature Reviews Molecular Cell Biology*, 18(8), 495–506.

Doudna, J. A., & Charpentier, E. (2014). The new frontier of genome engineering with CRISPR-Cas9. *Science*, 346(6213), 1258096.

Esvelt, K. M., & Gemmell, N. J. (2017). Conservation demands safe gene drive. *PLoS Biology*, 15(11), e2003850.

FDA. (2023). *FDA Approves First CRISPR-Based Gene-Editing Therapy*. U.S. Food and Drug Administration.

Frangoul, H., Altshuler, D., Cappellini, M. D., Chen, Y. S., Domm, J., Eustace, B. K., … Corbacioglu, S. (2021). CRISPR-Cas9 gene editing for sickle cell disease and β-thalassemia. *New England Journal of Medicine*, 384(3), 252–260.

Gaudelli, N. M., Komor, A. C., Rees, H. A., Packer, M. S., Badran, A. H., Bryson, D. I., & Liu, D. R. (2017). Programmable base editing of A T to G • C in genomic DNA without DNA cleavage. *Nature*, 551(7681), 464–471.

Gillmore, J. D., Gane, E., Taubel, J., Kao, J., Fontana, M., Maitland, M. L., … Lebwohl, D. (2021). CRISPR-Cas9 in vivo gene editing for transthyretin amyloidosis. *New England Journal of Medicine*, 385(6), 493–502.

Greely, H. T. (2019). CRISPR'd babies: Human germline genome editing in the 'He Jiankui affair'. *Journal of Law and the Biosciences*, 6(1), 111–183.

Jakočiūnas, T., Bonde, I., Herrgård, M., Harrison, S. J., Kristensen, M., Pedersen, L. E., … Keasling, J. D. (2015). Multiplex metabolic pathway engineering using CRISPR/Cas9 in *Saccharomyces cerevisiae*. *Metabolic Engineering*, 28, 213–222.

Jasanoff, S., & Hurlbut, J. B. (2018). A global observatory for gene editing. *Nature*, 555(7697), 435–437.

Jinek, M., Chylinski, K., Fonfara, I., Hauer, M., Doudna, J. A., & Charpentier, E. (2012). A programmable dual-RNA–guided DNA endonuclease in adaptive bacterial immunity. *Science*, 337(6096), 816–821.

Joung, J. K., & Sander, J. D. (2013). TALENs: A widely applicable technology for targeted genome editing. *Nature Reviews Molecular Cell Biology*, 14(1), 49–55.

Kim, Y. G., Cha, J., & Chandrasegaran, S. (1996). Hybrid restriction enzymes: Zinc finger fusions to Fok I cleavage domain. *Proceedings of the National Academy of Sciences*, 93(3), 1156–1160.

Komor, A. C., Kim, Y. B., Packer, M. S., Zuris, J. A., & Liu, D. R. (2016). Programmable editing of a target base in genomic DNA without double-stranded DNA cleavage. *Nature*, 533(7603), 420–424.

Mao, Z., Bozzella, M., Seluanov, A., & Gorbunova, V. (2008). DNA repair by nonhomologous end joining and homologous recombination during cell cycle in human cells. *Cell Cycle*, 7(18), 2902–2906.

Mitchell, L. A., Wang, A., Stracquadanio, G., Kuang, Z., Wang, X., Yang, K., … Boeke, J. D. (2017). Synthesis, debugging, and effects of synthetic chromosome consolidation: synVI and beyond. *Science*, 355(6329), eaaf4831.

Moscou, M. J., & Bogdanove, A. J. (2009). A simple cipher governs DNA recognition by TAL effectors. *Science*, 326(5959), 1501–1501.

Nayak, S. P., Sundararajan, S., Singha, D. L., & Ramalingam, S. (2026). CRISPR and functional RNAs: Identification, functional analysis, and delivery for plant stress tolerance. In Jen-Tsung Chen (Ed.), *Functional RNAs in Plants* (pp. 281–296). Academic Press.

Pausch, P., Al-Shayeb, B., Bisom-Rapp, E., Tsuchida, C. A., Li, Z., Cress, B. F., … Doudna, J. A. (2020). CRISPR-CasΦ from huge phages is a hypercompact genome editor. *Science*, 369(6501), 333–337.

Ran, F. A., Hsu, P. D., Lin, C. Y., Gootenberg, J. S., Konermann, S., Trevino, A. E., … Zhang, F. (2013). Double nicking by RNA-guided CRISPR Cas9 for enhanced genome editing specificity. *Cell*, 154(6), 1380–1389.

Stadtmauer, E. A., Fraietta, J. A., Davis, M. M., Cohen, A. D., Weber, K. L., Lancaster, E., … June, C. H. (2020). CRISPR-engineered T cells in patients with refractory cancer. *Science*, 367(6481), eaba7365.

Strecker, J., Ladha, A., Gardner, Z., Schmid-Burgk, J. L., Makarova, K. S., Koonin, E. V., & Zhang, F. (2019). RNA-guided DNA insertion with CRISPR-associated transposases. *Science*, 365(6448), 48–53.

Urnov, F. D., Rebar, E. J., Holmes, M. C., Zhang, H. S., & Gregory, P. D. (2010). Genome editing with engineered zinc finger nucleases. *Nature Reviews Genetics*, 11(9), 636–646.

Wang, Y., Cheng, X., Shan, Q., Zhang, Y., Liu, J., Gao, C., & Qiu, J. L. (2014). Simultaneous editing of three homoeoalleles in hexaploid bread wheat confers heritable resistance to powdery mildew. *Nature Biotechnology*, 32(9), 947–951.

Whitworth, K. M., Rowland, R. R., Ewen, C. L., Trible, B. R., Kerrigan, M. A., Cino-Ozuna, A. G., … Prather, R. S. (2016). Gene-edited pigs are protected from porcine reproductive and respiratory syndrome virus. *Nature Biotechnology*, 34(1), 20–22.

Zetsche, B., Gootenberg, J. S., Abudayyeh, O. O., Slaymaker, I. M., Makarova, K. S., Essletzbichler, P., … Zhang, F. (2015). Cpf1 is a single RNA-guided endonuclease of a class 2 CRISPR-Cas system. *Cell*, 163(3), 759–771.

2 Pre-CRISPR Genome Engineering Techniques (ZFNs and TALENs)

Sunil Kumar Sunani, Sandeep Kumar, Devendra Kumar Choudhary, Harvinder Kumar Singh, and Ramyashree Devi G. S.

2.1 INTRODUCTION

Before the targeted nuclease systems, the traditional technique of modifying genomic sequences was the gene targeting (GT) with the help of the homologous recombination (HR). But, in eukaryotes, the incidence of HR is much lower when naked DNA is introduced into cells, and the process in general is ineffective as compared to random integration, which impeded biomedical research. As an example, the insertion of a specific gene into a designated site in a human genome was also a difficulty, even though it could have been of great importance in the field of genetic medicine. Natural HR was so rare that routine GT applications became impossible in higher plants for 25 years after the initial success in tobacco. There have been efforts to enhance GT through the means such as overexpression of HR-related proteins or negative selectable markers. The greatest restriction which did not allow the effective and broadly applicable targeted modification was the lack of molecular reagents to cause double-strand breaks (DSBs) at defined positions throughout the genome. The invention of the first actually targetable reagents, the engineered nucleases, marked the beginning of the advance in the desired manipulation of genomes (Bhardwaj & Nain 2021).

The design of these chimeric proteins was to cause a DSB at a predefined genomic locus, which is essentially a significant step identified to increase at a crucial rate both HR and non-homologous end joining (NHEJ) repair at the site of the cut. The concept of combining a nonspecific DNA cleavage domain with a sequence-specific DNA-binding domain had been proposed earlier, notably in 1996 with the development of Splase, a restriction endonuclease created by genetically fusing the FokI DNA cleavage domain to the zinc-finger (ZF) DNA-binding domain of the Sp1 transcription factor. Splase had been identified as a rare cutter that was able to digest plasmid DNA containing Sp1 sites, and there was a possibility to develop other restriction enzymes by modulating the specificity of binding of the ZF recognition helix. The result of this work was the first significant programmable nuclease group, the Zinc-Finger Nucleases (ZFNs). This was then followed by the introduction of the ZFNs in 2005, which revealed that specific DSBs were capable of provoking high-efficiency gene repair in human cells. The second generation of programmable nucleases was developed in the form of Transcription Activator-Like Effector Nucleases (TALENs), which followed the identification of the TALE protein with a special code of DNA recognition (Chen & Gonçalves 2016).

The essence of the importance of ZFNs and TALENs is that these two technologies were the stepping stones on the path of instituting the paradigm of programmed genome editing. The work on ZFNs was done as an extension of the fundamental research on zinc-finger proteins (ZFPs) and FokI restriction enzyme wherein the authors established that by protein engineering, one could target any arbitrary DNA sequence to forced cleavage. It was demonstrated that one of the ZFNs capable of promoting HR in cells was a DSB induced by ZFN. This became much extended later

DOI: 10.1201/9781003735113-2

with the development of TALENs, that are bacterial TAL effectors fused to the FokI cleavage domain. These technologies placed scientists in a position of such extending capabilities to modify cells and organisms virtually, and with extensive ramifications on the fields of biology and medicine (Hirakawa et al. 2020).

Although ZFNs and TALENs had encountered several challenges to the form of complexity in protein engineering, limitations in TALEN delivery, and so forth, the initial functional successes of ZFNs and TALENs played a vital role in establishing the essential efficiency, specificity, and delivery benchmarks, which allowed developing the third-generation tool, the RNA-guided CRISPR-Cas9 system, within a remarkably short time (Khan 2019).

2.2 MECHANISM AND MOLECULAR ARCHITECTURE OF ZINC-FINGER NUCLEASES

2.2.1 Zinc-Finger DNA-Binding Domains Structure

Structure of zinc-finger DNA-binding domains appears in a variety of proteins. A DNA-binding domain is the functional center of a ZFN, and it is usually Cys_2His_2 ZF motifs. These motifs represent the most common DNA-binding motif which is often represented as tandem chains in most transcription factors. The ZF motifs have approximately 28–30 amino acids which are packed together into a small globular shape. This structure is characterized by a distinct helix which normally has an alpha-helix structure and a 3_{10} helix structure that is stacked between two beta-strands forming a hairpin structure. A zinc atom tightly complexed with two cysteine and two histidine ligands maintains the stability of this domain. The binding of the Zn (II) ion is required in the specific recognition of the DNA; the displacement of the Zn (II) will cause the release of an apoprotein that will not have the specific DNA recognition. Several ZFs are organized as tandem arrays or zinc finger array (ZFA) to bind long strands of DNA. As an example, the Sp1 transcription factor has three adjacent ZF motifs, while the myeloid zinc-finger gene 1 (MZF1) encodes a transcription factor containing 13 ZF motifs arranged into two distinct domains (Lee et al. 2016).

2.2.2 FokI Nuclease Domain and Requirement for Dimerization

ZFNs have their DNA cleavage activity conferred by the FokI nuclease domain which is derived as a restriction endonuclease FokI. This domain has a nonspecific cleavage activity in its own. A form of chimeric protein is a ZFN that is genetically engineered by a fusion of DNA-binding domain, ZFP, with the nonspecific cleavage domain (N) of the endonuclease FokI. More importantly, the FokI cleavage domain is only catalyzed when it is dimerized. Thus, an effective utility of chimeric nucleases in efficient double-strand cleavage needs two ZFN monomers to binding adjacent DNA sites in a close proximity, permitting the FokI domains to dimerize. Upon dimerization, both DNA strands are cleaved, typically producing staggered breaks separated by approximately 4 bp and generating 5′ overhangs characteristic of native FokI nuclease activity. Initial FokI versions used early glimpses of ZFNs used wild-type FokI, capable of forming cleavage-competent homodimers at undesired locations causing cytotoxicity and restricting its use as a means of control through off-target cleavage. It resulted in the structure-guided engineering of obligate heterodimeric ZFN scaffolds, with high affinity for cleaving DNA only in the presence of a heterodimer, a reduction in homodimer activity of several folds, and with highly reduced genotoxicity (Li et al. 2020).

2.2.3 DNA Recognition Principles and Specificity of the Target Site of DNA

DNA recognition by ZFNs relies on the specificity of individual zinc-finger domains, with each finger typically designed to recognize a 3-bp DNA subsite. The critical interaction is at the residues at three prominent positions in the alpha-helical reading head which is the dominant site of the recognition of the base. Initial directed mutagenesis screens based on sequence similarity between transcription factors such as Spl and Krox-20 were successful in finding amino acids that dictated binding specificity of DNA-binding proteins, and established that there were three nucleotides in the finger. Subsequently phage display libraries of zinc fingers selected by later years explained components of recognition rules which relate the amino acid sequence of a finger with a preferred DNA-binding site. Yet, as structural analysis of complexes such as the two zinc-finger domains complexed with DNA has shown, the recognition of ZFP is less straightforward as it was viewed to be at first, and occasionally the residues that are known to be outside of the three important positions are involved. To achieve adequate specificity to bind a unique genomic locus with large genomes, ZFP arrays of five to six distinct zinc fingers are usually fused together to bind a sequence of between 15 and 18 bp (Mussolino et al. 2011).

2.2.4 Plans for ZFN Design and Assembly

The architecture of ZFNs is designed to enable the reliable construction of ZFPs capable of specifically recognizing selected DNA target sites. Early strategies relied on selection and directed *in vitro* mutagenesis approaches, including phage display, to optimize DNA-binding specificity. Another widely explored strategy was modular assembly (MA), which was based on the assumption that individual ZF modules, each recognizing a 3-bp DNA subsite, could be readily combined to target longer DNA sequences. Although MA appeared theoretically straightforward, it proved difficult in practice and exhibited a high failure rate, largely due to context-dependent interactions between neighboring ZF domains that compromised binding specificity (Xiao et al. 2012). Because simple MA was inherently unreliable, more robust optimization and selection-based approaches were subsequently developed. These included Oligomerized Pool ENgineering (OPEN), which improved targeting efficiency compared with conventional MA, and Context-Dependent Assembly (CoDA), an open-source strategy that explicitly accounts for interactions between adjacent zinc fingers to ensure compatibility at finger–finger interfaces. Additional methods, such as the OverLap extension Polymerase Chain Reaction (PCR) and TA-cloning (OLTA) approach, employed overlap extension PCR and Thymine, Adenine (TA) cloning to enable rapid and efficient construction of customized DNA-binding domains. Beyond these assembly strategies, alternative architectural variants were explored to enhance specificity and expand targeting potential (Zhang et al. 2015).

2.2.5 Mechanism of DNA Cleavage and Cellular DNA Repair (NHEJ and HDR)

The initiation of the ZFN mechanism starts with FokI dimer-induced highly specific, sequence guided DSB. The cell endogenous DNA repair machinery subsequently does repair this DSB, and this is how the genome-editing process will conclude:

2.2.5.1 Non-Homologous End Joining

This is the most significant and mostly error-prone pathway in which DSBs repair with direct ligation of the broken ends. The pattern that is typical of NHEJ is small insertions or deletions (indels) at the site of break. When such indels are present in a gene coding region, they usually lead to a frameshift, resulting in functional disruption or gene knockout. NHEJ occurs through a major pathway via a Lig4-dependent mechanism, but an alternative microhomology-dependent mechanism is also activated, which Lig4 does not mediate, and is instructed by the disappearance of Lig4. Scientists reported pioneer evidence of somatic mutation that occurred due to NHEJ repair in *Drosophila* (Sahu et al. 2026).

2.2.5.2 Homologous Recombination/Homology-Directed Repair (HDR)

This is a pathway that is less common but is a more specific approach that needs a homologous DNA template (either endogenous or exogenous). DSB induced by ZFN symbolically promotes the

HDR allowing the specific integration of the preferred sequences of an extrachromosomal DNA donor into the genome. HDR plays a pivotal role in correcting genes and precise addition of genes (knock-in). This integration event takes place in a homology-guided fashion, and is usually due to synthesis-dependent strand annealing repair against the donor DNA as a template (Anis et al. 2025).

2.3 TALENS

2.3.1 Origin from Bacterial TAL Effectors

TALENs are a second generation of programmable nucleases, derived from the study of TALEs found in phytopathogenic *Xanthomonas* bacteria. TALEs are naturally occurring proteins secreted by these bacteria and injected into host plant cells where they localize to the nucleus. Their primary biological function is to act as transcription factors, binding to host DNA promoters to activate gene expression that facilitates pathogen growth. This ability of TALEs to bind specific DNA sequences based on their structure provided the critical conceptual foundation for developing TALENs as targeted genome-editing tools. The highly specific DNA recognition property of TAL effectors is defined by a central domain consisting of tandem repeats. The modular structure of these repeats establishes a simple, modular, one-module-to-one-base-pair recognition code (Ain et al. 2015).

2.3.2 Modular Repeat Domains and Repeat Variable Diresidues (RVDs)

The basis of TALE specificity resides in its central DNA-binding domain, which consists of tandem repeats, each typically 33–35 amino acids long. Within each repeat, the core element determining specificity is the RVDs, two hypervariable amino acids located at positions 12 and 13. This modular arrangement allows the protein to track along the DNA duplex's sense strand, with the RVDs contacting the major groove. Specifically, the 12th residue stabilizes the RVD loop, while the 13th residue makes the base-specific contact. The RVDs and their corresponding nucleotide targets generally follow a canonical code: NI typically specifies Adenine (A), HD typically specifies Cytosine (C), NN specifies Guanine (G) or Adenine (A), and NG specifies Thymine (T). This predictable RVD code enables the engineering of TALENs to target virtually any defined DNA sequence (Zess & Begemann 2021).

2.3.3 Fusion with FokI Nuclease and Mechanism of Action

TALENs are synthetic nucleases created by fusing the highly specific TALE DNA-binding domain to the nonspecific cleavage domain (N) of the FokI endonuclease. This fusion creates a targeted molecular scissor. The critical mechanistic feature is that the FokI nuclease domain is only active as a dimer. Therefore, cleavage requires the coordinated action of two TALEN monomers. These two TALENs are engineered to bind adjacent recognition sites on opposite DNA strands, separated by an appropriate "spacer" sequence. When correctly positioned, the two FokI domains dimerize, forming the active center that introduces a DSB in the spacer region. This induced DSB serves as a substrate for two main competing cellular DNA repair pathways: NHEJ, leading to gene disruption via insertions or deletions (indels), or HDR, enabling precise sequence replacement or insertion using a donor template. The development of the TALEN fusion concept was initially reported by Kim and Kim (2014), demonstrating their ability to direct DSBs to specific targeted sites.

2.3.4 Design Flexibility and Assembly Methods

TALENs offered substantial improvements in design flexibility compared with ZFNs because of the simple and modular RVD recognition code. This straightforward code enables researchers to readily engineer proteins capable of recognizing target DNA sequences of up to about 23 bases in length. However, despite the ease of conceptual design, the highly repetitive nature of TALE coding sequences created a major assembly bottleneck when using conventional cloning approaches, as extensive sequence homology often led to recombination errors in bacterial hosts. To overcome this limitation, several efficient and MA strategies were developed. Early MA approaches leveraged the RVD code to enable rapid construction of designer TALENs. Subsequently, Golden Gate cloning emerged as the most widely adopted method, using Type IIS restriction enzymes and standardized modular kits to facilitate seamless assembly of TALE repeat arrays from predefined building blocks. To further increase speed and scalability, high-throughput and automatable techniques were introduced, including solid-phase ligation-based methods, ligation-independent cloning strategies, and USER-based assembly approaches, all of which enabled rapid, reliable, and large-scale TALEN construction (González Castro et al. 2021).

2.3.5 Comparison of Efficiency and Specificity Relative to ZFNs

TALENs commonly demonstrated higher accessibility and reliability than ZFNs, mainly regarding their ease of construction and success rate. The RVD code made TALEN design conceptually simpler than the complex protein engineering required for ZFNs, leading to a success rate that was substantially greater than that for ZFNs in organisms like *Drosophila* and zebrafish. In terms of gene disruption activity, TALENs achieved similar efficiency to benchmark ZFNs in initial reporter assays in human cells. However, large-scale comparisons in organisms like zebrafish often favored TALENs. Scientists reported that TALENs were significantly more likely to be mutagenic and induced an average of tenfold more mutations than ZFNs generated using the CoDA method (Satheesh et al. 2019). For precise editing via HDR, both ZFNs and TALENs proved effective, stimulating targeted gene correction/knock-in; the efficiency ratio between HDR and NHEJ was found to be highly dependent on the gene locus, nuclease platform, and cell type. TALENs typically exhibited high specificity and significantly reduced nuclease-associated cytotoxicities compared to earlier ZFN architectures. The high toxicity associated with early ZFNs was due to off-target cleavage caused by FokI homodimerization, an issue largely mitigated in ZFNs only after the engineering of obligate heterodimers (Nakayama et al. 2014).

2.4 CHALLENGES IN DESIGN AND ENGINEERING OF ZFNS AND TALENS

2.4.1 Protein Engineering Complication for ZFNs Versus TALENs

The practical application of ZFNs was hindered by the substantial complexity involved in engineering proteins capable of reliable DNA recognition. ZFNs are composed of tandem arrays of Cys_2His_2 zinc-finger motifs, with each finger originally designed to recognize a 3-bp DNA subsite. However, the early strategy of MA, which assumed that individual ZF modules could be independently linked to target longer DNA sequences, frequently proved unsuccessful. This was largely due to context-dependent interactions, in which the DNA-binding specificity of one finger was influenced by adjacent fingers, leading to a high failure rate in ZFN design. To overcome these limitations, more sophisticated selection-based approaches were developed, including OPEN and CoDA. Although these strategies improved targeting success, they were technically demanding, labor-intensive, and required considerable expertise, thereby restricting the broader adoption of ZFN technology. In contrast, TALENs greatly simplified nuclease engineering. This improvement stemmed from the highly modular nature of the TALE DNA-binding domain, in which a straightforward one-module–to–one-base-pair recognition code is specified by RVDs. As a result, TALEN design became conceptually simpler and generally achieved a much higher success rate in generating functional nucleases than ZFNs. Nevertheless, TALENs introduced a distinct technical challenge: the high sequence similarity among TALE repeats promoted HR and instability in bacterial cloning (Shamshirgaran et al. 2022).

2.4.2 Context-Dependent Affects Impacting Targeting Fidelity

Targeting fidelity in both platforms is influenced by contextual factors that extend beyond simple sequence recognition. In ZFNs, a major limitation arises from inter-finger incompatibility, whereby the activity of ZFNs at endogenous chromosomal sites is strongly affected by local chromatin architecture and the ability of the protein to access its target DNA. In contrast, TALEN fidelity is constrained by the N-0 base requirement, as optimal activity generally requires the target sequence to be preceded by a thymine (T), thereby restricting the range of selectable target sites. Experimental studies have demonstrated that the absence of a 5′ T can reduce TALEN activity by more than tenfold in certain assays compared with targets containing a 5′ T. Additionally, TALEN activity is modulated by DNA methylation status; methylated cytosines have been shown to impair editing efficiency in rice, particularly in genomic regions exhibiting variable levels of cytosine methylation (Chen et al. 2013).

2.4.3 Off-Target Cleavage Issues and Methods for Detection

A major safety and efficacy challenge for early programmable nucleases was the presence of off-target cleavage leading to unintended and potentially mutagenic alterations. The principal mechanism for toxicity and genotoxicity, especially in ZFNs, was the formation of cleavage-competent FokI homodimers at unintended sites when nucleases employed wild-type FokI cleavage domains. Detecting these off-target effects accurately proved challenging. Initial ZFN designs showed that cleavage sites could not always be predicted accurately *in silico*. Furthermore, applying two different detection methods (an *in vitro* cleavage selection method and an *in vivo* lentiviral integration method) to the same ZFN pair often yielded largely nonoverlapping off-target sites, complicating comprehensive specificity profiling (Chen et al. 2013).

Methods for detecting mutations, such as the T7 Endonuclease I assay (T7E1), could be sensitive but often prone to false positives (e.g., from SNPs) and required substantial material. Gabriel et al. (2011) used integrase-defective lentiviral vectors (IDLVs) to map the actual *in vivo* cleavage activity of ZFNs genome-wide, confirming that cleavage required specific spatial arrangements and high homology to the intended target. While TALENs generally exhibited better specificity, even they could accommodate a small number of position-dependent mismatches while maintaining detectable activity at endogenous loci. Newer methods like deep sequencing and single molecule real-time (SMRT) sequencing have since been developed to quantify the frequency of editing events and provide a more comprehensive, single-molecule view of genome-editing outcomes for ZFNs, TALENs, and CRISPR-Cas9 systems (Shamshirgaran et al. 2022).

2.4.4 Strategies to Improve Specificity and Reduce Cytotoxicity

Extensive engineering efforts have been undertaken to improve nuclease specificity and to reduce cytotoxicity arising from off-target DNA cleavage. One of the most effective strategies for ZFNs was the development of obligate heterodimeric architectures, which prevent the formation of toxic FokI homodimers and thereby substantially reduce genome-wide off-target cleavage. Additional refinements to the FokI nuclease domain further enhanced cleavage fidelity, including variants with improved heterodimer performance and engineered substitutions that slowed cleavage kinetics, selectively suppressing activity at low-affinity off-target sites by several orders of magnitude. Similar architectural improvements were applied to TALENs, where obligate heterodimeric scaffolds reduced cellular toxicity while maintaining or even increasing on-target mutation frequencies. TALEN specificity was further fine-tuned through the use of nonconventional RVDs, which provide greater nucleotide discrimination and help minimize unintended cleavage events (Shamshirgaran et al. 2022).

Beyond nuclease architecture, strategies to regulate protein stability and expression were also explored to limit prolonged nuclease activity; controlled modulation of ZFN protein half-life was shown to preserve efficient GT while reducing cytotoxic effects, and transient stabilization of nuclease proteins enhanced editing efficiency when needed. A particularly effective safety-enhancing approach for both ZFNs and TALENs was their conversion into nickases by inactivating one cleavage domain, resulting in single-strand breaks rather than DSBs. This modification markedly reduced error-prone NHEJ, promoted error-free HDR, and enabled precise gene addition with minimal induction of unwanted insertions or deletions (Zubair et al. 2025).

2.5 DELIVERY AND CELLULAR IMPLEMENTATION

2.5.1 Delivery Systems for ZFNs and TALENs (Plasmid, mRNA, Viral Vectors)

Engineered nucleases such as ZFNs and TALENs require efficient delivery into target cells or organisms to enable successful genome modification, and the choice of delivery strategy is largely dictated by the target cell type and the intended application, particularly whether transient or stable expression is desired. Plasmid DNA delivery, in which ZFN or TALEN expression cassettes often together with a donor template are introduced by transfection, has been widely used in transformed cell lines and enables targeted transgenesis at relatively high frequencies; for example, transient ZFN plasmid transfection has achieved biallelic gene disruption at frequencies exceeding 1% in CHO cells, while TALENs are commonly delivered as paired plasmids followed by clonal isolation of edited cells. For *in vivo* applications and primary cells, delivery of *in vitro*-transcribed mRNA is generally preferred, as it ensures transient nuclease expression and minimizes the risk of insertional mutagenesis; ZFN and TALEN mRNAs have been extensively used via microinjection into one-cell embryos across a range of model organisms, including zebrafish, mice, Xenopus, and insects, as well as in large animals such as pigs and rats, to generate efficient gene knockouts. Viral vectors represent another powerful delivery platform, particularly for hard-to-transfect cell types such as hematopoietic stem and progenitor cells (HSPCs); IDLVs have been successfully employed to deliver ZFNs along with donor templates, supporting editing efficiencies of approximately 13%–39% in multiple cell types, while codon-swapped lentiviral constructs have enabled efficient single-vector delivery of ZFN pairs. Transient protein transduction using lentiviral particles loaded with ZFN or TALEN proteins further reduces prolonged nuclease activity and off-target effects. Adenoviral vectors, especially fiber-modified variants, can package and deliver large TALEN genes intact into both transformed and primary human cells, producing high levels of site-specific DSBs and enabling targeted genome correction when combined with donor templates. Finally, adeno-associated virus (AAV) vectors, particularly AAV6, are highly effective for delivering homologous donor DNA for HDR in combination with electroporated ZFN or TALEN mRNA, achieving high-efficiency targeted insertions in human HSPCs (Chen et al. 2013).

2.5.2 Challenges in Delivery Efficiency and Cellular Uptake

Despite the availability of diverse delivery strategies, achieving efficient delivery while controlling the activity and persistence of engineered nucleases remains a major challenge. Viral and nucleic acid–based delivery systems face issues related to vector integrity and targeting fidelity; for example, lentiviral vectors often fail to package the large, repetitive TALEN genes intact, resulting in heterogeneous proviruses, while codon swapping has been required to prevent HR during reverse transcription of ZFN-encoding lentiviruses. Conventional plasmid and viral vectors also pose risks of random genomic integration and prolonged nuclease expression, which can increase off-target effects. Balancing transient versus persistent expression is therefore critical, particularly in gene therapy, leading to the use of AAV vectors and IDLVs to achieve transient nuclease expression alongside sustained donor template delivery. Additional challenges arise at the cellular level, as many clinically relevant cell types, including primary cells and HSPCs, are difficult to transfect and often require viral vectors, although alternative approaches such as transferrin receptor–mediated delivery of ZFN proteins have shown promise. The large size of ZFN and TALEN proteins, TALEN monomers frequently exceeding 100 kDa, and the highly repetitive nature of TALEN DNA-binding domains further complicate nonviral uptake and viral packaging, prompting the development of direct protein delivery methods, such as conjugation with cell-penetrating poly-arginine peptides, to achieve efficient gene disruption. Finally, genome-editing outcomes are strongly influenced by cell cycle status, as HDR is most active during the G2/M phases; synchronizing cells such as human pluripotent stem cells in these phases has been shown to enhance ZFN- and TALEN-mediated HDR efficiency by three- to sixfold (Manikishore et al. 2025).

2.6 APPLICATIONS IN RESEARCH

The application of ZFNs and TALENs made one of the significant changes in research use, as they allow specific genome editing in a variety of models organism, plants, and cell system, by repairing DSBs. They were also the pioneers in developing trusted gene knockout and knock-in methodologies in creatures like Drosophila, zebrafish, Arabidopsis, rice, and maize, whereby standard genetic methodologies were constrained. These nucleases have been used to advantageously modify the characteristics of crops and polyploid plants in agriculture such as herbicide resistance, stacking functions in traits, enhanced oil quality, fragrance generation, and storage qualities and biomass optimization. ZFNs and TALENs provided the basis of modern functional genomics through reverse genetics, mass knockout generation, high-throughput nuclease library, and targeted chromosomal rearrangements. Their power to produce formatted deletions, inversions, translocations, and reporter lines revolutionized the study of gene functions and cancer biology, and laid the groundwork of fundamental approaches that formed a direct basis on the development of high-throughput genome-editing platforms (Shamshirgaran et al. 2022).

2.7 LIMITATIONS AND CONSTRAINTS

ZFNs and TALENs were limited in their technical and practical application despite their underlying usefulness to genome editing. Their size (especially when used in a TALEN size) posed significant delivery problems especially using viral vectors like lentiviruses, and this made it difficult to deliver the proteins efficiently using nonviral methods of delivery as well. The excessive repetitiveness of their DNA-binding domains is what made cloning unstable and subject to errors in assembling its parts, and the construction process was beginning to be time-consuming and unreliable. TALENs also had a limitation of a 5′ thymine requirement at the target site which makes targeting less flexible than CRISPR-Cas systems. Moreover, functional ZFNs and TALENs are complex to develop and demand lots of expertise including multifaceted protein engineering, lots of validation, and selective cloning techniques. All these restrictions in combination resulted in the necessity to create easier and more open genome-editing platforms and ultimately paved the way for simpler systems like CRISPR-Cas9 (Bhagtaney & Sundarrajan 2023).

2.8 COMPARATIVE PERSPECTIVE WITH CRISPR-CAS SYSTEMS

CRISPR-Cas systems are a diverse family of prokaryotic immune mechanisms that vary widely in their structure and function. The emergence of the CRISPR-Cas system marked a third major phase in genome editing, providing a necessary comparative perspective on the technical advantages and inherent limitations of the preceding ZFN and TALEN technologies (Shamshirgaran et al. 2022) (Table 2.1).

TABLE 2.1
Brief Comparisons of ZFNs, TALENs, and CRISPR-Cas9

Feature	ZFNs	TALENs	CRISPR-Cas9	References
Target recognition	Protein–DNA via zinc-finger arrays recognizing DNA triplets.	Protein–DNA via TALE repeats, each recognizing a single nucleotide.	RNA–DNA via guide RNA base-pairing to target plus PAM recognition by Cas9.	Shamshirgaran et al. (2022), Chang (2022), and Wani et al. (2023)
Double-strand break catalyst	FokI nuclease domain requiring dimerization.	FokI nuclease domain requiring dimerization.	Cas9 nuclease with two catalytic domains (RuvC, HNH) in a single protein.	
Ease of design	Difficult; each new target needs custom zinc-finger engineering and validation.	Easier than ZFNs; modular RVD code, but assembly of repetitive TALE arrays is labor-intensive.	Very easy; retargeting usually requires only changing the gRNA sequence.	
Typical target length	~9–18 bp recognized by zinc-finger array plus FokI spacing.	~30–40 bp recognized by TALE array plus spacer.	~20 bp protospacer plus PAM (e.g., NGG).	
Targeting constraints	Requires paired binding sites with proper spacing; context effects between zinc fingers; design less predictable.	Requires paired sites and suitable spacer; historically needed a 5′ T (N_0) at target; large construct size complicates delivery.	Requires presence of suitable PAM near target; otherwise broadly flexible.	
Editing efficiency (general)	Moderate; success strongly depends on good zinc-finger design.	Moderate to high; often better than ZFNs, but below or comparable to CRISPR in many systems.	High; frequently exceeds ZFNs and TALENs in side-by-side comparisons.	
Off-target profile	Can generate many off-targets if poorly designed; improved by obligate heterodimeric FokI variants.	Generally lower off-target activity than ZFNs and often than CRISPR, due to long recognition sequences.	Off-target risk depends on gRNA and nuclease variant; high-fidelity Cas9 and nickase designs reduce unwanted cuts.	
Chromatin sensitivity	Can cut in nucleosome-occupied and complex regions; useful in GC-rich or structurally constrained loci.	Performs well in heterochromatin; some studies show higher efficiency than Cas9 at tightly packed sites.	Activity reduced in nucleosome-dense or heterochromatic regions compared with some ZFN/TALEN designs.	
Construct size and delivery	Large protein dimers; delivery often via plasmid, mRNA, or protein; challenging for multiplexing.	Very large repetitive coding sequences; problematic for some viral vectors and multiplex delivery.	Single Cas9 plus small gRNA(s); well suited to multiplexing and diverse viral/nonviral delivery platforms.	

2.9 CURRENT STATUS AND FUTURE PROSPECTS

Regardless of the prevalence of CRISPR-Cas systems owing to their ease and affordability, ZFNs and TALENs remain confined to particular utility of genome engineering. These platforms have been demonstrated to work in over 40 organisms, which attests to the strength of the protein-based design. Under certain genomic conditions, ZFNs and TALENs have better performance compared to CRISPR-Cas9 especially in heterochromatin or nucleosome-bounded areas, where Cas9 activity is suppressed frequently. They are highly specific (and therefore more attractive) to therapeutic uses, as well as GMP-compatible TALEN-mediated correction of the sickle cell mutation, as well as insertions targeted by ZFNs in livestock genomes. Future development is also being made to enhance their accuracy, including refined FokI domain engineering to minimize off-target cleavage, conversion into nickases to prefer HDR, and building TALE-based base editors, which can be used to make a precision nucleotide replacement without inducing a DSB (Wani et al. 2023). Optimization of systems of nonviral and non-DNA-free delivery is also improving their clinical and regulatory acceptability (Tariq et al. 2026).

2.10 CONCLUSION

The pre-CRISPR era of genome engineering was fundamentally established and advanced by two pivotal technologies: ZFNs and TALENs, which paved the way for the breakthrough in targeted genome manipulation by confirming the power of the targeted DSB paradigm. The development of ZFNs represented the crucial first step, demonstrating that arbitrary DNA sequences could be addressed for cleavage through sophisticated protein engineering. Later, the development of TALENs leveraged the natural modularity of bacterial TALEs. TALENs dramatically simplified the design process compared to ZFNs, owing to their simple, modular, one-module-to-one-base-pair recognition code encoded by RVDs. Collectively, both ZFNs and TALENs proved the general efficacy of targeted cleavage, successfully achieving genome modification in more than 40 different organisms and cell types. By establishing that an induced DSB is the critical, rate-limiting step for stimulating cellular repair pathways like HDR and NHEJ, these technologies provided the conceptual and technical foundation for all subsequent genome-editing efforts.

REFERENCES

Ain, Q. U., Chung, J. Y., & Kim, Y. H. (2015). Current and future delivery systems for engineered nucleases: ZFN, TALEN and RGEN. *Journal of Controlled Release*, 205, 120–127.

Anis, A., Dash, S., Karmakar, S., & Sahoo, J. P. (2025). Genetic engineering of cis-regulatory elements by CRISPR/Cas technology for crop improvement. In J.-T. Chen (Ed.), *Genome and Epigenome Editing for Stress-Tolerant Crops* (pp. 157–177). Wiley. https://doi.org/10.1002/9781394280049.ch10

Bhagtaney, L., & Sundarrajan, P. (2023). An overview of tools for genome editing: ZFNs, mega nucleases, and TALENs. In Pandita, D., & Pandita, A. (Eds.), *CRISPR/Cas-Mediated Genome Editing in Plants* (pp. 37–64) CRC Press. https://doi.org/10.1201/9781003331759

Bhardwaj, A., & Nain, V. (2021). TALENs—An indispensable tool in the era of CRISPR: A mini review. *Journal of Genetic Engineering and Biotechnology*, 19(1), 125.

Chang, A. Y. (2022). Genome engineering with CRISPR/Cas9, ZFNs, and TALENs. In Tsang, S. H. (Ed.), *CRISPR Genome Surgery in Stem Cells and Disease Tissues* (pp. 39–45). Academic Press.

Chen, X., & Gonçalves, M. A. (2016). Engineered viruses as genome editing devices. *Molecular Therapy*, 24(3), 447–457.

Chen, S., Oikonomou, G., Chiu, C. N., Niles, B. J., Liu, J., Lee, D. A., … Prober, D. A. (2013). A large-scale in vivo analysis reveals that TALENs are significantly more mutagenic than ZFNs generated using context-dependent assembly. *Nucleic Acids Research*, 41(4), 2769–2778.

Gabriel, R., Lombardo, A., Arens, A., Miller, J. C., Genovese, P., Kaeppel, C., & Von Kalle, C. (2011). An unbiased genome-wide analysis of zinc-finger nuclease specificity. *Nature Biotechnology*, *29*(9), 816–823.

González Castro, N., Bjelic, J., Malhotra, G., Huang, C., & Alsaffar, S. H. (2021). Comparison of the feasibility, efficiency, and safety of genome editing technologies. *International Journal of Molecular Sciences*, 22(19), 10355.

Hirakawa, M. P., Krishnakumar, R., Timlin, J. A., Carney, J. P., & Butler, K. S. (2020). Gene editing and CRISPR in the clinic: Current and future perspectives. *Bioscience Reports*, *40*(4), BSR20200127.

Khan, S. H. (2019). Genome-editing technologies: Concept, pros, and cons of various genome-editing techniques and bioethical concerns for clinical application. *Molecular Therapy Nucleic Acids*, 16, 326–334.

Kim, H., & Kim, J. S. (2014). A guide to genome engineering with programmable nucleases. *Nature Reviews Genetics*, 15(5), 321–334.

Lee, C. M., Cradick, T. J., Fine, E. J., & Bao, G. (2016). Nuclease target site selection for maximizing on-target activity and minimizing off-target effects in genome editing. *Molecular Therapy*, 24(3), 475–487.

Li, H., Yang, Y., Hong, W., Huang, M., Wu, M., & Zhao, X. (2020). Applications of genome editing technology in the targeted therapy of human diseases: Mechanisms, advances and prospects. *Signal Transduction and Targeted Therapy*, 5(1), 1.

Manikishore, M., Maurya, S. K., Rathee, S., & Patil, U. K. (2025). Genome editing approaches using Zinc Finger Nucleases (ZFNs) for the treatment of motor neuron diseases. *Current Pharmaceutical Biotechnology*, 26(10), 1514–1531.

Mussolino, C., Morbitzer, R., Lütge, F., Dannemann, N., Lahaye, T., & Cathomen, T. (2011). A novel TALE nuclease scaffold enables high genome editing activity in combination with low toxicity. *Nucleic Acids Research*, 39(21), 9283–9293.

Nakayama, T. J., Borém, A., Chiari, L., Molinari, H. B. C., & Nepomuceno, A. L. (2014). Precision genetic engineering. In Borém, A., and Fritsche-Neto, R., (Eds.), *Omics in Plant Breeding* (pp. 187–205). John Wiley & Sons, Inc. https://doi.org/10.1002/9781118820971.ch10

Sahu, A., Bebarta, A., Jena, A. K., Mohanty, M., Patra, S., Sahoo, J. P., & Sahu, N. (2026). CRISPR-edited plants for resistance to insect herbivores: Methods and applications. In Abd-Elsalam, K. A., and Chen, J. T. (Eds.), *CRISPR Technology for Combating Plant Pests and Pathogens* (pp. 69–94). Academic Press.

Satheesh, V., Zhang, H., Wang, X., & Lei, M. (2019). Precise editing of plant genomes–prospects and challenges. In Jain, A. & Rai, V. (Eds.), *Seminars in Cell & Developmental Biology* (Vol. 96, pp. 115–123). Academic Press.

Shamshirgaran, Y., Liu, J., Sumer, H., Verma, P. J., & Taheri-Ghahfarokhi, A. (2022). Tools for efficient genome editing; ZFN, TALEN, and CRISPR. In Verma, P. J., Sumer, H., Liu, J., (Eds.), *Applications of Genome Modulation and Editing* (pp. 29–46). Humana New York, NY.

Tariq, H., Ali, A., Chen, Y. M., & Zang, L. S. (2026). In-depth exploration of CRISPR/Cas9 and other genome editing tools for pest management. In Ali, H. & Xiao, Y. (Eds.), *Genome Editing for Pest Management* (pp. 39–58). CRC Press.

Wani, A. K., Akhtar, N., Singh, R., Prakash, A., Raza, S. H. A., Cavalu, S., … Hashem, N. M. (2023). Genome centric engineering using ZFNs, TALENs and CRISPR-Cas9 systems for trait improvement and disease control in Animals. *Veterinary Research Communications*, 47(1), 1–16.

Xiao, A., Wu, Y., Yang, Z., Hu, Y., Wang, W., Zhang, Y., & Zhang, B. (2012). EENdb: A database and knowledge base of ZFNs and TALENs for endonuclease engineering. *Nucleic Acids Research*, *41*(D1), D415–D422.

Zess, E., & Begemann, M. (2021). CRISPR-Cas9 and beyond: What's next in plant genome engineering. *In Vitro Cellular & Developmental Biology-Plant*, 57(4), 584–594.

Zhang, X. H., Tee, L. Y., Wang, X. G., Huang, Q. S., & Yang, S. H. (2015). Off-target effects in CRISPR/Cas9-mediated genome engineering. *Molecular Therapy Nucleic acids*, *4*. e264

Zubair, A., Ali, M., Ahmad, F., & Althobaiti, S. A. (2025). Unlocking the role of transcription activator-like effector nuclease (TALENs) and zinc finger nuclease (ZFN) in the treatment of HIV. *Molecular Biology Reports*, 52(1), 948.

3 The CRISPR Toolbox
Expanding the Universe of Genome Editing

Devendra Kumar Choudhary, Harvinder Kumar Singh, Sunil Kumar Sunani, and Sandeep Kumar

3.1 INTRODUCTION

Genome editing is now a key technology in contemporary molecular biology, and has extensive implications in the fields of basic research, therapeutic intervention, agriculture and synthetic biology. These techniques allow fine-scale modification of the DNA sequence of an organism-targeted inactivation of genetic material, repair or expression of genetic material, and therefore can be used directly to study the function of genes, recapitulate disease phenotype and design traits of agronomic or clinical utility. These capabilities are based on the use of programmable nucleases (that cause specific DNA lesions and can trigger cellular repair pathways) and can be translated into the capabilities of various experimental and translational applications (Cloney, 2016).

By knowing the history of genome editing tools, one can clearly understand the evolution of the tool to overcome previous limitations and influence the current possibilities. The history of targeted manipulation of DNA is in the exploration and use of restriction endonucleases in the early 1970s. These bacterial enzymes cut the DNA at given sequences and recombinant DNA techniques and molecular cloning became a possibility with such enzymes making it possible to cut and ligate pieces of DNA accurately. The limited understanding of the conceptual and technical principles offered by restriction enzymes paved the way to in vivo editing approaches by showing that DNA could be edited reproducibly and predictably (Cloney, 2016).

The discovery of programmable protein-based nucleases was the initial steps to targeted editing into living cells. In the late 1990s, zinc-finger nucleases (ZFNs) were created by conjugating modular zinc-finger DNA-binding motifs with the FokI nuclease domain to create site-specific double-strand breaks (DSBs) upon binding adjacent sequences by pairs of ZFNs. ZFNs proved that the heritable sequence alterations in mammalian genomes could be induced using the DSB, making genome engineering a feasible technology (Damian and Porteus, 2013). However, ZFNs showed practical limitations, such as context-dependent recognition of DNA, design requirements and the issue of off-target cleavage (Cloney, 2016). Design predictability and target flexibility were then enhanced by transcription activator-like effector nucleases (TALENs). They are recognized by arrays of repeats in TALENs using defined repeat-variable diresidue codes, and together with FokI result in specific targeted DSBs with relatively enhanced specificity and reduced cytotoxicity than early ZFNs. Nevertheless, repetitions of TALE arrays make molecular assembly more complex and have the potential to hinder effective delivery, especially when using a viral vector that has a limited cargo capacity (Osborn et al., 2013).

The discovery of the CRISPR-Cas systems as genome engineering tools represented a paradigm shift in that protein design was replaced by RNA-programmable target recognition. CRISPR-Cas effectors, in particular, *Streptococcus pyogenes* Cas9, utilize short guide RNAs that base-pair with complementary locations in genomes to guide nuclease action. This RNA-directed process made the retargeting process easier, facilitated multiplexed editing and increased the availability of genome

 DOI: 10.1201/9781003735113-3

editing methods to any laboratory in the world (Wang et al., 2016). Even CRISPR systems have high diversity, and the distinction between Class 1 and Class 2 architectures has different effector constructions and engineering opportunities. Class 1 systems use multi-protein effector complexes, whereas Class 2 systems (such as Cas9 and Cas12 families) utilize single effector proteins, which have the target recognition and nuclease activity abilities combined, which enabled their rapid adaptation to eukaryotic cell editing (Crawley et al., 2018). Moreover, the classification and functional overview of major CRISPR-Cas systems are indexed in Table 3.1.

In addition to nuclease-dependent DSB approaches, base editors (BEs) and prime editors have been developed, and their use has broadened the toolbox by allowing the conversion of nucleotides and the insertion of small sequences without creating large-scale indels and rearrangements (Roy and Pokale, 2025). In addition to these examples, further expansion of editing modalities with RNA-targeting effectors and transposon-based integrases provides transient RNA editing and site-specific delivery of larger cargos of DNA with no canonical DSB repair responses (Xie et al., 2014). Locating modern practices in this historical development puts into focus enduring issues in delivery efficiency, specificity control and ethical translation, that continue to guide the research and clinical development. This introduction puts the technological advances in context and prepares the reader to a close examination of applications, mechanisms, ethical concerns and translational concerns in the future.

TABLE 3.1
Classification and Functional Overview of Major CRISPR-Cas Systems

System/Nuclease	Class/Type	Key Mechanism	Applications	References
CRISPR-Cas9	Class 2, Type II	RNA-guided DNA cleavage (DSBs), often resulting in knockouts via non-homologous end joining (NHEJ). Requires a single-guide RNA (sgRNA).	Most widely used system for generating targeted insertions/ deletions (indels) for gene knockout. Used for multiplex editing, DNA-free editing via ribonucleoproteins, and developing various resistance traits.	Ma et al. (2015)
CRISPR-Cas12a (formerly Cpf1)	Class 2, Type V	RNA-guided DNA cleavage. Uses a CRISPR RNA (crRNA).	Used for generating targeted DNA modifications, including multiplex editing (often with tRNA-crRNA arrays). Shown to be highly specific.	Hu et al. (2020)
CRISPR-Cas13 (e.g., Cas13b, Cas13d, CasRx)	Class 2, Type VI	RNA-guided RNA cleavage (ribonuclease activity).	Used for transcriptome editing (RNA knockdown/silencing) and conferring virus resistance. Cas13 subtypes include VI-A (LwaCas13a), VI-B (PbuCas13b), VI-D (RfxCas13d), VI-X and VI-Y.	Wang et al. (2025)
Base editors (Cas9-derived)	Modified Cas9	Facilitates targeted nucleotide substitution without generating DSBs, typically fusing a nuclease-deficient Cas9 (or a nickase Cas9) with a deaminase.	Used for precision genome editing, generating specific point mutations (e.g., C to T or G to A) and base substitution.	Liu et al. (2021)

3.2 CRISPR EFFECTORS IN PLANT SYSTEMS

3.2.1 Historical Development and Significance of CRISPR Effectors

CRISPR effectors developed as an adaptive immune response of prokaryotes into multifunctional tools of plant genome engineering. In archaea and bacteria, CRISPR loci take pieces of invading nucleic acids and encode guide RNAs which guide Cas proteins to complementary sequences, resulting in cleavage and acquired immunity. Recycling this RNA-guided nuclease action to eukaryotic systems revolutionized functional plant genomics and speeded up crop enhancement. The implementation of *S. pyogenes* Cas9 (SpCas9) heralded the introduction of routine and programmable editing of DNA in plants due to its relative ease and efficiency. The toolbox has since been extended to a variety of different effectors with different biochemical characteristics to further increase flexibility, specificity and the number of edits that can be achieved in crops and model plants. A significant part of current plant genome modification work is grounded in the historical patterns of microbial defense against agricultural biotechnology (Tyumentseva et al., 2023).

3.2.2 Classification of CRISPR-Cas Systems in Plants

The CRISPR-Cas systems are classified by their architecture and mechanism as per their effector and classification. The Class II systems have single multidomain effector proteins that are used in plant applications since they are simple to design and deliver guides. Type II (Cas9) and Type V (Cas12) of the Class II target a double strand of DNA, whereas Type VI (Cas13) targets a single strand of RNA. These systems vary in catalytic domain, preference of target (DNA or RNA) and recognition sequences, i.e., PAMs of DNA-targeting nucleases and PFSs (Protospacer Flanking Sequences) of RNA-targeting enzymes. The choice of an effector will therefore be based on the need to change genomic sequence, transiently regulate transcripts or create diagnostics; mechanistic differences will inform decisions (Nie et al., 2023).

3.2.3 SpCas9 and Engineered Cas9 Versions: Mechanism, PAM Limitations, Size, Fidelity

SpCas9 is an RNA-guided endonuclease. A single-guide RNA (sgRNA), formed by the fusion of crRNA and tracrRNA, directs sequence-specific binding of SpCas9 to complementary target DNA adjacent to a 5′-NGG-3′ protospacer adjacent motif (PAM). Recognition induces cleavage of both strands of DNA, and a blunt DSB is repaired by the endogenous pathways in the plant non-homologous end joining (NHEJ), rather than homology-directed repair (HDR) in the presence of a template. NHEJ generally induces indels that cause loss-of-function in genes, a simple path to loss-of-function alleles in crops. This mechanism is so clear and predictable that it is now being actively used in rice, maize, tomato, *Arabidopsis* and several other plants (Wang et al., 2016).

The canonical 5′-NGG-3′ PAM requirement of SpCas9 limits target selection in PAM-poor genomic regions. To overcome this limitation, engineered variants such as xCas9 and SpCas9-NG have been developed to recognize a broader range of PAM sequences, thereby enabling access to genomic loci that were previously inaccessible. These modified specificities of PAMs extend editing opportunities in regulatory regions, introns and genes containing small numbers of NGG sites. Nevertheless, optimized PAM relaxation can have implications on kinetics and specificity, and empirical optimization is required in every plant species and tissue. When a specific nucleotide modification is necessary, as is the case with crop traits, these variants can greatly enhance the probability of having a good target site to match the repair-based strategies (Li et al., 2023).

The large size of SpCas9 makes certain delivery methods more challenging and has led to the search of smaller or split systems, as well as methods that leave the transgene in a smaller footprint (e.g., ribonucleoprotein [RNP] delivery). High-fidelity mutants (e.g., eSpCas9, HF-Cas9) make amino acid changes that minimize off-target interactions with DNA and, consequently, off-target activity. Such fidelity improvements are useful in breeding scenarios where they would problematically lead to undesirable mutations in related agronomic loci. However, fidelity increases with on-target cleavage efficiency and necessitate guide selection and potentially increased nuclease expression. In practice, the weight that researchers give the significance of maximum specificity and the efficiency of practical editing rates should be at the cost of selecting the type of wild and engineered Cas9 (Naeem et al., 2020).

3.2.4 Cas12 Family: Collateral Cleavage, Smaller Orthologs, Single-Stranded DNase Activity

The cas12 nucleases (Type V) are different in PAM preference and cleavage pattern compared to Cas9. Cas12a (Cpf1) identifies T-rich PAMs, including 5′-TTTV-3′, and cleaves with staggered ends that have 5′ overhangs instead of blunt ends. The staggered ends have the potential to make more accurate insertions and can enhance the success of HDR-mediated edits in the plants. Strikingly, collateral single-stranded DNase activity is also a feature of most Cas12 enzymes, which are triggered by target recognition; this activity has been repackaged to detect nucleic acids with high sensitivity, such as fast plant pathogen surveillance (Paul and Montoya, 2020). Cas12 has orthologs including AsCas12a, LbCas12a and FnCas12a, all of which have different temperature optima, activity profile and size. Certain orthologs are more thermophilic, which can be important in a thermophilic crop system or in an in vitro assay. Smaller Cas12 orthologs have the potential to reduce delivery requirements and potentially achieve even greater editing efficiencies in monocasts and dicasts when temperature, expression timing and guide architecture are optimized (Mahas et al., 2021).

Even empirical testing of model plants (e.g., *Arabidopsis*) and crops (e.g., rice) often demonstrates that there are significant differences among orthologs, and that matching nuclease choice to species and growth conditions is essential (Wada et al., 2022). The collateral ssDNase activity of Cas12 has been paired with isothermal amplification (RPA: Recombinase Polymerase Amplification, LAMP: Loop-Mediated Isothermal Amplification) to produce field-based diagnostic tests of plant pathogens (Malzahn et al., 2019). Recognition of reporters causes reporter cleavage, which produces fluorescent or colorimetric signals that can be used to quickly detect viral, bacterial or fungal sequences within crop tissue. The staggered cuts introduced by Cas12 are useful in the approach to insertion and multiplexed knockouts in the editing of applications (Karimi et al., 2025).

3.2.5 Cas13 Family: RNA Targeting, Transcriptome Editing and Diagnostics

The unique characteristics of Cas13 (Type VI) effectors include dealing with single-stranded RNA through guide RNAs, and the HEPN (Higher Eukaryotes and Prokaryotes Nucleotide-binding domain) domains contain RNase activity. There are several subtypes (Cas13a–d), which vary in size and specificity with Cas13d being compact and effective and demonstrated potential in plant applications. Since Cas13 cuts RNA and not DNA, it allows temporary, reversible gene expression repression and direct targeting of RNA viruses, which is an appealing antiviral defense mechanism (Gupta et al., 2022).

3.2.6 Editing of Transcriptomes, Gene Regulation and Resistance to Virus

Cas13 systems promote post-transcription regulating genes: specific mRNA knockdown, control of splicing, or translation regulation in a way that is programmable. They have uses where it is good to be able to interrogate genes in a functional manner that permanent mutation would be inadvisable, and where there is a route to engineering resistance to viruses by degrading viral genomes in planta. Cas13-based methods present a quick, reversible defense approach to crops at risk due to RNA viruses, which can be delivered transiently or expressed stably based on regulatory and breeding purposes (Xue et al., 2022). A comparison of various CRISPR toolboxes is indexed in Table 3.2.

TABLE 3.2
Comparison Table of Various CRISPR Toolboxes

Feature	Cas9	Cas12a (Cpf1)	Base Editors (BEs)	Cas13
Target nucleic acid	DNA	DNA	DNA (non-cleaving/ nickase)	RNA
Primary output	Indels (knockout)	Indels (knockout)/gene replacement	Single-nucleotide substitution	RNA silencing (knockdown)
Break type	Double-strand break (DSB, blunt ends)	DSB (5′-protruding ends)	No DSB, nick	RNA cleavage
PAM requirement	G-rich (e.g., NGG)	T-rich	Varies (often NGG or NNG)	None (targets RNA)
Advantages	High efficiency, versatility, maturity	Optimized multiplexing architecture (crRNA processing), distinct PAM	Precise, predictable point mutations	Nonpermanent editing, useful for viral resistance/gene knockdown

3.3 PRECISION EDITING IN PLANT SYSTEMS: BASE EDITORS AND PRIME EDITORS

3.3.1 Base Editors: Cytosine and Adenine Base Editors

BEs can be modified to allow accurate C-T (G-A) or A-G (T-C) replacements without generating a DSB. Cytosine base editors (CBEs) complex, a cytidine deaminase (e.g., APOBEC1, PmCDA1) with a Cas9 nick to deaminate cytosines in a single-stranded piece of DNA is opened by R-loop formation; adenine base editors (ABEs) use modified versions of TadA, which inosinize adenine in DNA. The editing window (typically positions of about 4–8 with many CBEs and 4–7 with ABEs, but the choice of deaminase can change the range; A3A/Y130F has a wider window) is of great importance in editing efficiency and outcome. Bystander editing into adjacent bases in that window is still a viable constraint; solutions to shorten windows or change deaminase specificity (e.g., V106W mutations in some TadCBEs) have minimized undesired conversions in rice and tomato (Ren et al., 2021). The choice of promoters (viral or endogenous like RPS5a), choice of vectors and multiplex tRNA–gRNA architecture-significantly influence the results in *Arabidopsis*, rice and other crops (Li et al., 2022).

3.3.2 Prime Editing: pegRNA Structure, Reverse Transcriptase Fusion, Strengths and Limitations

Prime editing (PE) is a fusion of nCas9-reverse transcriptase and a prime editing guide RNA (pegRNA) that encodes the edit of interest using a primer binding site (PBS) and RT (Reverse Transcription) template. The RT introduces the edit at the pegRNA nick into the target strand; the flap is replaced by a downstream repair. PE is capable of installing all the 12 base conversions and small indels without DSBs and exogenous donors and is versatile compared to BEs. Nonetheless, plant applications tend to be characterized by low efficiency that depends on the state of chromatin and pegRNA design (PBS length, RT template) and pegRNA stability. Other methods to enhance efficiencies in rice and other species (epegRNAs, RT engineering, codon optimization, NLS (Nuclear Localization Signal) additions) have been made, but PE optimization is still more difficult in many dicots, including tomato (Jiang et al., 2022).

3.3.3 Design Advances: Evolved Fidelity Mutations, pegRNA Deaminases, Evolved Deaminases

Upgrades on both platforms will enhance activity and accuracy. Better purity is provided by evolved deaminases (e.g., TadA-8e); A3A/Y130F is better at targeting cytosines in plants (Fan et al., 2024). The off-target deamination can be mitigated by using high-fidelity Cas9 variants (SpCas9-HF1, HypaCas9) when modified to work in plants (Hassan et al., 2021). It has been demonstrated that PE can be enhanced by pegRNA stabilization (epegRNAs), truncated/optimized RTs and transient silencing of mismatch repair (dominant-negative MLH1) (Gao et al., 2022).

3.3.4 Uses and Impetus in Agriculture

In practice, BEs have been used to produce agronomically useful alleles, e.g., ALS and EPSPS mutations imparting resistance to herbicides in cereals including rice and maize, showing practical trait engineering (Butt et al., 2020; Jiang et al., 2022). PE has already generated narrow-focused replacements, deletions and minuscule insertions in rice, wheat, tomato and peanut, such as polyploid-context edits, suggesting a future optimism on trait stacking and precision breeding (Li et al., 2022; Ni et al., 2023). The two modalities support functional genomics by facilitating finer forms of allelic alteration without broad indels, which helps in regulatory tolerance.

3.4 CRISPR DIAGNOSTICS, IMAGING AND READOUTS IN PLANT SYSTEMS

The CRISPR-based technologies have developed further than the direct gene editing, to be employed as flexible platforms in diagnostics, imaging and various readout strategies in plant science. These capabilities are particularly useful in plant pathology: such a sequence-specific identification of pathogens allows making decisions in the field as quickly as possible; visualization and quantitative indicators allow the dynamics of the infection process to be tracked and the effectiveness of control measures to be evaluated. The key modalities of readout are summarized below, along with their combination with amplification and sample-preparation protocols appropriate to agricultural applications, plant-specific applications and practical implementation considerations.

3.4.1 Readout Modalities Fluorescence, Lateral Flow, Colorimetric and Electrochemical Readouts

The essence of most CRISPR diagnostics is collateral cleavage activated Cas nuclease (such as Cas12 or Cas13) cleaves reporter molecules in a nonspecific manner upon recognizing its specific target sequence. The collateral activity is couplable with a number of reporter chemistries. The most commonly used are fluorescence readouts in development and validation. Reporters based on fluorophore–quencher pairs give a fast rise in fluorescence on cleavage by Cas, allowing sensitive, usually quantitative, detection in combination with a portable fluorometer or a simple blue-light viewer (Nguyen et al., 2020). Plant RNA viruses, e.g., sorghum mosaic virus and rice stripe mosaic virus, have been detected using fluorescent readouts and have a detection time below 30 minutes (Wang et al., 2022). In the lateral flow (immunochromatographic) strips, molecular recognition is translated into an instrument-free visual band that is easy to understand like a pregnancy test. CRISPR collateral cleavage releases or cleaves a labeled reporter that migrates on the strip creating an interpretable pattern of lines. Lateral flow assays are specifically appealing to on-farm applications where simplicity and sturdiness are important factors (Nguyen et al., 2020).

There are also adaptations to colorimetric readouts, i.e., enzyme-linked or nanoparticle-based. These systems generate a visible color change on target recognition and are applicable when the fluorescence readers and lateral flow strips are not available, albeit at the expense of some sensitivity. CRISPR activity can also be translated into an electrical response by emerging electrochemical readouts; these methods can be quantitative and sensitive with minimal power usage, which is why they have been adopted in portable diagnostic systems, but are not generally used in contemporary plant diagnostics literature. The tradeoffs are sensitivity versus quantification versus cost versus field-friendliness in each modality, selection of the surveillance task (early detection, routine screening, strain discrimination) and the resources at the disposal of the user (Nguyen et al., 2020).

3.4.2 Combination with Isothermal Amplification and Low Volumes of Sample Preparation

Due to the potential low copy numbers of the pathogen nucleic acid in the plant tissue or vectors, common CRISPR diagnostics of plants often include isothermal amplification techniques like RT-RPA or RT-LAMP preceding CRISPR detection (Li et al., 2021; Mahas et al., 2021). These amplification cycles use a constant temperature and do away with thermocyclers, allowing single pot or sequential protocol that can be performed using simple heat blocks or chemical heaters. In most plant-targeted experiments, the amplification and CRISPR detection have been merged into a single tube to minimize the risk of contamination and to minimize steps taken by the user (Wang et al., 2023b). Fieldwork has also been simplified in sample preparation. The use of crude extraction buffers, short-term mechanical homogenization and little centrifugation have been demonstrated to be sufficient to downstream RT-RPA/CRISPR detection of a variety of plant viruses and bacteria, enabling the turnaround of a leaf or sap to readout less than an hour (Wang et al., 2022, 2023a). Cas enzymes have been reported to be stabilized by buffer optimization and additives to enhance signal strength in conditions that are not ideal (Wan et al., 2024).

3.4.3 In Vivo Imaging and Spatial Readings in Plants

In addition to endpoint detection, CRISPR-based reporters can be used in imaging and spatially resolved readouts of plant tissues. Fluorescent reporters activated by nucleases can be utilized either in microscopic or bench-top imaging systems to map a pathogen nucleic acid localization in infected tissue and study both infection foci and tissue tropism. Although the majority of published plant applications focus on molecular detection, a continually growing literature adapts CRISPR sensors to in situ visualization, such as fluorescence reporters that show the location of viruses in the leaf or the root, and the application of CRISPR reporters to measure the load of pathogens in the tissues of vectors (Wang et al., 2022). These imaging methods can be useful in a research scenario, whereby the space data can be used to guide epidemiology or host–pathogen interaction studies.

The above elements are useful in crop systems as can be illustrated by practical examples. Cas12a and Cas13a-based assays have quickly and sensitively sensed RNA viruses, including sorghum mosaic virus and rice stripe mosaic virus, and workflows that can be deployed in the field have paired RT-RPA amplification with lateral flow or fluorescence readouts to generate results within 30 minutes (Mahas et al., 2021). There are also applications of CRISPR tools in the detection of bacterial pathogens in plants, e.g., the speedy identification of *Xylophilus* spp. in plant matter through isothermal amplification and CRISPR-mediated readout (Wang et al., 2023a). Bar-Cas12a assays are used in seed certification and GMO (Genetically Modified Organism) screening to barcode regions and transgenic constructs (such as T-nos) to verify the identity of species and to detect transgenic constructs with a sensitivity and speed superior to electrophoretic techniques (Buddhachat et al., 2021; Wang et al., 2023b).

3.5 DELIVERY OF CRISPR-CAS SYSTEMS IN PLANTS

The effective genome editing of plants is equally dependent on the delivery method of the editing reagents, as well as on the nuclease or editing method utilized. Plants present special barriers: a rigid cell wall, species and genotype receptivity that is not constant, and regeneration that is often difficult. In this way, the editing efficiency, the probability of stable integration of a transgene, the probability of an editing experiment producing transgene-free progeny, and finally the viability of translating an editing experiment into a useful cultivar all depend on the method of delivery. An array of tools has been developed and streamlined to various crops, tissues and aims in the past decade; these tools are *Agrobacterium*-mediated transfer, PEG-mediated protoplast transfection, particle bombardment, viral vectors, RNP delivery, nanoparticle carriers, pollen-mediated systems and physical techniques, including electroporation and microinjection. These methods are outlined in Table 3.3, along with their mechanisms, uses in plants and practical limitations.

TABLE 3.3
Overview of CRISPR Delivery Methods: Mechanisms, Applications and Limitations

Delivery Method	Short Description	Mechanism/Typical Uses	Practical Limitations	References
Agrobacterium-mediated delivery	T-DNA transfer of Cas and sgRNA cassettes into the plant nuclear genome via *Agrobacterium tumefaciens*.	Binary vectors with Cas (e.g., Cas9/Cas12a) and sgRNA under plant promoters (35S, ubiquitin, actin); co-culture with explants (leaf discs, cotyledons, immature embryos). Favored for stable, low-copy insertions in many dicots; with strain optimization and morphogenic regulators used in some monocots.	Genotype dependence; multi-month tissue culture and regeneration; regulatory issues from foreign DNA in final product. Work-arounds include transient expression, excisable "suicide" vectors or combining *Agrobacterium* sgRNA delivery with DNA-free Cas sources to reduce transgene persistence.	Tang et al. (2018)
PEG-mediated protoplast transfection	PEG-$CaCl_2$ facilitates uptake of plasmid DNA, RNA transcripts or preassembled RNPs into enzymatically generated protoplasts.	Rapid transient assays to verify guide activity and on-target editing; useful for DNA-free editing with RNPs; reported successes in rice, wheat, lettuce, tobacco and *Arabidopsis*. Enables multiplexing by co-delivery of multiple RNPs.	Major bottleneck is regeneration from protoplasts—somatic embryogenesis and organogenesis are species- and genotype-dependent, limiting use to species with robust protoplast-to-plant protocols.	Woo et al. (2015)
Particle bombardment (biolistics)	High-velocity microcarrier particles (gold/ tungsten) deliver DNA or RNPs into cells by physical penetration.	Widely used in monocots (maize, rice, barley, wheat); tolerates large constructs, mixed cargo (selectable markers, donor templates), and can deliver preassembled RNPs for DNA-free edits. Useful when *Agrobacterium* is inefficient.	Can cause tissue damage, random multi-copy DNA integration, and genomic rearrangements from high-energy impacts; still broadly applicable across many crops.	Svitashev et al. (2015)

(Continued)

TABLE 3.3 (*Continued*)
Overview of CRISPR Delivery Methods: Mechanisms, Applications and Limitations

Delivery Method	Short Description	Mechanism/Typical Uses	Practical Limitations	References
Ribonucleoprotein (RNP) delivery	Direct delivery of preassembled Cas-sgRNA complexes to produce transgene-free edits.	Rapid action and transient presence reduce long-term nuclease exposure and off-target activity; delivery via PEG transfection, biolistics or electroporation; enables multiplex editing. Reported transgene-free, heritable edits in key crops (rice, wheat, maize, barley).	Cost and stability issues for purified Cas proteins; technical challenges to achieve efficient uptake into intact, regenerable tissues.	Liang et al. (2017)
Viral vector–mediated delivery (VIGE)	Use of engineered plant viruses to systemically deliver sgRNAs (and in some cases compact Cas variants) without stable integration.	Fast, tissue-culture-free functional genomics and screening; used with TRV, ALSV, FoMV, BSMV in hosts like *Nicotiana benthamiana*, *Arabidopsis*, soybean, rice. Often deliver sgRNAs to plants that already express Cas or are combined with transient/ other Cas delivery.	Cargo-size limits (full-length Cas usually too large); host range and viral stability constraints; regulatory and biosafety considerations.	Wang et al. (2025)
Nanoparticle-assisted delivery	Nanocarriers (carbon nanotubes, mesoporous silica, layered double hydroxide nanosheets, chitosan, lipid nanoparticles) ferry DNA, RNA or RNPs across cell walls/membranes.	Surface functionalization enables controlled release and targeting; demonstrated plasmid delivery to *Nicotiana* leaves and delivery of guide RNAs/RNPs with LDH nanosheets for localized editing; low tissue damage and repeatable application.	Challenges in tissue compatibility, scale-up to whole-plant/field level, and unknown environmental/ biosafety impacts of nanoparticles in agriculture.	Zhi et al. (2022)
Pollen-mediated/ pollen tube pathway	Introduction of DNA, RNPs or magnetic nanoparticles into pollen grains or via pollen tube to reach the embryo sac during fertilization.	Potential to produce edited seeds in a single generation without tissue culture, promising for high-throughput, non-tissue-culture editing.	Species-dependent efficiency; methods need further optimization before routine, large-scale use.	Zhao et al. (2017)
Electroporation and microinjection	Temporary permeabilization (electroporation) or direct injection of reagents into cells/ zygotes.	Electroporation has delivered RNPs into wheat and maize protoplasts; microinjection into single cells or zygotes demonstrated in rice and tobacco for specialized projects.	Electroporation typically limited to isolated cells/ protoplasts; microinjection is low-throughput, technically demanding, suitable only for specialized, small-scale experiments.	Subburaj and Agapito-Tenfen (2023)

3.6 FUTURE AND CONCLUSION

The future of plant genome editing can be seen in the incessant improvement of the tools and their conversion into viable agricultural products to sustain world food security, sustainability and climate resistance. This progress is still being propelled by technical innovation. Recent discoveries include high-fidelity Cas variants, base and prime editors, and RNA-guided transposases (CASTs: CRISPR-Associated Transposases), and are increasing the precision and breadth of genome manipulation and minimizing unwanted genomic changes (Ryu et al., 2023). With the development of the multiplex editing systems, it is now possible to modify multiple loci at once, which allows the engineering of complex phenotypes, including stress resistance, yielding growth and nutritional remediation (Vats et al., 2019). Likewise, entirely in vivo PE and CAST systems offer to insert DNA or make corrections in DNA and do not produce DSBs that lead to genotoxicity and aggravate genomic stability. Combinatory strategies that unite the CRISPR technology with epigenome editing, RNA interference, and nanotechnology-based delivery further add to this technological ecosystem and offer reversible and precise control over gene expression (Alok et al., 2021).

Scalable, efficient and species-independent delivery strategies, such as viral replicons, nanoparticles and DNA-free transient systems, are the new elements of translational development (Ahmar et al., 2021). These platforms should enable multiplex and accurate editing and at the same time be biosafe and field-compatible. Safety data must be considered over the long term to determine the stability of genome, the integrity of the ecosystem and the confidence of regulation (Peterson et al., 2016). Mass field tests will also be essential to validate laboratory results, test agronomic performance and will guarantee the consistency of the performance of edited crops to work under varying environmental conditions (Razzaq et al., 2021).

In the future, overcoming the chronic issues that include immune reactivity, rare side effects and disparate access to genome editing technology would be the main issue in achieving their potential. Genome editing must be disseminated fairly, controlled openly and involve smallholder farmers in a manner that is inclusive of developing regions to make the application of the technology assist more in the overall practice of sustainable agriculture (Pixley et al., 2019).

3.7 CONCLUSION

The CRISPR toolbox has rapidly evolved from a simple gene-editing system into a versatile and expanding platform that is redefining the boundaries of genome engineering. Advances such as base editing, PE, epigenome editing, RNA-targeting CRISPR systems and refined Cas variants have significantly improved precision, efficiency and target range while minimizing off-target effects. Together, these innovations enable not only targeted mutagenesis but also programmable regulation of gene expression, chromatin state and cellular function without permanent DNA alteration. As delivery methods, computational design tools and regulatory frameworks continue to mature, the CRISPR toolbox is poised to accelerate breakthroughs across agriculture, biotechnology and medicine. Ultimately, the expanding universe of CRISPR technologies offers unprecedented opportunities to understand, redesign and responsibly harness biological systems for sustainable and transformative applications.

REFERENCES

Ahmar, S., Mahmood, T., Fiaz, S., Mora-Poblete, F., Shafique, M. S., Chattha, M. S., & Jung, K. H. (2021). Advantage of nanotechnology-based genome editing system and its application in crop improvement. *Frontiers in Plant Science,* 12, 663849.

Alok, A., Chauhan, H., Upadhyay, S. K., Pandey, A., Kumar, J., & Singh, K. (2021). Compendium of plant-specific CRISPR vectors and their technical advantages. *Life,* 11(10), 1021.

Buddhachat, K., Paenkaew, S., Sripairoj, N., Gupta, Y. M., Pradit, W., & Chomdej, S. (2021). Bar-Cas12a, a novel and rapid method for plant species authentication: A case of *Phyllanthus* species. *Scientific Reports,* 11, 20888.

Butt, H., Rao, G. S., Sedeek, K., Aman, R., Kamel, R., & Mahfouz, M. (2020). Engineering herbicide resistance via prime editing in rice. *Plant Biotechnology Journal*, 18(12), 2370.

Cloney, R. (2016). Patent law and genome engineering: A short guide to a rapidly changing landscape. *Molecular Therapy,* 24(3), 419–421. https://doi.org/10.1038/mt.2016.31

Crawley, A. B., Henriksen, J. R., & Barrangou, R. (2018). CRISPRdisco: An automated pipeline for the discovery and analysis of CRISPR-Cas systems. *The CRISPR Journal,* 1(2), 171–181. https://doi.org/10.1089/CRISPR.2017.0022

Damian, M., & Porteus, M. H. (2013). A crisper look at genome editing: RNA-guided genome modification. *Molecular Therapy,* 21(4), 720–722.

Fan, T., Cheng, Y., Wu, Y., Liu, S., Tang, X., He, Y., … Zhang, Y. (2024). High performance TadA-8e derived cytosine and dual base editors with undetectable off-target effects in plants. *Nature Communications,* 15(1), 5103.

Gao, Z., Ravendran, S., Mikkelsen, N. S., Haldrup, J., Cai, H., Ding, X., … Bak, R. O. (2022). A truncated reverse transcriptase enhances prime editing by split AAV vectors. *Molecular Therapy,* 30(9), 2942–2951.

Gupta, R., Ghosh, A., Chakravarti, R., Singh, R., Ravichandiran, V., Swarnakar, S., & Ghosh, D. (2022). Cas13d: A new molecular scissor for transcriptome engineering. *Frontiers in Cell and Developmental Biology,* 10, 866800.

Hassan, M. M., Zhang, Y., Yuan, G., De, K., Chen, J. G., Muchero, W., ... & Yang, X. (2021). Construct design for CRISPR/Cas-based genome editing in plants. *Trends in Plant Science*, *26*(11), 1133–1152.

Hu, K. D., Zhang, X. Y., Yao, G. F., Rong, Y. L., Ding, C., Tang, J., Yang, F., Huang, Z. Q., Xu, Z. M., Chen, X. Y., Li, Y. H., Hu, L. Y., & Zhang, H. (2020). A nuclear-localized cysteine desulfhydrase plays a role in fruit ripening in tomato. *Horticulture Research, 7*(1), Article 211.

Jiang, Y., Chai, Y., Qiao, D., Wang, J., Xin, C., Sun, W., … Chen, Q. J. (2022). Optimized prime editing efficiently generates glyphosate-resistant rice plants carrying homozygous TAP-IVS mutation in EPSPS. *Molecular Plant,* 15(11), 1646–1649.

Karimi, M., Ghorbani, A., Niazi, A., Rostami, M., & Tahmasebi, A. (2025). CRISPR-Cas13a as a next-generation tool for rapid and precise plant RNA virus diagnostics. *Plant Methods,* 21, Article 83. https://doi.org/10.1186/s13007-025-01401-9

Li, H., Zhu, Z., Li, S., Li, J., Yan, L., Zhang, C., … Xia, L. (2022). Multiplex precision gene editing by a surrogate prime editor in rice. *Molecular Plant,* 15(7), 1077–1080.

Li, T., Yang, Y., Qi, H., Cui, W., Zhang, L., Fu, X., He, X., Liu, M., Li, P., & Yu, T. (2023). CRISPR/Cas9 therapeutics: Progress and prospects. *Signal Transduction and Targeted Therapy.* https://doi.org/10.1038/s41392-023-01309-7

Li, Z., Zhao, W., Ma, S., Li, Z., Yao, Y., & Fei, T. (2021). A chemical-enhanced system for CRISPR-based nucleic acid detection. *Biosensors and Bioelectronics,* 192, 113493.

Liang, Z., Chen, K. L., Li, T. D., Zhang, Y., Wang, Y. P., Zhao, Q., Liu, J. X., Zhang, H. W., Liu, C. M., Ran, Y. D., & Gao, C. X. (2017). Efficient DNA-free genome editing of bread wheat using CRISPR/Cas9 ribonucleoprotein complexes. *Nature Communications,* 8, Article 14261.

Liu, T. L., Zeng, D. C., Zheng, Z. Y., Lin, Z. S., Xue, Y., Li, T., Xie, X. R., Ma, G. L., Liu, Y. G., & Zhu, Q. L. (2021). The ScCas9++ variant expands the CRISPR toolbox for genome editing in plants. *Journal of Integrative Plant Biology,* 63(9), 1611–1619. https://doi.org/10.1111/jipb.13164

Ma, X. L., Zhang, Q. Y., Zhu, Q. L., Liu, W., Chen, Y., Qiu, R., Wang, B., Yang, Z. F., Li, H. Y., Lin, Y. R., Xie, Y. Y., Shen, R. X., Chen, S. F., Wang, Z., Chen, Y. L., Guo, J. X., Chen, L. T., Zhao, X. C., Dong, Z. C., & Liu, Y. G. (2015). A robust CRISPR/Cas9 system for convenient, high-efficiency multiplex genome editing in monocot and dicot plants. *Molecular Plant,* 8(8), 1274–1284. https://doi.org/10.1016/j.molp.2015.04.007

Mahas, A., Hassan, N., Aman, R., Marsic, T., Wang, Q., Ali, Z., & Mahfouz, M. M. (2021). LAMP-coupled CRISPR–Cas12a module for rapid and sensitive detection of plant DNA viruses. *Viruses,* 13(3), 466. https://doi.org/10.3390/v13030466

Malzahn, A. A., Tang, X., Lee, K., Ren, Q., Sretenovic, S., Zhang, Y., Chen, H., Kang, M., Bao, Y., Zheng, X., Deng, K., Salcedo, V., Wang, K., Zhang, Y., & Qi, Y. (2019). Application of CRISPR-Cas12a temperature sensitivity for improved genome editing in rice, maize, and Arabidopsis. *BMC Biology,* 17, Article 9. https://doi.org/10.1186/s12915-019-0629-5

Naeem, M., Majeed, S., Hoque, M. Z., & Ahmad, I. (2020). Latest developed strategies to minimize the off-target effects in CRISPR-Cas-mediated genome editing. *Cells, 9*(7), Article 1608. https://doi.org/10.3390/cells9071608

Nguyen, L. T., Smith, B. M., & Jain, P. K. (2020). Enhancement of trans-cleavage activity of Cas12a with engineered crRNA enables amplified nucleic acid detection. *Nature Communications,* 11(1), 4906.

Ni, P., Zhao, Y., Zhou, X., Liu, Z., Huang, Z., Ni, Z., … Zong, Y. (2023). Efficient and versatile multiplex prime editing in hexaploid wheat. *Genome Biology,* 24(1), 156.

Nie, X., Wang, D., Pan, Y., Hua, Y., Li, P., & Yang, Y. (2023). Discovery, classification and application of the CRISPR-Cas13 system. *Technology and Health Care.* https://doi.org/10.3233/THC-230258

Osborn, M. J., Starker, C. G., McElroy, A. N., Webber, B. R., Riddle, M. J., Xia, L., … Tolar, J. (2013). TALEN-based gene correction for epidermolysis bullosa. *Molecular Therapy,* 21(6), 1151–1159.

Paul, B., & Montoya, G. (2020). CRISPR-Cas12a: Functional overview and applications. *Biochimica et Biophysica Acta (BBA) – Gene Regulatory Mechanisms.* https://doi.org/10.1016/j.bbagrm.2019.194449

Peterson, B. A., Haak, D. C., Nishimura, M. T., Teixeira, P. J., James, S. R., Dangl, J. L., & Nimchuk, Z. L. (2016). Genome-wide assessment of efficiency and specificity in CRISPR/Cas9 mediated multiple site targeting in Arabidopsis. *PloS One,* 11(9), e0162169.

Pixley, K. V., Falck-Zepeda, J. B., Giller, K. E., Glenna, L. L., Gould, F., Mallory-Smith, C. A., … Stewart Jr, C. N. (2019). Genome editing, gene drives, and synthetic biology: Will they contribute to disease-resistant crops, and who will benefit?. *Annual Review of Phytopathology,* 57(1), 165–188.

Razzaq, A., Kaur, P., Akhter, N., Wani, S. H., & Saleem, F. (2021). Next-generation breeding strategies for climate-ready crops. *Frontiers in Plant Science,* 12, 620420.

Ren, Q., Sretenovic, S., Liu, G., Zhong, Z., Wang, J., Huang, L., … Zhang, Y. (2021). Improved plant cytosine base editors with high editing activity, purity, and specificity. *Plant Biotechnology Journal*, 19(10), 2052–2068.

Roy, A. M., & Pokale, R. (2025). Synergizing CRISPR-Cas9 with advanced artificial intelligence and machine learning for precision drug delivery: Technological nexus and regulatory insights. *Current Gene Therapy,* 25(4), 467–496. https://doi.org/10.2174/0115665232342293241120033251

Ryu, J., Adashi, E. Y., & Hennebold, J. D. (2023). The history, use, and challenges of therapeutic somatic cell and germline gene editing. *Fertility and Sterility,* 120(3), 528–538.

Subburaj, S., & Agapito-Tenfen, S. Z. (2023). Establishment of targeted mutagenesis in soybean protoplasts using CRISPR/Cas9 RNP delivery via electro– transfection. *Frontiers in Plant Science,* 14, 1255819.

Svitashev, S., Young, J. K., Schwartz, C., Gao, H. R., Falco, S. C., & Cigan, A. M. (2015). Targeted mutagenesis, precise gene editing, and site-specific gene insertion in maize using Cas9 and guide RNA. *Plant Physiology,* 169(2), 931–945. https://doi.org/10.1104/pp.15.00793

Tang, X., Liu, G. Q., Zhou, J. P., Ren, Q. R., You, Q., Tian, L., Xin, X. H., Zhong, Z. H., Liu, B. L., Zheng, X. L., Zhang, D. W., Malzahn, A., Gong, Z. Y., Qi, Y. P., Zhang, T., & Zhang, Y. (2018). A large-scale whole-genome sequencing analysis reveals highly specific genome editing by both Cas9 and Cpf1 (Cas12a) nucleases in rice. *Genome Biology,* 19, Article 84. https://doi.org/10.1186/s13059-018-1458-5

Tyumentseva, M., Tyumentsev, A., & Akimkin, V. (2023). CRISPR/Cas9 landscape: Current state and future perspectives. *International Journal of Molecular Sciences,* 24(22), 16077.

Vats, S., Kumawat, S., Kumar, V., Patil, G. B., Joshi, T., Sonah, H., … Deshmukh, R. (2019). Genome editing in plants: Exploration of technological advancements and challenges. *Cells,* 8(11), 1386.

Wada, N., Osakabe, K., & Osakabe, Y. (2022). Expanding the plant genome editing toolbox with recently developed CRISPR–Cas systems. *Plant Physiology,* 188(4), 1825–1837.

Wan, Y., Li, S., Xu, W., Wang, K., Guo, W., Yang, C., … Wang, J. (2024). Terminal chemical modifications of crRNAs enable improvement in the performance of CRISPR-Cas for point-of-care nucleic acid detection. *Analytical Chemistry,* 96(41), 16346–16354.

Wang, C. L., Zhou, Y. Y., Wang, Y. M., Jiao, P., Liu, S. Y., Guan, S. Y., & Ma, Y. Y. (2025). CRISPR-Cas9-mediated editing of ZmPL1 gene improves tolerance to drought stress in maize. *GM Crops & Food,* 16(1), 1–16. https://doi.org/10.1080/21645698.2024.2448869

Wang, H., La Russa, M., & Qi, L. S. (2016). CRISPR/Cas9 in genome editing and beyond. *Annual Review of Biochemistry,* 85(1), 227–264.

Wang, J., Huang, X., Chen, S., Chen, J., Liang, Z., Chen, B., … Zhang, T. (2023a). On-site and visual detection of sorghum mosaic virus and rice stripe mosaic virus based on reverse transcription-recombinase-aided amplification and CRISPR/Cas12a. *Frontiers in Genome Editing,* 5, 1124794.

Wang, X., Wang, L. F., Cao, Y. F., Yuan, Y. Z., Hu, J., Chen, Z. H., … Wang, X. Z. (2022). *Bursaphelenchus xylophilus* detection and analysis system based on CRISPR–Cas12. *Frontiers in Plant Science,* 13, 1075838.

Wang, Y., Peng, C., Ding, L., Su, Z., Chen, X., Wang, X., … Xu, J. (2023b). An accurate, rapid and cost-effective method for T-nos detection based on CRISPR/Cas12a. *Foods,* 12(3), 615.

Woo, J. W., Kim, J., Kwon, S. I., Corvalán, C., Cho, S. W., Kim, H., … Kim, J. S. (2015). DNA-free genome editing in plants with preassembled CRISPR-Cas9 ribonucleoproteins. *Nature Biotechnology,* 33(11), 1162–1164.

Xie, F., Chang, J. C., & Ye, L. (2014). Seamless gene correction of β-thalassemia mutations in patient-specific iPSCs using CRISPR/Cas9 and piggyBac. *Genome Research,* 24(8), 1526–1533.

Xue, Y., Chen, Z., Zhang, W., & Zhang, J. (2022). Engineering CRISPR/Cas13 system against RNA viruses: From diagnostics to therapeutics. *Bioengineering,* 9(7), Article 290.

Zhao, X., Meng, Z., Wang, Y., Chen, W., Sun, C., Cui, B., … Cui, H. (2017). Pollen magnetofection for genetic modification with magnetic nanoparticles as gene carriers. *Nature Plants,* 3(12), 956–964.

Zhi, H., Zhou, S., Pan, W., Shang, Y., Zeng, Z., & Zhang, H. (2022). The promising nanovectors for gene delivery in plant genome engineering. *International Journal of Molecular Sciences,* 23(15), 8501. https://doi.org/10.3390/ijms23158501

4 Next-Generation Functional Genomics

Methodologies for CRISPR-Based Screening

K. Keerthana and Asha Velayudhan Nair

4.1 INTRODUCTION

The inception of CRISPR (clustered regularly interspaced short palindromic repeats) screening and editing tool marked the initiation of a new era of functional genomics. It all started in 1987 when Japanese scientist Yoshizumi Ishino found strange repetitive sequences in *Escherichia coli* DNA, whose function was unknown at the time (Ishino et al. 2018). Almost for a two-decade span, these sequences, later named CRISPR, lingered as an intriguing but unexplained biological phenomenon until Francisco Mojica's research in 2005 revealed something remarkable: these same sequences appeared consistently across vastly different bacterial species, suggesting they served an important evolutionary purpose (Mojica et al. 2005). In 2007, the French scientists, Rodolphe Barrangou and Philippe Horvath, made a groundbreaking discovery while studying the bacteria used in yogurt production (Horvath & Barrangou 2010), where CRISPR functioned as a sophisticated bacterial immune system, essentially with a molecular memory that allowed bacteria to recognize and combat viral threats. The next turning point arrived in 2012 when researchers like Jennifer Doudna and Emmanuelle Charpentier successfully adapted this bacterial defence system into a precise editing tool for human genes (Jinek et al. 2012). This study focused on the CRISPR-Cas9 redesigning to target and alter the chosen DNA sequences in living cells. Further, in 2020, the Nobel Prize in Chemistry was given to Doudna and Charpentier for their successful development of the genome editing tool: CRISPR-Cas9.

CRISPR is all over the life sciences research map right now, with its uses running the spectrum from omics studies, through gene expression tweaks, disease modelling, cancer studies, developing strategies for tailor-made medicine, etc. The focus of CRISPR research is on fixing single-gene inherited disorders, coming up with new cancer treatments and immunotherapies, and characterizing disease mechanics at a whole lot faster pace than was thought possible before. And it's making strides in developing medicinal protocols to treat genetic disorders and cancer, and perform high-throughput genomics at an incredible rate. Meanwhile, it's also being used to develop diagnostic tools for plant diseases and pests in the agricultural sector to keep crops healthy. CRISPR has also opened up new avenues to fight viral infections like AIDS, hepatitis B, human papillomavirus, etc., which could significantly improve the public health sector (Singh et al. 2024b).

4.2 CRISPR AND FUNCTIONAL GENOMICS

With the CRISPR-Cas system in the forefront, the last 25 years have shown a tremendous growth in the domain of functional genomics. The new methodologies and applications of CRISPR have become the support system of all omics-based studies. The gene targeting and editing capabilities of CRISPR have aided in the development of treatments for different congenital disorders and genetic

DOI: 10.1201/9781003735113-4

abnormalities. The CRISPR-Cas system has two necessary components: a Cas nuclease enzyme and a guide RNA (gRNA). The two primary methods for screening with CRISPR technology are positive and negative selection. Positive selection screens pick out the good ones – cells with useful mutations, like drug resistance, because their single-guide RNAs (sgRNA) levels are enhanced. On the other hand, negative selection screens look for the critical genes that the cell needs to survive and grow, by identifying the loss of function of these genes (cell cycle regulators and transcription factors) through diminished sgRNA representation (Sage et al. 2019).

Researchers from the Department of Medicine (University of Florida, USA) have developed the MEDS system – multiple expression and dual selection, which works with CRISPR to make precise and effective single-base genome edits. This method utilizes two enzymes named NPT-II (neomycin phosphotransferase-II) and CD/UPRT (cytosine deaminase/uracil phosphoribosyltransferase) – with opposing functions (one for positive selection and the other for negative) to provide a finer result (Sharma et al. 2025). CRISPR is re-moulding the genetic studies in complex systems and disease models, enabling the researchers to pace the studies from identifying the role of a gene to establishing its function with respect to health and diseases. New dimensions of genetic regulation and cellular pathways are detected with every single cell line or animal model edited, helping us bridge the gap between the lab-based molecular characterizations and clinical studies., is enabling the global pool of researchers to address the most relevant health questions.

4.3 CRISPR-BASED SCREENING METHODOLOGIES

The four steps in the CRISPR screening are library building, delivery, selection, and analysis. For delivery into cells expressing Cas9, a library of sgRNAs encompassing either the complete genome or a particular group of genes is first created and cloned into a lentiviral vector. To guarantee one sgRNA per cell, cells are transduced at a low multiplicity of infection. The cell population is then subjected to a specific selection pressure, like chemical treatment or stress, which makes it possible to identify the genes that affect behaviour or survival in that situation. Ultimately, the genes causing the observed phenotype are identified by using next-generation sequencing (NGS) to monitor variations in sgRNA abundance across experimental and control groups. CRISPR-based screening methodologies are indexed in Figure 4.1.

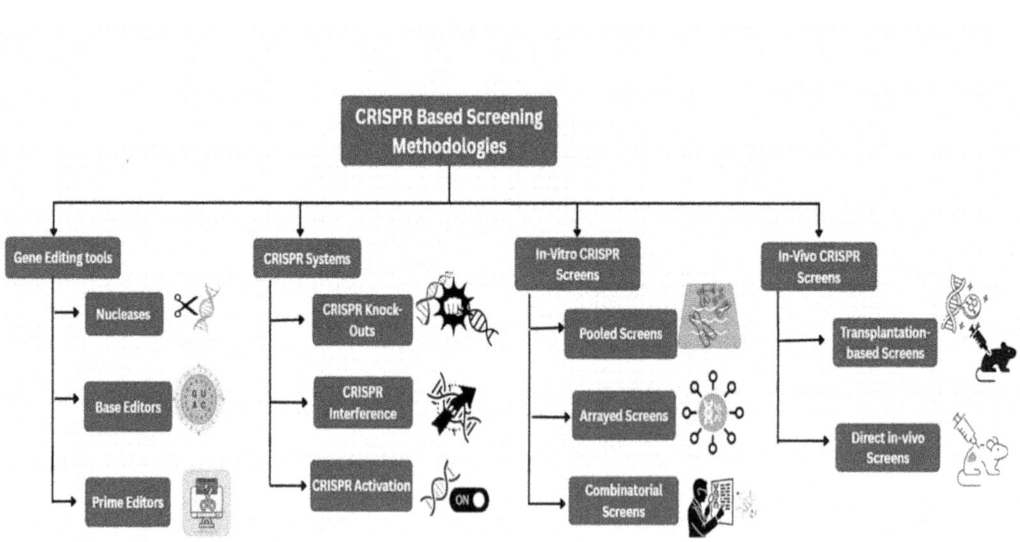

FIGURE 4.1 CRISPR-based screening methodologies – comprising gene editing tools, CRISPR systems, in vitro CRISPR screens, and in vivo CRISPR screens.

4.3.1 Gene Editing Tools

In human mutation databases, point mutations are the most prevalent ones, making up over half (50%) of all mutations (Kim et al. 2024). It involves the alteration of one base for another or adding or removing a single base, which results in either no change or total loss of gene function. They occur during cell division or due to exposure to mutagens. Their occurrence is inevitable, but their effect can be alleviated using gene editing tools, although certain gene editing approaches are hazardous, as in off-target effects like activation of oncogenes, silencing of regulatory genes, and chromosomal rearrangements leading to chromothripsis (Anderson et al. 2018). The three classes of gene editing tools discovered are nucleases, base editors, and prime editors (PEs).

4.3.1.1 Nucleases

Nucleases are the hydrolases that cut the nucleic acids by hydrolysing the phosphodiester bonds. Particularly, among the nucleases, endonucleases are used in genetic modifications owing to their internal incision abilities and specificity. Zinc finger nucleases (ZFNs), transcription activator-like effector nucleases (TALENs), mega-nucleases, and CRISPR-associated nucleases (Cas) are the four types of nucleases. Among these, CRISPR-Cas nucleases possess the ability to multi-target when tagged with multiple RNA sequences to edit more than one target gene site. Also, CRISPR-Cas is cost-effective, simple, and reprogrammable, unlike ZFN and TALEN (Wang et al. 2016). The two-component system of CRISPR-associated nucleases includes the Cas9 nuclease protein and sgRNA. The CRISPR-Cas regulates its target specificity through two factors – the protospacer adjacent motif (PAM) in the gene of interest and the complementary spacer sequence of the gRNA (Anderson et al. 2018). The PAM is defined as a short, conserved nucleotide sequence that is located adjacent to the target sequence, serving as the recognition signal for the identification of the target gene. The spacer sequence of gRNA binds complementarily to the target DNA, without which the Cas protein fails to recognise the target for cleavage. Changing the spacer sequence in gRNAs alters the target site of CRISPR-Cas, enabling versatility for functional genomic studies.

Cas nucleases function by recognizing the target and performing the molecular incision activity via a double-strand break (DSB), which is then repaired using homology-directed repair (HDR) and/or non-homologous end joining (NHEJ). In the HDR pathway, cells repair the DSB accurately using a DNA template or a single-stranded oligo-deoxy-nucleotide sequence as a reference strand. HDR applies only during the dividing phases: synthesis (S) and growth-2 (G2) phase. In the NHEJ pathway, there are small insertions and deletions (indels) made by the cell repair mechanism at the cut site, resulting in frameshift mutations. Contrary to HDR, NHEJ acts through all the stages of cell division and growth. HDR is used for the insertion of a desired sequence, while NHEJ is preferred for the knockout of a gene. Cas nucleases are currently used in infectious viral therapies such as coronavirus (Najafi et al. 2022) and cancer treatments (Far et al. 2025).

4.3.1.2 Base Editors

These are gene editing tools that function in both dividing and nondividing cells via the HDR mechanism. Base Editors (BE are developed by the integration of nickase Cas9 protein or catalytically dead Cas9 (dCas9) with the deaminase enzyme. The Cas protein, along with the gRNA, recognizes the target site and hybridizes to form a single-stranded DNA R-loop. Using BEs, base conversions occur without a DSB in the DNA strands through pyrimidine-to-pyrimidine or purine-to-purine conversions. The deaminase enzyme performs the base edits following Chargaff's rule. Based on the deaminases, BEs are classified as cytosine base editor (CBE), adenine base editor (ABE), and dual base editor (DBE). The CBE substitutes cytosine:guanine (C:G) to thymine:adenine (T:A), the ABE substitutes adenine:thymine (A:T) to guanine:cytosine (G:C) (Gaudelli et al. 2018), and the ACEs simultaneously edit both adenine and cytosine nucleotide bases (Cross 2020). For reducing off-target effects in base editors, improvements are introduced by integrating the deaminase domain within the Cas9 structure and by introducing mutations in the deaminase domain (Liu et al. 2020).

BEs can introduce disease-suppressive mutations in human cells effectively with fewer off-target effects than the Cas nucleases. Additionally, they generate a fewer indels (<0.1%) with higher product purity (~99.9%) (Gaudelli et al. 2018). BEs are also equipped to introduce stop codons (Billon et al. 2017) or to alter the RNA splicing sites (Yuan et al. 2018), thus encouraging reliable genome editing.

4.3.1.3 Prime Editors

PEs are an improved form of the "find and substitute" gene editing method, extensively used to edit genes without forming a DSB with precision (Chen & Liu, 2023). PEs are constituted by an impaired dCas9 endonuclease and a reverse transcriptase, which are guided via a prime editing-guide RNA (pegRNA) to the point of excision. The dCas9 in PE is a nickase enzyme variant that targets the DNA and produces a single-stranded nick instead of a DSB. The reverse transcriptase then inserts the newly synthesized DNA edit onto the target strand. The pegRNA has the ability to recognize PAM along with the property to guide Cas9 and to encode a new DNA sequence. The pegRNA consists of three parts: guide sequence, template sequence, and primer sequence. The guide sequence aims the dCas9 to the target site; the template sequence modifies the gene; and the primer sequence binds the primer to the target DNA before incision by the nickase to produce a DNA overhang.

The incorporation of the modified sequence into the site of DNA overhang is then carried out by the cell's natural repair mechanism. PEs function based on HDR, but with much lower off-target owing to its features like stabilization of pegRNAs, reduction in the cell's mismatch repairs, and modification of the reverse transcriptase domain. The versatile editing features enable PEs to perform all 12 base conversions (including both transitions and transversions), unlike the BEs that can only do C-to-T or A-to-G conversions. PEs also exhibit high specificity and accuracy due to the dual precise hybridization from Cas9 binding and reverse transcription fusion. Prime editing is used to introduce epitopes and tags into the desired loci and to treat sickle cell anaemia and Tay–Sachs disease (Anzalone et al. 2019).

4.3.2 CRISPR Systems

Approximately 40% of bacterial species and 90% of archaeal species possess endogenous CRISPR-Cas systems (Ghavami & Pandi 2021). Through CRISPR-Cas, bacteria acquired immunity to the viral "spacer" DNAs, which were further transcribed into CRISPR RNAs (crRNAs). These crRNAs guided the Cas nucleases to specifically recognize and cleave the invading viral DNA, thus protecting the bacterial genetic integrity. This motivated researchers to implement it on human cells for genome editing (Wang et al. 2024). They used a Cas nuclease guided by RNA to target specific DNA sequences for editing through gene knockouts, activation, or repression pathways. Based on the approach to genetic modifications, CRISPR systems are broadly classified into CRISPR knockout (CRISPR KO), CRISPR interference (CRISPRi), and CRISPR activation (CRISPRa). These systems offer scalable, accurate, and versatile methods for understanding gene function and developing therapies.

4.3.2.1 CRISPR Knockout

4.3.2.1.1 Purpose

Among the CRISPR systems' approaches, gene knockout is the most common strategy. CRISPR KO focuses on studying the gene annotations and functions by reversibly or irreversibly disrupting the gene function. This programmable disruption of gene function enables systematic loss of function to investigate gene essentiality, dissect signalling pathways, and to observe disease-associated alleles in model systems. It works by permanently correcting deleterious base mutations or disrupting problematic genes with remarkable accuracy and efficiency. CRISPR KOs sometimes result in discrepancies such as gene redundancies, post-transcriptional gene silencing, and genetic compensations (Wang et al. 2024).

4.3.2.1.2 Structure

CRISPR gene knockout employs a range of Cas nucleases. Cas9 is the most widely used nuclease. There are two major components of the knockout CRISPR Cas9 system – the Cas9 nuclease and an sgRNA. The Cas9 nuclease is also referred to as spCas. The Cas9 nuclease can make DSBs and can work as a multi-modular enzyme. It is the protein isolated from the bacteria *Streptococcus pyogenes.* Cas9 consists of two lobes, the nuclease lobe (NUC) and recognition lobe (REC). The NUC lobe acts as a molecular scissor to cleave Double stranded DNA (dsDNA), whereas REC lobe and sgRNA recognize and bind to the target DNA, forming the RNA–DNA complex. The NUC lobe hosts three main domains that include a protected NHN domain, a PAM-interacting (PI) domain, and a split RuvC domain. The NUC lobe domains have the following functions: NHN domain cuts the complementary strand; PI domain is responsible for PAM recognition and target validation before NUC cleavage of the strand, and RuvC domain cuts the template DNA strand, creating a DSB in the target gene sequence. The REC section has three parts – REC1, REC2, and REC3; with their collective action, the nuclease domain gets activated only in the presence of the actual target, which helps to reduce off-target effects. The sgRNA portion of CRISPR is not a natural genome. It is a man-made RNA sequence that consists of about 20 nucleotides. The sgRNA is made of a trans-activating CRISPR RNA (tracrRNA) and a crRNA. A crRNA is a short sequence that is complementary to a target DNA, whereas the tracrRNA is the scaffold that binds the crRNA and Cas9 together.

4.3.2.1.3 Mechanism

The CRISPR-Cas9 components are delivered into the cultured cells via transduction using viral particles or via plasmid transfection. Plasmid DNA is commonly used as a vector for delivering the CRISPR system because it is stable, can be produced in large amounts, and is easy to modify. Once inside the cell, the plasmid carrying the CRISPR-Cas9 components enters the nucleus with the help of nuclear localization signals, where, through transcription, it produces the mRNA for Cas9 protein and the sgRNA, which together enable targeted gene editing. Viral transduction brings about permanent expression of the Cas, while transfection by plasmids results in prolonged Cas nuclease activity, often leading to off-target events (Van Campenhout et al. 2019).

CRISPR KO mechanism involves Cas9–gRNA complex assembly and RNA–DNA hybrid formation. Cas9 attaches to sgRNA that allows the "spacer" region of 20 nt length to precisely bind to the target DNA. When Cas9 and a gRNA target a particular DNA sequence, a DSB occurs close to a PAM sequence. To fix these breaks, the cell uses NHEJ, which frequently results in frameshift mutations, early stop codons, or indels that permanently impair gene function. It operates by removing a part of the genomic sequence (gene deletion), making the gene inoperative by introducing "indels" or introducing malfunctioning mutations.

KO is not a completely efficient gene turn-off system, as there are residual proteins formed during the gene knockouts that rescue the cells from total loss of gene function. This results in pseudo-phenotypes and interrupted gene function investigations, which is termed as "knockout escaping" (Wang et al. 2024). To avoid the interference of the residual proteins in CRISPR KOs, two DNA breaks are simultaneously introduced to cut off the gene permanently from the genome (Wang et al. 2024).

However, this could not stop translational re-initiation or alternative splicing due to the presence of persistent exons; hence, physical removal of these exons is required through lengthy genomic deletions (Dolan et al. 2019). CRISPR KO efficacy can be enhanced by optimized sgRNA expression strategies using multiple RNA polymerase III promoters or Cpf1 crRNA arrays, by enabling multiplex targeting, primarily for combinatorial interaction mapping and noncoding deletions. The knockout efficiency can also be improved by single-gene disruption through concentrated Cas activity (Fleiss et al. 2019). Further, to optimize multiplex CRISPR KOs and to validate and reanalyse the existing data with comparative screening, sgRNA selection, and dual-expression algorithms are developed (Morchikh et al. 2017).

4.3.2.1.4 Applications

In functional genomics, CRISPR KO has helped in unveiling gene function by gene deletion, gene ablation, or creation of a null allele. The KO approach has helped to reveal the role of TGFβ receptor 2 (Tgfbr2) gene in lung tumours (Dhainaut et al. 2022). Knocking out the PCSK9 gene in liver cells (Lebeau et al. 2019) has led to reduced levels of Low density cholesterol (LDL) cholesterol, which has guided the development of cholesterol-lowering therapies and drugs aimed at preventing cardiovascular diseases in the field of drug development. In agricultural biotechnology, CRISPR KO has enabled knockouts in rice genes, enhancing flood tolerance and disease resistance. Through CRISPR KO of genes like Slc2a4 in neural stem cells uncovered mechanisms of ageing-related cognitive decline in neuroscience (Ruetz et al. 2024). In immunotherapies, CRISPR KO of immune checkpoint genes, for example, Programmed Death-11(PD-1) in T cells, has boosted anti-cancer immune responses (Zhao et al. 2017). CRISPR KO also enabled scientists to create models of diseases by turning off specific genes, such as using the BCL11A gene to develop treatments for sickle cell disease based on syndrome models and disease simulations (Javaid & Choi 2021).

4.3.2.2 CRISPR Interference

4.3.2.2.1 Purpose

CRISPRi is the gene "knockdown" (suppression) approach, which is different from the regular gene "knockout" (removal), wherein gene expression modulation is achieved for functional or epigenetic studies without permanent DNA changes. It has fine-tuned genetic mapping and gene regulatory pathways to differentiate essential and nonessential genes. It helps to study the importance of noncoding regions, including silenced promoters, enhancers, and regulatory RNAs, and their corresponding biological roles. It can potentially reverse the suppressed gene expression invariably, working to temporarily silence the expression of the gene through the gene repression mechanism in order to study its role.

4.3.2.2.2 Structure

It comprises dCas9 fused to an additional repressor domain, such as KRAB (Krüppel-associated box), Mxi1 (mammalian transcriptional repressor domain), and MeCP2 (methyl-CpG-binding protein 2). The additional epigenetic domains bring more stability and efficiency to CRISPRi's Cas9. KRAB works together with chromatin modifiers that induce hetero-chromatinization, resulting in a stronger gene silencing effect. The fusion protein dCas9-Mxi1, containing the Mxi1 repressor domain, blocks transcriptional activators (Wensing et al. 2019). When fused with MeCP2, it knocks down the specific gene activity via the enhancement of DNA methylation and histone modification (Duke et al. 2020).

4.3.2.2.3 Mechanism

The dCas9:gRNA complex binds promoter or regulatory regions, physically blocking transcription machinery and recruiting repressive chromatin modifiers. A catalytically dead Cas9 (dCas9) is used by CRISPRi, which can only bind DNA at the target gene but is not able to cut it. When a repressor domain is fused to the dCas9, that directs it towards the promoter or the early coding sequence of the gene, the RNA polymerase and transcriptional machinery are blocked, leading to the reversible suppression of gene expression without any DNA sequence modification. Further, if the dCAs9 binds in the coding region, the transcription is terminated at the elongation stage itself. In CRISPRi, the gene expression can be restored by deactivating dCas9, as there is no alteration in the genome sequence. Another modified dCas, named dCas12a (also called dCpf1), exhibits better transcriptional suppression properties with a unique capability to cut a single long RNA transcript to different sgRNAs. This helps researchers to easily target multiple genes at once for transcriptional repression without extra steps or components, making multiplex gene regulation much simpler (Karlson et al. 2021).

The CRISPRi has three systems based on the variations in the target DNA, crRNA, and the PAM. In type I CRISPR systems, the invading DNA is recognized by a multi-protein complex called

Cascade, bound to crRNA, which identifies foreign DNA sequences via PAM. This recognition recruits the Cas3 nuclease that unwinds and degrades the target DNA. Type II systems differ by requiring only the Cas9 protein for interference; Cas9 forms a complex with tracrRNA and pre-crRNA, which RNase III processes into mature crRNAs. The Cas9-tracrRNA:crRNA complex then binds to the invading DNA and introduces blunt-ended DSBs. Type III systems rely on a multi-protein complex and Cas6 to process pre-crRNA into mature forms. This complex, guided by crRNA, targets either invading DNA or RNA to induce degradation, highlighting their dual targeting capability (Richter et al. 2013).

4.3.2.2.4 Applications

In drug target discovery, CRISPRi is used to knock down genes temporarily to test their role in the various diseases. CRISPRi is successfully employed to treat retinal diseases that are dominantly inherited (Peddle et al. 2020). Due to the recoverable gene activity, CRISPRi is readily used for silencing disease-causing oncogenes and viral genes for potential therapeutic investigations. CRISPRi has the ability to silence the *inhA* gene in *Mycobacterium,* thereby increasing its susceptibility to isoniazid and rifampicin (Singpanomchai & Ratthawongjirakul 2024). It is reversible, programmable, and nonlethal, hence, suitable for knockdowns in essential genes, where knockouts are life-threatening. CRISPRi doesn't involve cleavage of DNA and is scalable; therefore, it can be applied for wide-genome screening.

4.3.2.3 CRISPR Activation

4.3.2.3.1 Purpose

CRISPRa is widely used in eukaryotes and less in prokaryotes owing to the versatile nature of bacterial promoters. In bacteria, developing CRISPRa systems to regulate multiple genes was challenging due to the lack of activation domains that can reliably activate gene expression (Fontana et al. 2020). Similar to CRISPR KO and CRISPRi, CRISPRa also uses dCas9, but fused to a transcriptional activator domain, which enhances the transcription of the target gene.

4.3.2.3.2 Structure

The core component of the CRISPRa system is the dCas9, which lacks endonuclease activity but has DNA-binding abilities. It also consists of an sgRNA that has a spacer sequence and a secondary motif (that binds to dCas9, guiding it to the target site) (Karlson et al. 2021). The machinery of CRISPRa consists of the key effector modules called transcriptional activation domains, which include proteins such as VP64, p65, or Rta. The first CRISPRa module identified was the dCas9-VP64 extracted from herpes simplex virus, which has a low to moderate efficiency level. This led to the designing of second-generation CRISPRa variants like VPR, constituted of VP64, p65, and Rta, to boost gene expression (Becirovic 2022).

In the recent CRISPRa systems, sgRNA tetraloops and stem loops are modified with RNA aptamers to recruit trans-activating proteins in addition to dCas9 fusions. The dCas9-SAM (synergistic activator mediator) system illustrates this approach, where the system binds to modified sgRNA loops through the MS2 coat protein domain and promotes transcription via p65 and heat shock transcription factor-1 (HSF1) activator domains, thus combining dCas9-VP64 with an MS2-p65-HSF1 fusion protein (Valdivieso-Rivera et al. 2023).

4.3.2.3.3 Mechanism

The dCas9 is fused to transcriptional activation domains like VP64 (Karlson et al. 2021) and p65, and is guided to the site of promoter by an extensively designed gRNA. Direct attachment or recruitment of activation domains to the dCas9 can promote interactions with RNA polymerase and different transcription factors, which can lead to the enhanced transcription initiation and hence, enhanced gene expression. Additionally, expression of several distinct sgRNAs that target multiple regulatory elements can simultaneously activate a range of genes through multiplexed transcriptional activation (Karlson et al. 2021). The activation domains can either be directly fused to dCas9 or can be brought to the target site through engineered recruitment systems like MS2, SunTag, or SAM. Among the three Cas9 activators, SAM has more potential to activate transcription at specific genomic locations (Chavez et al. 2016). In this approach, the sgRNA is modified by the addition of a small hairpin aptamer sequence to its tetra-loop and stem loop 2 regions. This aptamer specifically binds to MS2 bacteriophage coat proteins, creating an MS2-mediated sgRNA. When these MS2 proteins are fused to transcriptional activators like p65 and HSF1, the resulting SAM complex enhances the recruitment of transcription factors to the target promoters, leading to a significant increase in endogenous gene expression (Karlson et al. 2021).

4.3.2.3.4 Applications

CRISPRa enables robust gene activation in target cells for functional investigation or therapeutic protein production. CRISPRa has been used to study antiviral interferon stimulated gene expression, wherein the CRISPRa system is used to turn on specific antiviral genes and is tested for their ability to protect cells from infection by several viruses, including Dengue, Zika, and coronavirus (Kirby et al. 2024). CRISPRa is used to study the osimertinib resistance of lung cancer cells. It is also used to study the effect of drugs like venetoclax, paclitaxel, and temozolomide on double-hit large B cell lymphoma, oesophageal squamous carcinoma, and malignant glioblastoma-like brain tumours, respectively (La Marca et al. 2024). CRISPRa has improved the molecular delivery of nanoparticles into tumour cells (La Marca et al. 2024). These systems offer great flexibility in their modular design, allowing for the tailoring of different components to meet specific experimental needs.

4.4 IN VITRO CRISPR SCREENS

In vitro CRISPR screens are a transformative, high-throughput method for studying gene function in lab-cultured cells. By leveraging the CRISPR-Cas9 genome editing system, these screens allow scientists to systematically disrupt, repress, or activate genes using comprehensive gRNA libraries. This approach enables us to study how certain genetic changes impact complex cellular behaviours, such as growth, response to drugs, and the progression of diseases, thereby offering an in-depth understanding of gene function, cellular pathways, and potential therapeutic targets. CRISPR screening is simple and more direct in haploid cells due to their single set of chromosomes; however, it is also effectively used in diploid cells, which have a mix of single-copy (heterozygous) and double-copy (homozygous) gene knockouts. The in vitro CRISPR screens have transformed gene function analysis in mammalian cells owing to their scalability, precision, and flexibility, establishing them as a crucial tool in the fields of precision medicine and targeted therapeutics development (La Marca et al. 2024).

4.4.1 Pooled Screens

Pooled CRISPR screens are a high-throughput technique for studying gene function, wherein multiple genes are simultaneously targeted by a huge library of gRNA in a single cell population. Pooled screens are particularly ideal for loss-of-function or gain-of-function experiments where population-level data is sufficient, such as identifying essential genes or factors contributing to drug resistance. A recent study using pooled CRISPR screens targeted the genes involved in tRNA creation and identified valyl-tRNA synthetase as a key player in protein synthesis in T-cell acute lymphoblastic leukaemia, emphasizing aminoacyl-tRNA synthetases as cancer therapy targets (Yin et al. 2024). Unique autophagy regulators are also discovered using the pooled T. spiezzo screening library.

4.4.1.1 Mechanism

Pooled CRISPR screens employ a cell line exhibiting robust phenotypic responses to selective conditions, such as pharmacological perturbation, followed by transduction with a diverse sgRNA library and Cas9 to generate genome-wide knockouts. Next, selective pressure is applied, which enriches the fit sgRNAs and depletes the detrimental ones. Then with deep sequencing, sgRNA abundance is quantified to assess gene essentiality. Bioinformatic analysis identifies candidate genes via differential enrichment metrics (e.g., MAGeCK), followed by orthogonal validation of hits in arrayed formats (Walton et al. 2022). The generation of gene knockouts in the population occurs due to the stable expression of both Cas9 nuclease and sgRNAs in the cells. But it's very hard to find the correlation between a particular genetic change and the corresponding phenotype in these screens, as all the changes occur within a single pooled cell culture. Owing to this disadvantage, the use of pooled screens is restricted to binary assays that demonstrate cell survival or proliferation, which can distinguish between the cells that express the phenotype and the ones that do not. Hence, pooled screens are considered less effective in their role in complex phenotyping.

4.4.2 Arrayed Screens

Arrayed screens provide an error-free and targeted CRISPR screening approach through the introduction of a single gene-targeting gRNA in every single well of a multi-well plate. Therefore, these screens demonstrate the gene disruption effects of single genes that can be easily correlated to the particular phenotypes, without any complex sequencing studies.

Arrayed screens assist various assays, like flow cytometry, imaging experiments, and detailed investigations of cellular responses. One key use of arrayed transposon screens is to find Mtb mutants that cannot survive well inside macrophages, helping to understand bacterial factors needed for infections. This approach enhances the signal-to-noise ratio and improves the sensitivity of pooled screening experiments (De Wet et al. 2020). Arrayed lentiviral production prevents template switching artefacts during gRNA library pooling by enabling individual preparation of gRNAs paired with matched barcodes under RNA polymerase II promoters, ensuring reliable single-cell RNA-seq detection (De Wet et al. 2020).

4.4.2.1 Mechanism

In arrayed screens, individual sgRNAs or small groups of sgRNAs are introduced into individual wells or samples, allowing each well to contain a single genetic perturbation. Cells expressing Cas9 or a compatible CRISPR system are plated into wells containing pre-arranged gRNAs, followed by delivery of gene editing components via transfection or lentiviral transduction, to ensure high efficiency and minimal cytotoxicity. After an appropriate incubation period, phenotypic assays are performed directly in each well using automated system for high-throughput processing, with appropriate controls incorporated to maintain data quality, allowing direct correlation of observed phenotypes with targeted gene perturbations without additional sequencing (Walton et al. 2022). The in vitro CRISPR screening – pooled versus arrayed screens – is shown in Figure 4.2.

4.4.3 Combinatorial Screens

Combinatorial CRISPR screens are a cutting-edge method that lets researchers explore how multiple genes work together by targeting two or more genes in the same cells at once so that the combined effect of targeting multiple genes (or elements) can be measured simultaneously across a population. In combinatorial pooled screens, vectors encode multiple gRNAs for interaction testing, while arrayed approaches physically isolate perturbations per well. Unlike single-gRNA pooled screens, combinatorial scoring in multiplexed systems is complicated by gRNA design factors like gRNA count per construct, positional effects, and features influencing expression, processing, or capture (Najm et al. 2017).

Transcriptomic profiling of combinatorial perturbations offers enhanced resolution for dissecting intricate signalling pathways (Najm et al. 2017). Combinatorial CRISPR screening was applied to a knockout library targeting 73 genes consisting of tumour suppressor genes and oncogenic targets to

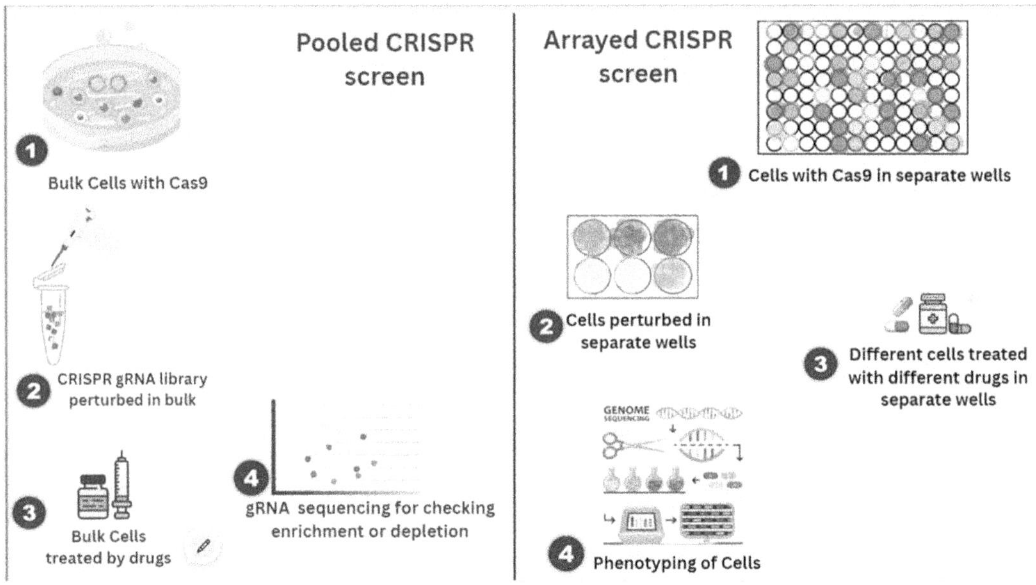

FIGURE 4.2 In vitro CRISPR screening – pooled versus arrayed screens.

comparatively study gene–gene and gene–drug interactions. A systematic approach was developed to map human genetic networks using combinatorial CRISPR-Cas9 perturbations coupled with robust growth kinetics analysis, in which pairs of 73 cancer genes were targeted with dual-guide RNAs across three cell lines, including 141,912 interaction-based tests (Li et al. 2022).

4.4.3.1 Mechanism

Combinatorial CRISPR screening involves the simultaneous perturbation of two or more genes within the same cell using multiple sgRNAs. This approach can be conducted in either pooled or arrayed formats with libraries specifically designed to target gene pairs or combinations. Based on library construction and strategy adaptation, combinatorial screens are classified into four formats: orthologous Cas9 systems, dual-gRNA systems, hybrid gRNA systems, and higher-order multiplexing systems. Orthologous Cas9 systems pair different Cas9 versions from bacteria like *S. pyogenes* (SpCas9) and *Staphylococcus aureus* (SaCas9), which have distinct tracrRNA parts that stop gRNAs from mixing up during library making, allowing cells to get hit by two separate gene targets at once in big pooled experiments. Dual-gRNA systems employ vectors with two gRNAs targeting distinct genes, paired with Cas9 variants like spCas9–saCas9 or engineered tracrRNAs such as VCR1-WCR3 to reduce recombination and ensure precise readout. Hybrid gRNA platforms, exemplified by CHyMErA, integrate Cas9 and Cas12a gRNAs that Cas12a processes into separate guides for recombination-free multiplexing. Higher-order multiplexing methods, including 3Cs (covalently closed circular synthesis) and extensible assembly, produce libraries for trigenic or greater combinations with minimal skew, lowering required screening coverage (Chow & Chen 2018).

4.5 IN VIVO CRISPR SCREENS

In vivo CRISPR screens extend the power of genome-wide screening technologies by enabling the analysis of gene function directly within living organisms. Unlike in vitro screens, in vivo screens allow the exploration of gene functions in the complex physiological environment of intact tissues, organs, and whole organisms, capturing the effects of the native microenvironment and systemic interactions. This method has revolutionized fields like cancer research, immunology, developmental biology, etc., by discovering the genes involved in tumour progression, immune evasion, and drug response under clinically relevant conditions. Generally, in vivo CRISPR screening involves the delivery of sgRNA libraries into genetically engineered animals that express Cas9, or transplanting engineered cells back into host organisms, followed by phenotypic monitoring of outcomes like tumour growth or immune cell activity. Continuing progress in delivery methods and sequencing techniques is further expanding the capabilities of in vivo CRISPR screens. The two types of in vivo CRISPR screens are transplantation-based and direct delivery-based screens.

4.5.1 Transplantation-Based In Vivo Screens

Transplantation-based in vivo CRISPR screens involve the genetic editing of cells in vitro, which are then transplanted into host animals. This approach provides exact control over the initial genetic modifications and enables expansion of the edited cells before implantation. It enables focused investigations of specific gene functions in disease models such as cancer (Santinha et al. 2025).

4.5.1.1 Mechanism

A two-step process is followed in transplantation-based in vivo CRISPR screens, wherein first, the cells undergo an in vitro editing via the introduction of a library of sgRNAs that generate a diverse population of genetically modified cells. These cells are then transplanted into immune compromised or genetically compatible animal models, such as mice. Once in the physiological environment, the transplanted cells undergo selective pressures and are monitored for specific phenotypes to study gene function in a context that mimics natural tissue conditions. After the selection process, genomic DNA is extracted from the cells, and the abundance of each sgRNA is determined using NGS, which identifies genes that affect the observed phenotype. This approach brings in the scalability of pooled CRISPR screens with the biological complexity of in vivo models, offering insights that better reflect real disease states and tissue interactions.

4.5.1.2 Advantages

The transplantation method offers controlled cell preparation, the ability to expand cells, and precise timing of implantation. These features make it particularly useful for studying dynamic processes such as tumour growth, metastasis, and immune cell functions under well-controlled experimental conditions. However, this method may not fully capture the complexities of tissue-specific gene regulation that occur naturally within intact organs.

4.5.2 Direct In Vivo Screens

Direct in vivo CRISPR screens, unlike transplantation-based screens, deliver CRISPR components, including Cas9 and gRNAs, directly into the tissues of living animals using viral vectors, nanoparticles, or electroporation. This method facilitates gene perturbation within their natural environment, preserving physiological contexts such as immune interactions and microenvironmental influences.

4.5.2.1 Mechanism

A library of sgRNAs is packaged into vectors such as adeno-associated virus, lentivirus, or plasmid DNA and administered directly to target tissues in Cas9-expressing animals. This method induces gene mutations on-site, preserving the microenvironment and immune system. After allowing sufficient time for gene editing effects to occur, tissues are collected for analysis. The abundance of sgRNAs is quantified by sequencing, identifying genes that influence the observed phenotype.

4.5.2.2 Advantages

Direct in vivo CRISPR screens maintain endogenous tissue architecture and cellular interactions, providing insights with higher physiological relevance. They are used to integrate cancer genetics

TABLE 4.1
Comparison of CRISPR-Based Screening Methodologies

Methodology	Features	In Vitro Screens	In Vivo Screens	Advantages	Limitations
CRISPR knockout (KO)	Uses Cas9, saCas9, Cas12a; turns OFF gene expression by removing the gene	Pooled KOs: for preliminary discovery of KO phenotypes; arrayed KOs: for targeted validation of specific KOs	Transplantation-based: cells with target gene KOs are introduced into model organisms; direct-based: Cas9 components (Cas9, mRNA, sgRNA) are delivered directly into tissues	Helps discover gene functions; removal of disease-causing and mutating genes	Permanent and irreversible; cannot study translational re-initiation or alternative splicing; occurrence of KO escaping (Wang et al. 2024)
CRISPR interference	Uses dCas9, dCas9-KRAB, Cas13; turns down gene expression through repressor molecules	Pooled interferences: for large-scale mapping of regulatory genes; arrayed interferences: for specialized cell types based on subcellular responses	Transplantation-based: mutagenized cells integrated into the transplant; direct-based: transgenic Cas9 models receive sgRNAs	Temporary and reversible; useful for studying essential genes; no DNA cleavage	PAM targeting constraints; unintended transcriptional repression
CRISPR activation	Uses dCas9-SAM, dCas9-SPH, dCas9-MS2; turns ON gene expression using activator molecules	Pooled activation: for activating >1,000 sgRNAs in exploratory studies; arrayed activation: for activating >1,000 sgRNAs in integrated studies	Transplantation-based: engineered cells are transferred and injected into models; direct-based: dCas9-sgRNA activators are delivered into tissues	Temporary and reversible; enhanced transcription of target genes; enables activation of native genes (Kirby et al. 2024)	Spillover activation of neighbouring cells; reduced specificity to broad genomic regions; may induce squelching due to high activator levels (Maddineni et al. 2024)

with mouse models to reveal how tumour-associated mutations dictate immunotherapy responses and nominate novel targets. Challenges in the direct in vivo approach include delivery efficiency, tissue specificity, and the complexity of tracking and analysing gene edits in a living organism. However, comparison of CRISPR-based screening methodologies is indexed in Table 4.1.

4.6 ADVANCEMENTS AND FUTURE PROSPECTS

CRISPR is revolutionizing genomics and transcriptomics, with direct implications for many genetic disorders, infectious diseases, and malignant tumours. SARS-CoV-2 DETECTR with less processivity was developed from CRISPR-Cas12. CRISPR is now an FDA-approved therapy that follows strict regulatory guidelines which has a higher benefits-to-risk ratio (Singh et al. 2024a). From restoring dystrophin protein in muscular dystrophy to mitigating optical congenital amaurosis type 10, in vivo gene editing has succeeded. Human-induced pluripotent stem cells were combined with ex vivo CRISPR genome editing to manage type 1 diabetes (Leite et al. 2020).

Studies focussing on inheritable hypercholesterolemia are currently ongoing. In future, researchers are looking forward to develop ultra-precise Cas enzymes, reversible epigenetic editing tools, more powerful multiplex CAR-T cell therapies, and scalable nonviral delivery strategies aimed to tackle complex conditions such as HIV and solid tumours. Novel genome editing approaches are being formulated using retro-transposons and RNA-targeting technologies as well. Insertion of large, lengthy gene sequences into postmitotic cells, with impaired HDR, remains a challenge in genome editing. With its wide range of benefits and existing challenges, CRISPR is evolving towards a bright future that it can further revolutionize modern medicine, healthcare, and the agriculture sector.

4.7 CONCLUSION

Next-generation functional genomics has been profoundly advanced by CRISPR-based screening methodologies, which provide powerful, scalable, and precise tools to interrogate gene function across diverse biological systems. The integration of CRISPR knockout, interference, and activation approaches with pooled and arrayed in vitro and in vivo screening platforms enables systematic dissection of gene regulatory networks, identification of essential genes, and functional validation of disease-associated loci. These technologies offer unmatched flexibility, allowing both permanent and reversible genetic perturbations while accommodating complex physiological contexts. Despite challenges such as off-target effects, delivery constraints, and context-dependent responses, continuous improvements in CRISPR enzymes, gRNA design, and screening analytics are steadily enhancing specificity and translational relevance. Collectively, CRISPR-based screening stands as a cornerstone of next-generation functional genomics, driving deeper biological insights and accelerating applications in biomedical research, agriculture, and therapeutic discovery.

REFERENCES

Anderson, K. R., Haeussler, M., Watanabe, C., Janakiraman, V., Lund, J., Modrusan, Z., Stinson, J., Bei, Q., Buechler, A., Yu, C., Thamminana, S. R., Tam, L., Sowick, M., Alcantar, T., O'Neil, N., Li, J., Ta, L., Lima, L., Roose-Girma, M., … Warming, S. (2018). CRISPR off-target analysis in genetically engineered rats and mice. *Nature Methods*, 15(7), 512–514. https://doi.org/10.1038/s41592-018-0011-5

Anzalone, A. V., Randolph, P. B., Davis, J. R., Sousa, A. A., Koblan, L. W., Levy, J.M., Chen, P. J., Wilson, C., Newby, G. A., Raguram, A., & Liu, D. R. (2019). Search-and-replace genome editing without double-strand breaks or donor DNA. *Nature*, 576(7785), 149–157. https://doi.org/10.1038/s41586-019-1711-4

Becirovic, E. (2022). Maybe you can turn me on: CRISPRa-based strategies for therapeutic applications. *Cellular and Molecular Life Sciences: CMLS*, 79(2), 130. https://doi.org/10.1007/s00018-022-04175-8

Billon, P., Bryant, E. E., Joseph, S. A., Nambiar, T. S., Hayward, S. B., Rothstein, R., & Ciccia, A. (2017). CRISPR-mediated base editing enables efficient disruption of eukaryotic genes through induction of STOP codons. *Molecular Cell*, 67(6), 1068–1079.e4. https://doi.org/10.1016/j.molcel.2017.08.008

Chavez, A., Tuttle, M., Pruitt, B. W., Ewen-Campen, B., Chari, R., Ter-Ovanesyan, D., Haque, S. J., Cecchi, R. J., Kowal, E. J., Buchthal, J., Housden, B. E., Perrimon, N., Collins, J. J., & Church, G. (2016). Comparison of Cas9 activators in multiple species. *Nature Methods*, 13(7), 563–567. https://doi.org/10.1038/nmeth.3871

Chow, R. D., & Chen, S. (2018). Cancer CRISPR screens in vivo. *Trends in Cancer*, 4(5), 349–358. https://doi.org/10.1016/j.trecan.2018.03.002

Chen, P. J., & Liu, D. R. (2023). Prime editing for precise and highly versatile genome manipulation. *Nature Reviews Genetics*, 24(3), 161–177. https://doi.org/10.1038/s41576-022-00541-1

Cross, R. (2020, June 3). Making 2 DNA edits at once. *Chemical & Engineering News,* 98(22). https://cen.acs.org/biological-chemistry/gene-editing/21-base-editors-make-two/98/i22

De Wet, T. J., Winkler, K. R., Mhlanga, M., Mizrahi, V., & Warner, D. F. (2020). Arrayed CRISPRi and quantitative imaging describe the morphotypic landscape of essential mycobacterial genes. *eLife*, *9*. https://doi.org/10.7554/elife.60083

Dhainaut, M., Rose, S. A., Akturk, G., Wroblewska, A., Nielsen, S. R., Park, E. S., Buckup, M., Roudko, V., Pia, L., Sweeney, R., Berichel, J. L., Wilk, C. M., Bektesevic, A., Lee, B. H., Bhardwaj, N., Rahman, A. H., Baccarini, A., Gnjatic, S., Pe'er, D., … Brown, B. D. (2022). Spatial CRISPR genomics identifies regulators of the tumor microenvironment. *Cell*, 185(7), 1223–1239.e20. https://doi.org/10.1016/j.cell.2022.02.015

Dolan, A. E., Hou, Z., Xiao, Y., Gramelspacher, M. J., Heo, J., Howden, S. E., ... & Zhang, Y. (2019). Introducing a spectrum of long-range genomic deletions in human embryonic stem cells using type I CRISPR-Cas. *Molecular Cell*, 74(5), 936–950. https://doi.org/10.1016/j.molcel.2019.03.014

Duke, C. G., Bach, S. V., Revanna, J. S., Sultan, F. A., Southern, N. T., Davis, M. N., Carullo, N. V. N., Bauman, A. J., Phillips, R. A., & Day, J. J. (2020). An improved CRISPR/DCAS9 interference tool for neuronal gene suppression. *Frontiers in Genome Editing*, 2, 9. https://doi.org/10.3389/fgeed.2020.00009

Far, B. F., Akbari, M., Habibi, M. A., Katavand, M., & Nasseri, S. (2025). CRISPR technology in disease management: An updated review of clinical translation and therapeutic potential. *Cell Proliferation*, 58(11), e70099. https://doi.org/10.1111/cpr.70099

Fleiss, A., O'Donnell, S., Fournier, T., Lu, W., Agier, N., Delmas, S., ... & Fischer, G. (2019). Reshuffling yeast chromosomes with CRISPR/Cas9. *PLoS Genetics*, 15(8), e1008332. https://doi.org/10.1371/journal.pgen.1008332

Fontana, J., Sparkman-Yager, D., Zalatan, J. G., & Carothers, J. M. (2020). Challenges and opportunities with CRISPR activation in bacteria for data-driven metabolic engineering. *Current Opinion in Biotechnology*, 64, 190–198. https://doi.org/10.1016/j.copbio.2020.04.005

Gaudelli, N. M., Komor, A. C., Rees, H. A., Packer, M. S., Badran, A. H., Bryson, D. I., & Liu, D. R. (2018). Publisher correction: Programmable base editing of AT to G•C in genomic DNA without DNA cleavage. *Nature*, 559(7714), E8. https://doi.org/10.1038/s41586-018-0070-x

Ghavami, S., & Pandi, A. (2021). CRISPR interference and its applications. Progress in Molecular Biology and *Translational Science*, 180, 123–140. https://doi.org/10.1016/bs.pmbts.2021.01.007

Horvath, P., & Barrangou, R. (2010). CRISPR/CAS, the immune system of bacteria and archaea. *Science*, 327(5962), 167–170. https://doi.org/10.1126/science.1179555

Ishino, Y., Krupovic, M., & Forterre, P. (2018). History of CRISPR-Cas from encounter with a mysterious repeated sequence to genome editing technology. *Journal of Bacteriology*, 200(7). https://doi.org/10.1128/jb.00580-17

Javaid, N., & Choi, S. (2021). CRISPR/CAS system and factors affecting its precision and efficiency. *Frontiers in Cell and Developmental Biology*, 9, 761709. https://doi.org/10.3389/fcell.2021.761709

Jinek, M., Chylinski, K., Fonfara, I., Hauer, M., Doudna, J. A., & Charpentier, E. (2012). A programmable Dual-RNA–Guided DNA endonuclease in adaptive bacterial immunity. *Science*, 337(6096), 816–821. https://doi.org/10.1126/science.1225829

Karlson, C. K. S., Mohd-Noor, S. N., Nolte, N., & Tan, B. C. (2021). CRISPR/DCAS9-based systems: Mechanisms and applications in plant sciences. *Plants*, 10(10), 2055. https://doi.org/10.3390/plants10102055

Kim, H. S., Kweon, J., & Kim, Y. (2024). Recent advances in CRISPR-based functional genomics for the study of disease-associated genetic variants. *Experimental & Molecular Medicine*, 56(4), 861–869. https://doi.org/10.1038/s12276-024-01212-3

Kirby, E. N., Montin, X. B., Allen, T. P., Densumite, J., Trowbridge, B. N., & Beard, M. R. (2024). CRISPR activation as a platform to identify interferon stimulated genes with anti-viral function. *Innate Immunity*, 30(2–4), 40–54. https://doi.org/10.1177/17534259231225611

La Marca, J. E., Diepstraten, S. T., Kelly, G. L., & Herold, M. J. (2024). Awakening the genome: CRISPR activation to unravel the origin and vulnerabilities of cancer. *Annual Review of Cancer Biology*, 9(1), 141–159. https://doi.org/10.1146/annurev-cancerbio-062624-104149

Lebeau, P. F., et al. (2019). PCSK9 knockout exacerbates diet-induced non-alcoholic steatohepatitis, fibrosis and liver injury in mice. *Metabolism: Clinical and Experimental*. https://doi.org/10.1016/j.metabol.2019.01.012

Leite, N. C., Sintov, E., Meissner, T. B., Brehm, M. A., Greiner, D. L., Harlan, D. M., & Melton, D. A. (2020). Modeling type 1 diabetes in vitro using human pluripotent stem cells. *Cell Reports*, 32(2), 107894. https://doi.org/10.1016/j.celrep.2020.107894

Li, R., Klingbeil, O., Monducci, D., Young, M. J., Rodriguez, D. J., Bayyat, Z., Dempster, J. M., Kesar, D., Yang, X., Zamanighomi, M., Vakoc, C. R., Ito, T., & Sellers, W. R. (2022). Comparative optimization of combinatorial CRISPR screens. *Nature Communications*, 13(1), 2469. https://doi.org/10.1038/s41467-022-30196-9

Liu, Y., Zhou, C., Huang, S., Dang, L., Wei, Y., He, J., Zhou, Y., Mao, S., Tao, W., Zhang, Y., Yang, H., Huang, X., & Chi, T. (2020b). A Cas-embedding strategy for minimizing off-target effects of DNA base editors. *Nature Communications*, 11(1), 6073. https://doi.org/10.1038/s41467-020-19690-0

Maddineni, A., Liang, Z., Jambardi, S., Roy, S., Tycko, J., Patil, A., Manzano, M., & Gottwein, E. (2024). *Cytotoxicity of activator expression in CRISPR-based transcriptional activation systems*. bioRxiv. https://doi.org/10.1101/2024.09.23.614524

Mojica, F. J. M., Díez-Villaseñor, C., García-Martínez, J., & Soria, E. (2005). Intervening sequences of regularly spaced prokaryotic repeats derive from foreign genetic elements. *Journal of Molecular Evolution*, 60(2), 174–182. https://doi.org/10.1007/s00239-004-0046-3

Morchikh, M., Cribier, A., Raffel, R., Amraoui, S., Cau, J., Severac, D., … Benkirane, M. (2017). HEXIM1 and NEAT1 long non-coding RNA form a multi-subunit complex that regulates DNA-mediated innate immune response. *Molecular Cell*, 67(3), 387–399.e5. https://doi.org/10.1016/j.molcel.2017.06.030

Najafi, S., Tan, S. C., Aghamiri, S., Raee, P., Ebrahimi, Z., Jahromi, Z. K., Rahmati, Y., Nahand, J. S., Piroozmand, A., Jajarmi, V., & Mirzaei, H. (2022). Therapeutic potentials of CRISPR-Cas genome editing technology in human viral infections. *Biomedicine & Pharmacotherapy*, 148, 112743. https://doi.org/10.1016/j.biopha.2022.112743

Najm, F. J., Strand, C., Donovan, K. F., Hegde, M., Sanson, K. R., Vaimberg, E. W., Sullender, M. E., Hartenian, E., Kalani, Z., Fusi, N., Listgarten, J., Younger, S. T., Bernstein, B. E., Root, D. E., & Doench, J. G. (2017). Orthologous CRISPR–Cas9 enzymes for combinatorial genetic screens. *Nature Biotechnology*, 36(2), 179–189. https://doi.org/10.1038/nbt.4048

Peddle, C. F., Fry, L. E., McClements, M. E., & MacLaren, R. E. (2020). CRISPR interference-potential application in retinal disease. *International Journal of Molecular Sciences*, 21(7), 2329. https://doi.org/10.3390/ijms21072329

Richter, H., Randau, L., & Plagens, A. (2013). Exploiting CRISPR/CAS: Interference mechanisms and applications. *International Journal of Molecular Sciences*, 14(7), 14518–14531. https://doi.org/10.3390/ijms140714518

Ruetz, T. J., Pogson, A. N., Kashiwagi, C. M., Gagnon, S. D., Morton, B., Sun, E. D., Na, J., Yeo, R. W., Leeman, D. S., Morgens, D. W., Tsui, C. K., Li, A., Bassik, M. C., & Brunet, A. (2024). CRISPR-Cas9 screens reveal regulators of ageing in neural stem cells. *Nature*, 634(8036), 1150–1159. https://doi.org/10.1038/s41586-024-07972-2

Sage, C. L., Lawo, S., & Cross, B. C. (2019). CRISPR: A screener's guide. *SLAS Discovery*, 25(3), 233–240. https://doi.org/10.1177/2472555219883621

Santinha, A. J., Strano, A., & Platt, R. J. (2025). Methods and applications of in vivo CRISPR screening. *Nature Reviews Genetics*, 26(10), 702–718. https://doi.org/10.1038/s41576-025-00873-8

Sharma, I., Hall, K., & Moonah, S. (2025). CRISPR genome editing using a combined positive and negative selection system. *PLoS ONE*, 20(5), e0321881. https://doi.org/10.1371/journal.pone.0321881

Singh, A., Irfan, H., Fatima, E., Nazir, Z., Verma, A., & Akilimali, A. (2024a). Revolutionary breakthrough: FDA approves CASGEVY, the first CRISPR/Cas9 gene therapy for sickle cell disease. *Annals of Medicine and Surgery (2012)*, 86(8), 4555–4559. https://doi.org/10.1097/MS9.0000000000002146

Singh, A., Kumar, R., & Sharma, P. (2024b). CRISPR-Cas systems in antiviral therapy. *International Research Journal of Medicine and Surgery, 1*(3), 20–23. https://www.irjmeds.com/wp-content/uploads/2024/11/Manuscript_IRJMedS_015_WS.pdf

Singpanomchai, N., & Ratthawongjirakul, P. (2024). The CRISPR-dCas9 interference system suppresses inhA gene expression in *Mycobacterium smegmatis*. *Scientific Reports*, 14(1), 26116. https://doi.org/10.1038/s41598-024-77442-2

Valdivieso-Rivera, F. B., De Oliveira Furino, V., & Sponton, C. H. (2023). Investigation of beige fat biology and metabolism using the CRISPR SunTag-P65-HSF1 activation system. *Journal of Visualized Experiments*, 191. https://doi.org/10.3791/64849

Van Campenhout, C., Cabochette, P., Veillard, A., Laczik, M., Zelisko-Schmidt, A., Sabatel, C., Dhainaut, M., Vanhollebeke, B., Gueydan, C., & Kruys, V. (2019). Guidelines for optimized gene knockout using CRISPR/CAS9. *BioTechniques*, 66(6), 295–302. https://doi.org/10.2144/btn-2018-0187

Walton, R. T., Singh, A., & Blainey, P. C. (2022). Pooled genetic screens with image-based profiling. *Molecular Systems Biology*, 18(11), e10768. https://doi.org/10.15252/msb.202110768

Wang, M., Glass, Z. A., & Xu, Q. (2016). Non-viral delivery of genome-editing nucleases for gene therapy. *Gene Therapy*, 24(3), 144–150. https://doi.org/10.1038/gt.2016.72

Wang, Y., Zhai, Y., Zhang, M., Song, C., Zhang, Y., & Zhang, G. (2024). Escaping from CRISPR–Cas-mediated knockout: The facts, mechanisms, and applications. *Cellular & Molecular Biology Letters*, 29(1), 48. https://doi.org/10.1186/s11658-024-00565-x

Wensing, L., Sharma, J., Uthayakumar, D., Proteau, Y., Chavez, A., & Shapiro, R. S. (2019). A CRISPR interference platform for efficient genetic repression in *Candida albicans*. mSphere, 4(1), e00002–19. https://doi.org/10.1128/mSphere.00002-19

Yin, J., Frick, L., Scheidmann, M. C., Liu, T., Trevisan, C., Dhingra, A., Spinelli, A., Wu, Y., Yao, L., Vena, D. L., Knapp, B., Guo, J., De Cecco, E., Ging, K., Armani, A., Oakeley, E. J., Nigsch, F., Jenzer, J., Haegele, J., … Aguzzi, A. (2024). Arrayed CRISPR libraries for the genome-wide activation, deletion and silencing of human protein-coding genes. *Nature Biomedical Engineering*, 9(1), 127–148. https://doi.org/10.1038/s41551-024-01278-4

Yuan, J., Ma, Y., Huang, T., Chen, Y., Peng, Y., Li, B., Li, J., Zhang, Y., Song, B., Sun, X., Ding, Q., Song, Y., & Chang, X. (2018). Genetic modulation of RNA splicing with a CRISPR-guided cytidine deaminase. *Molecular Cell*, 72(2), 380–394.e7. https://doi.org/10.1016/j.molcel.2018.09.002

Zhao, Z., Shi, L., Zhang, W., Han, J., Zhang, S., Fu, Z., & Cai, J. (2017). CRISPR knock out of programmed cell death protein 1 enhances anti-tumor activity of cytotoxic T lymphocytes. *Oncotarget*, 9(4), 5208–5215. https://doi.org/10.18632/oncotarget.23730

5 Multiplex Genome Editing
Concepts, Strategies, and Applications

Alok Ranjan Nayak, Abhijit Ratha, Anshuman Sarangi, and Neelanjana Choudhury

5.1 INTRODUCTION

Some fundamental genetic engineering methods, such as homologous recombination and mutagenesis, contributed to the beginning of the genome editing era. However, these techniques were imprecise and ineffective. These techniques made it challenging to regulate the location and mode of genetic change. In the early 2000s, a major turning point resulted from the advancement of another genetic tools, zinc finger nucleases (ZFNs). It was the first truly engineered programable nuclease. ZFNs contain zinc finger array (that targets 9–18 bp sequence) and folk nuclease domain. These tools allowed researchers to modify DNA at specific location. Using this method, gene editing was more accurate and precise (Urnov et al. 2005). After few years (in 2011), scientists developed another tool named transcription activator-like effector nucleases (TALENs). TALENs were easier to form and more flexible for targeting different DNA sequences (Boch et al. 2009). In 2012, the real revolution began in genome editing with CRISPR-Cas9 system. CRISPR sequence refers to clustered regularly interspaced short palindromic repeats, and was first discovered in *Escherichia coli* in 1987.

This was identified as the major component of bacterial adaptive system used to defend against virus by storing the viral DNA fragments (spacers). That spacer was used as a guide RNA (gRNA) to identify the target location with the help of Cas protein. The Cas protein will only cut the DNA if it finds a PAM sequence right next to the target. The PAM is a tiny DNA segment, that contain two to six base pairs, for example, in Cas9, the PAM is generally "NGG" where "N" is any base. This PAM sequence is essential because it helps the Cas protein differentiate viral DNA from the cell's own DNA, which prevent it from cutting its own genome (Jinek et al. 2012). After publishing their work on CRISPR, Emmanuelle Charpentier and Jennifer Doudna transformed it into a potent genome editing tool. They streamlined the natural system from *Streptococcus pyogenes* by uniting the crRNA and a second RNA (tracrRNA) into a single gRNA. They revealed that this CRISPR-Cas9 system could be programmed with any sgRNA sequence to cut matching DNA in vitro (Jinek et al. 2012).

A new era has been brought most recently by the establishment of more precise editing tools such as base editors and prime editors. By bypassing the error-prone repair pathways associated with double-strand DNA cleavage these technologies reduce off-target effects by direct modification of genetic information at the single base level or by small insertions or deletions introduced with catalytically compromised Cas enzymes fused to effector protein (Anzalone et al. 2019; Komor et al. 2016). A very precise genetic engineering technique called multiplex genome editing means that it is possible to simultaneously alter several genetic locations (loci) in one organism or a cell. This method, primarily enabled by multiplex genome-editing systems, utilizes multiple guide RNAs (gRNAs) to direct a Cas nuclease to simultaneously target and cleave multiple DNA sequences.

DOI: 10.1201/9781003735113-5

As compared to sequential editing, this ability has greatly improved the speed of genetic research and engineering by making it possible to introduce multiple mutations or gene modifications during a single experimental step, saving money, time, and labor. Multiplex genome editing enables complex genetic editing such as multigene knockouts, pathway editing, and studying the interaction of genes by delivering several gRNAs in single or polycistronic expression cassettes together with Cas proteins. It is extensively utilized across a wide range of organisms, including microorganisms, plants, and animals, for diverse applications such as crop improvement, therapeutic development, and functional genomics (Adiego-Pérez et al. 2019).

5.2 STRATEGIES FOR MULTIPLEX GENOME EDITING

Strategies for expressing multiplex gRNA encompass various techniques to effectively express multiple gRNAs simultaneously for CRISPR-based genome editing. The primary strategies are described in the following subsections:

5.2.1 Polycistronic tRNA-gRNA (tRNA-gRNA) Arrays

Using the cell's natural tRNA processing system, polycistronic tRNA-gRNA (tRNA-gRNA) (PTG) arrays are a sophisticated multiplex genome editing technique that effectively expresses several gRNAs from a single transcript. Utilizing this technique, a polycistronic RNA molecule transcribed from single Pol III promoter contains different gRNAs arranged in tandem, each separated by a tRNA sequence. Individual mature gRNAs are obtained when endogenous RNases (RNaseP and RNaseZ) identify and precisely cleave the tRNA sequences. These gRNAs can then direct the Cas9 protein to multiple genomic target sites at once (Dong et al. 2017; Bindal & Rath 2025). The processing of PTG arrays is indexed in Figure 5.1.

5.2.2 The Csy4 Ribonuclease System

Multiple gRNAs are co-expressed from a single transcript using the potent Csy4-based system. It avoids the requirement for numerous, repetitive terminators and promoters, which are the main cause of genetic instability in conventional multi-cassette vectors. The Csy4 endoribonuclease

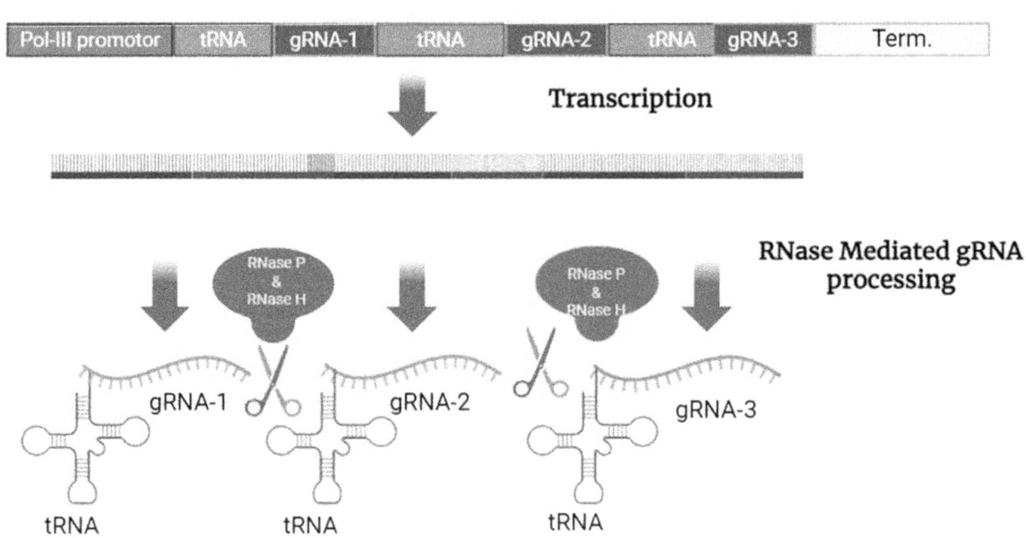

FIGURE 5.1 Processing of polycistronic tRNA-gRNA arrays.

from *Pseudomonas aeruginosa* is utilized by the system. In this bacterium, Csy4 is a part of the CRISPR-Cas system and is essential for converting the sgRNA transcript into mature sgRNAs. Its primary characteristic is that, in heterologous systems such as human cells, it produces a sequence-specific cut with remarkably high efficiency and accuracy. The Csy4 protein is a 25-kDa endoribonuclease enzyme that specifically recognizes and cleaves a 28-nucleotide RNA sequence known as the Csy4 Recognition Site (CRS), thereby enabling precise RNA processing and regulated gene expression.

There is a unique, highly structured 28-base sequence (5′-GTTT-N20-TTC-3′), to which Csy4 binds with high affinity and breaks at the 3′ end of the GTTT motif. N20 is the variable region that forms a stem-loop. It produces a single polycistronic gRNA transcript rather than several distinct U6-gRNA cassettes. The U6 promoter generates a single, lengthy RNA transcript. Every single Csy4 recognition site (CRS) in the transcript is bound by the Csy4 protein with a very high affinity. Csy4 cleaves the RNA exactly after its recognition site's fourth nucleotide (GTTT↓). The upstream gRNA sequence is released by this cleavage event. The cleavage product is a functional, mature gRNA with the exact 5′ end required for Cas9 binding and targeting. The cleavage site's short, 4-nt overhang (GTTT) at the 3′ end does not affect the function of gRNA (Haurwitz et al. 2010; Nissim et al. 2014). The processing of sgRNA by Csy4 ribonuclease system is indexed in Figure 5.2.

5.2.3 The Cas12a (Cpf1) System

The Class 2, Type V CRISPR-associated endonuclease enzyme Cas12a (Cpf1) has a structure that looks like two lobes, one for recognition (REC) and one for nuclease (NUC). It doesn't have the HNH domain that Cas9 has, but it does have a RuvC-like domain that cuts DNA. Cas12a has two different nuclease activities: one RNA-guided DNase that cleaves target DNA and another pre-crRNA processing endoribonuclease that helps its own gRNA maturation. Cas12a detects the thymine-rich PAM site, which is located 5′ upstream of the target sequence. This is a big plus when trying to find AT-rich areas of the genome (Bandyopadhyay et al. 2020). The RuvC domain contains the catalytic site that mediates the cleavage of both DNA strands in a single active site. About 18–23 nucleotides downstream of the PAM (TTTV) is where DNA cleavage takes place (Paul & Montoya 2020). The Cas12a/cpf1-assisted genome editing is indexed in Figure 5.3.

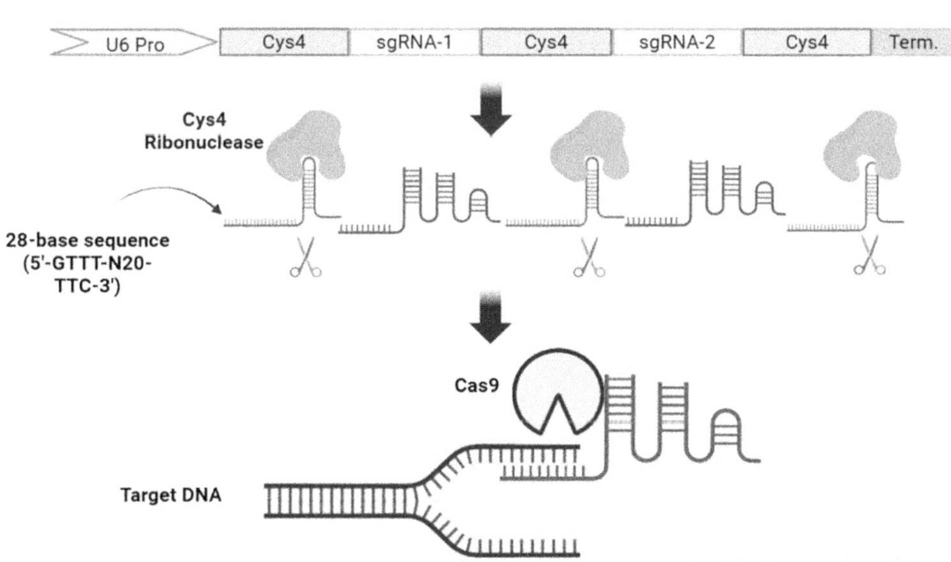

FIGURE 5.2 Processing of sgRNA by Csy4 ribonuclease system.

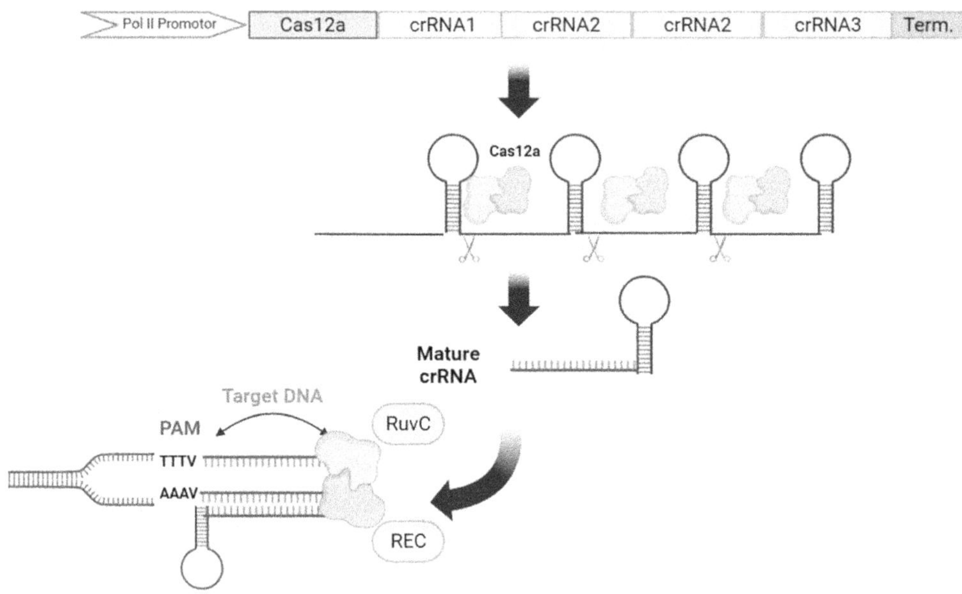

FIGURE 5.3 The Cas12a/cpf1-assisted genome editing.

5.2.4 Base and Prime Editing in Multiplex Genome Editing

5.2.4.1 Base Editing

Base editors are fusion proteins composed of a deaminase enzyme with either a completely inactive Cas9 (dCas9) or a catalytically impaired Cas9 (Cas9 nickase/nCas9). They do not cleave the backbone of DNA; instead, they function via a "search-and-replace" mechanism. The base editors are guided toward the target sequence by the gRNA. The Cas9 domain binds as well as unwinds DNA and expose a narrow window of single-stranded DNA within the R-loop structure, known as the "editing window" (within 4–8 nts of protospacer region). Within this window, the linked deaminase enzyme acts on a particular base. Cytosine base editors deaminate cytosine to uracil to modify GC base pairing to AT. U is subsequently treated as T by the cellular machinery, resulting in a permanent conversion during replication. Important elements uracil glycosylase inhibitor will be added to stop unwanted repair and stop the deaminase enzyme activity. Adenine base editors deaminate adenine to inosine to change an AT base pairing to GC. Polymerases recognize inosine as guanine. A transfer RNA-specific adenosine deaminase (TadA) is engineered to produce adenine base editors (ABEs) (Gaudelli et al. 2017).

5.2.4.2 Prime Editing

Prime editors are flexible fusion proteins that can mediate small insertions and deletions as well as direct base switching without the requiring for either donor DNA templates or double-strand breaks. The three main parts of the system are a reverse transcriptase enzyme, pegRNA, which is a prime editing gRNA that identifies the target site and has the required modifications within its 3′ extension, and a Cas9 nickase, for DNA binding and nicking the non-edited strand. The pegRNA directs the prime editor to the target DNA, where the nCas9 nicks the PAM sequence to create a DNA opening, initiating the multistep process. This nicked strand's exposed 3′ end then anneal to the pegRNA's primer binding site, allowing the reverse transcriptase to use the pegRNA's extension as a template to create a new DNA flap with the desired edit. By integrating the modified region into the genome, cellular repair mechanisms ultimately correct this structure. This process can be improved by creating a second nick on the opposing strand to skew repair in the desired direction (Anzalone et al. 2019).

5.3 DELIVERY METHODS FOR MULTIPLEX EDITING TOOLS

5.3.1 Viral Vector

Viral and nonviral delivery systems that carry the required gene-editing components into cells are the main techniques for delivering multiplex editing tools. Microinjection, viral vectors (like adenoviruses), and plasmid or RNA delivery systems encoding Cas nucleases and multiple gRNAs are common delivery methods for multiplex genome editing, which intends to modify several locations at once. Viral vectors are among the most favored and effective methods for delivering multiplex genome editing tools, especially those utilizing CRISPR-Cas systems. The main types of viral vectors employed are adenoviral vectors, adeno-associated viral (AAV) vectors, and lentiviral vectors, each possessing unique features that suit different delivery needs.

5.3.1.1 Adenoviral Vectors–Based Delivery Method

Adenoviral vectors are double-stranded DNA viruses are beneficial for delivering Cas nucleases and several gRNAs at once because they can carry large genetic components (7.5–30 kb). Their use is restricted by their relatively high immunogenicity, despite their high transduction yield and capability of infecting both proliferative and quiescent cells. More sophisticated AdV generations with deleted viral genes enhance editing accuracy and minimize cytotoxicity. AdVs are useful for delivering homology donor constructs in vivo and can be combined with AAV vectors to enable precise gene editing (Ates et al. 2020).

5.3.1.1.1 Construction of Adenoviral Vectors for Multiplex Genome Editing

The helper-dependent adenovirus is the recommended system for multiplex editing. Only the necessary inverted terminal repeat (ITR) and packaging signal remain after all viral genes (E1, E2, E3, and E4) are eliminated. This results in a massive ~30 kb capacity that can hold multiple gRNA expression cassettes with a U6 promoter and gRNA scaffold (each ~0.3 kb) and an A Cas9 nuclease (e.g., ~4.2 kb for SpCas9). Ten or more gRNAs can be carried by a single vector. A specific helper cell line, such as HEK293, is transfected with the HD-AdV genome, which contains the editing machinery. In order to facilitate viral replication and inadequate packaging, these cells supply the absent E1, E2, and E4 genes in trans. As a result, the multiplex editing machinery is present in a high-titer stock of purified, replication-incompetent AdV virions. By attaching to the Coxsackie and Adenovirus Receptor and co-receptors ($\alpha v\beta 3/5$ integrins), the purified AdV virions infect target cells through receptor-mediated endocytosis. When the endosomal membrane is broken by the AdV capsid, the viral capsid can be released into the cytoplasm. The viral DNA is actively moved into the cell nucleus, where it stays as a non-integrating episome while still being connected to core proteins.

The expression of the Cas9 mRNA, which is exported to the cytoplasm and translated into the Cas9 protein, is driven by mammalian RNA Polymerase II promoter inside the nucleus. The expression of several gRNAs in a single vector is controlled by RNA Polymerase III promoters, e.g., U6. In the nucleus and nucleoplasm, the Cas9 protein and gRNAs combine to create active ribonucleoprotein (RNP) complexes. Every unique RNP complex (Cas9 with its own gRNA) looks for the complementary DNA target site. When Cas9 binds the DNA, a double-strand break occurs. The mechanisms for cell repair are triggered by non-homologous end joining and homology-directed repair (Lee et al. 2017).

5.3.1.2 Adeno-Associated Viral Vectors–Based Delivery Method

AAV vectors are used extensively for accurate and effective genome editing in a range of mammalian cells despite having a smaller packaging capacity (~4.7 kb) and being less immunogenic. In multiplex editing techniques, they are excellent donor DNA carriers for homologous recombination or homology-mediated end joining. To supplement Cas9 delivery by other vectors, single-stranded and self-complementary AAVs can be designed to efficiently deliver donor templates (Ates et al. 2020).

5.3.1.2.1 Construction of Adeno-Associated Viral Vectors for Multiplex Genome Editing

This process involves creating the DNA plasmid, or vector genome, that will be contained within the AAV capsid. The most widely used alternative is SaCas9 from *Staphylococcus aureus*, which is roughly 3.2 kb in size and saves roughly 1 kb over SpCas9. Another choice is Cas12f, which enables even larger payloads due to the fact that these minuscule effectors (~1.5–2 kb) are presently being developed for high efficiency in human cells. Use minimal viral promoters (e.g., mini-CMV, ~0.2–0.3 kb) or short synthetic promoters to drive Cas9 expression. H1/U6 type RNA Polymerase III promoters will be taken into consideration with regard to the gRNAs (Friedland et al. 2015). The HEK293T cells are co-transferred with three distinct plasmids.

This is called triple-transfection method. The engineered "all-in-one" genome found in the vector plasmid includes the Cas gene, gRNA code, and promotor enclosed by ITRs, the only cis-elements needed for packaging. The AAV rep (replication) and cap (capsid protein) genes are supplied by the Rep/Cap plasmid. Adenovirus helper functions (E1A, E1B, E2A, E4, VA RNA) needed for AAV replication and packaging are provided by the adenovirus helper plasmid. The ITR-flanked vector genome from the plasmid is replicated inside the cell by the Rep proteins. The Cap proteins form the capsid, and the replicated vector genome is "packaged" into these empty capsids (Friedland et al. 2015).

5.3.1.3 Lentiviral Vectors–Based Delivery Method

Lentiviral vectors are appropriate for the persistent expression of CRISPR components, including multiple gRNAs in multiplex editing, because of their stable integration into the host genome and moderate packaging capacity (~8 kb). They have clinical trial status for gene therapy techniques and are extensively utilized in ex vivo editing applications.

5.3.1.3.1 Construction of Lentiviral Vector for Multiplex Genome Editing

The initial process of developing a lentiviral vector for genome editing is to engineer a transfer plasmid with the genetic payload flanked by crucial viral components, such as the Rev-Responsive Element for nuclear export of the full-length viral transcript, the packaging signal for RNA encapsidation, and the 5′ and 3′ long terminal repeats for integration (Dull et al. 1998). This plasmid is customized with a multiplexed gRNA system, typically a PTG array, where individual gRNAs are separated by tRNA sequences before being processed by endogenous cellular RNases to create an all-in-one multiplex editing vector. This allows for the concurrent targeting of multiple genomic loci from a single transcriptional unit (Xie et al. 2014). In HEK293T cells, a triple-transfection technique is applied to generate infectious but replication-incompetent viral particles.

The transfer plasmid is co-transfected with a packaging plasmid encoding the structural and enzymatic proteins Gag and Pol, along with an envelope plasmid typically expressing the VSV-G glycoprotein to pseudotype the virions, thereby broadening their host range and enhancing particle stability (Stewart et al. 2011). After transfection, the viral particle-containing supernatant is collected, concentrated, and used to transduce target cells. During this process, the viral capsid is introduced into the cytoplasm, where the viral RNA undergoes reverse transcription to become DNA, and the VSV-G protein facilitates cell entry through endocytosis. The viral integrase imports this DNA as part of a pre-integration complex into the nucleus and randomly integrates it into the host genome, that results stable and long-term expression of the editing machinery and permanent genetic modifications (Naldini et al. 1996). The comparison of different viral delivery methods for multiplex genome editing is indexed in Table 5.1 (Wang et al. 2020).

5.3.2 Nonviral Vector

Nonviral vectors for multiplex genome editing include a range of delivery methods aimed at introducing CRISPR-Cas elements (like multiple sgRNAs and Cas9 protein or mRNA) into cells without the use of viral particles. These include synthetic carriers like lipid nanoparticles, polymers, cell-penetrating peptides, and gold nanoparticle, or physical methods like electroporation and

TABLE 5.1
Comparison of Different Viral Delivery Methods for Multiplex Genome Editing

Feature	Adenovirus	Adeno-Associated Virus	Lentivirus
Packaging capacity	Very high (~30 kb)	Low (~4.7 kb)	High (~8 kb)
Integration profile	Non-integrating (episomal)	Mostly non-integrating (episomal)	Integrating (random)
Editing duration	Long-term transient	Long-term transient	Permanent/stable
In vivo application	Excellent	Excellent	Limited (immune response)
Ex vivo application	Good	Good	Excellent
Multiplexing potential	Single vector can carry large editors + many gRNAs	Requires small effectors (SaCas9) and compact gRNA arrays (tRNA)	Single vector can carry Cas9 + multiple gRNAs
Ideal use case	In vivo delivery of large cargo (e.g., base/prime editors + gRNAs)	In vivo delivery for nondividing cells with compact editors	Stable engineering in ex vivo settings (e.g., cell therapies, screens)

nucleofection that deliver plasmid DNA, mRNA, or RNP complexes. Co-delivering several gRNAs, either as individual molecules or encoded in polycistronic arrays that are processed into distinct guides within the cell, is how multiplex editing is performed. Benefits of nonviral techniques include decreased immunogenicity, no risk of viral recombination, increased packaging capacity for large payloads, and comparatively simple large-scale manufacturing (Li et al. 2018).

5.3.2.1 Nucleofection

A cell suspension is exposed to a regulated electrical pulse, which causes transient holes in the cell membrane. This makes it possible for negatively charged molecules to enter cells directly, such as DNA, mRNA, or RNPs. Nucleofection is a more sophisticated method that enables direct delivery to the nucleus by utilizing particular electrical parameters and solutions. It is possible to efficiently co-deliver several different RNP complexes that each target a different genomic locus at the same time. For ex vivo editing of difficult-to-transfect primary cells like T-cells, NK cells, and hematopoietic stem cells, this is considered the ultimate procedure (Kim et al. 2014).

5.3.2.2 Lipid Nanoparticles

A biodegradable lipid bilayer encloses the editing content (RNPs or mRNA encoding Cas9 with sgRNAs). These Lipid Nanoparticles (LNPs) are either taken up by endocytosis or fuse with the cell's plasma membrane. They are designed with ionizable lipids that, in an acidic endosome, acquire a positive charge, causing vascular release of the content into the cytoplasm. Multiplexed editing in vivo is made possible by the co-encapsulation of mRNA for Cas9 and synthetic gRNAs (or multiple RNP complexes). LNPs are being developed for affinity to other tissues and are very successful at targeting the liver (Cullis & Hope 2017).

5.3.2.3 Microinjection

A precise volume of editing substance is injected directly into the cytoplasm or nucleus of a single cell applying a glass capillary needle with an incredibly fine tip (diameter 0.5–5 μm) that mechanically penetrates the plasma membrane (and optionally to nuclear envelope). A preferred Cas9 RNP complex is one of the most popular and efficient substances. To create active RNP complexes, recombinant Cas9 protein is combined with synthetic, chemically altered gRNAs (sgRNAs or crRNAs/tracrRNAs) and incubated. The administered mixture contains a number of different RNP complexes, each of which is programmed with a distinct gRNA to target a different genomic locus (Wang et al. 2013).

5.3.2.4 Cell-Penetrating Peptides

Positively charged peptides are bound to the RNP complex or the Cas9 protein. Through direct translocation or endocytosis, these peptides promote cellular uptake by facilitating interaction with the negatively charged cell membrane. RNPs can be combined and delivered with unique gRNAs. Its efficiency is generally lower and less consistent across cell types, despite being simpler than LNPs (Gaj et al. 2014).

5.4 ADVANCED APPLICATIONS OF MULTIPLEX EDITING

5.4.1 Functional Genomics and Genetic Interaction Mapping

The systematic, large-scale screening of genetic interactions such as synthetic toxicity (a novel concept for cancer therapy where the simultaneous disruption of two genes causes cell death while targeting either alone does not) is made possible by multiplex editing, a fundamental application in functional genomics and genetic interaction mapping. To ensure that most cells receive just single set of genetic alterations, researchers use pooled libraries of lentiviral vectors, each containing a construct with multiple gRNAs (such as 2 or 3). These vectors are then introduced into a cell population at a low multiplicity of infection (Han et al. 2017).

5.4.2 Modeling Complex Human Diseases

Researchers are able to explain some complicated human diseases occurring as a result of mutations in many genes, such as cancer and neurodegenerative disorders, by using multiplex genome editing. As it is difficult to reproduce the effects of more than one mutation combined, these conditions cannot be modeled with traditional methods of research. By allowing precise and simultaneous changes to more than one gene per single human pluripotent stem cell or organoid, the technique of multiplex editing got around this problem. This implies that laborious cross-breeding is no longer required by the researchers (González et al. 2014). For example, in one such experiment, researchers can turn on an oncogene such as KRASG12D, turn off a stability gene such as PTEN, and turn off a tumor suppressor gene such as P53, to get a similar genotype as cancer. Researcher used the technique called, CRP-Ras9 and the human intestinal organoids to model the development of colorectal cancer by introducing a series of mutations progressively (APC, P53, KraS, and SMAD4). The process can now be accelerated by making all of these mutations at once using modern multiplex tools. This technique thus allows researchers to trial the functions of genetic variants and develop very accurate models of diseases on normal cells (Drost et al. 2015).

5.4.3 Metabolic Engineering and Synthetic Biology

Multiplex CRISPR in microbes is a very powerful method that helps scientists to make multiple genetic changes at the same time as opposed to editing genes one at a time. Instead of tweaking genes one by one, scientists will be able to make multiple genetic alterations simultaneously in microbes with the powerful technique known as multiplexing, or the acronym for the whole method – CRISPR. This approach is what is being used in metabolic engineering to alternately switch on (activate) crucial biosynthetic genes, and switch off (knock out) genes that interfere or block a desired chemical pathway. By means of modulating the metabolism of the cell, more resources are allocated to the production of beneficial compounds as opposed to by-products, such as biofuels, medications, or certain chemicals (Behler et al. 2018).

Recently, a study showed the cyanobacterium *Synechococcus* sp. was manipulated through a multiplex editing approach to produce an important high-value product, the ketocarotenoid named astaxanthin, by expressing recombinant beta-ketolase (bKT) and a beta-hydroxylase enzymes (CtrZ) (Victoria et al. 2024). By altering a number of genes simultaneously, multiplexing, or using a technology known as multiplex, a much faster and more efficient process is achieved with the use of the technique known as CRISPR. The establishment and latest developments in technologies related to multiplex automated genome engineering enable the production of useful strains of microorganisms with demanded phenotype for high yields of product or desired phenotype in just one experiment, which can save a lot time and effort at comparing to difficulties in using classic methodology with high number of cycles of mutation and selection (Gao et al. 2024). This procedure has been widely used in bacteria and yeast to enhance yields for industrial bioproduction and to study the interaction of genes in a metabolic network. This approach has also been very widely used in yeast and bacteria for the purposes of enhancing yields for industrial bioproduction and studying the interactions between genes in metabolic networks (Bindal & Rath 2025).

5.4.4 Applications in Agriculture Sector

In agriculture sector, the use of multiplex genome editing is an incredibly powerful tool that has allowed scientists to precisely modify multiple genes simultaneously. Compared to traditional breeding, which takes an average of years to improve crops, this is extremely fast-paced. Farmers can boost rice production through the modification of several genes important for such traits as grain size, numbers, and weight. Recently, researchers alter 36 genetic loci simultaneously in polyploid crops (Miroshnichenko et al. 2023). Similarly, other crops such as cucumbers build up long-lasting resistance against viruses by removing multiple genes that make the plants prone to infection by diseases. By modifying a number of genes that control the synthesis of vitamins and reduce toxic substances this way, this technique also contributes to enhancing the nutritional value of crops. In order to produce plants that are more suited to dense farming, it can alter plant growth characteristics such as height and flowering time. Crucially, these enhancements are safe and effective, since they do not require foreign DNA to add up on advantageous mutations (Prajapati & Tyagi 2024). Programmable nuclease systems, such as CRISPR/Cas12i3, are employed in precision molecular breeding for the development of herbicide-resistant rice plants (Lv et al. 2024).

5.5 CONCLUSION

Multiplex editing of multiple target sites from a single construct is enabled by strategies such as tRNA–gRNA arrays and the intrinsic ability of Cas12a to process multiple guide RNAs simultaneously. Autochthonous genome editing is becoming revolutionary because it goes beyond single gene editing to address the complexity of biological systems. The exact manipulation of genetic networks is made possible by new base and prime editing technologies as well as sophisticated delivery techniques. The systematic mapping of genetic interactions, the establishment of models for the treatment of polygenic disorders, the rewiring of metabolic pathways for synthetic biology, and the speed up of the development of cell and gene therapy are just some of the many and revolutionary applications of multiplex genome editing. It enables the study and de novo design of complex gene networks, thereby driving major biological discoveries and the development of novel therapeutic strategies. However, there are still issues that need to be solved, such as effectiveness of delivery, off-target effects, cytotoxicity of cells, and potential undesired changes to the genome (e.g., chromosomal rearrangement and epigenetic modifications). In order to minimize these risks and improve editing reliability, ongoing research is focused on optimizing both genome-editing tools and their delivery systems.

REFERENCES

Adiego-Pérez, B., Randazzo, P., Daran, J. M., Verwaal, R., Roubos, J. A., Daran-Lapujade, P., & Van Der Oost, J. (2019). Multiplex genome editing of microorganisms using CRISPR-Cas. *FEMS Microbiology Letters*, 366(8). https://doi.org/10.1093/femsle/fnz086

Anzalone, A. V., Randolph, P. B., Davis, J. R., Sousa, A. A., Koblan, L. W., Levy, J. M., Chen, P. J., Wilson, C., Newby, G. A., Raguram, A., & Liu, D. R. (2019). Search-and-replace genome editing without double-strand breaks or donor DNA. *Nature*, 576(7785). https://doi.org/10.1038/s41586-019-1711-4

Ates, I., Rathbone, T., Stuart, C., Hudson Bridges, P., & Cottle, R. N. (2020). Delivery approaches for therapeutic genome editing and challenges. *Genes*, 11(10). https://doi.org/10.3390/genes11101113

Bandyopadhyay, A., Kancharla, N., Javalkote, V. S., Dasgupta, S., & Brutnell, T. P. (2020). CRISPR-Cas12a (Cpf1): A versatile tool in the plant genome editing tool box for agricultural advancement. *Frontiers in Plant Science*, 11. https://doi.org/10.3389/fpls.2020.584151

Behler, J., Vijay, D., Hess, W. R., & Akhtar, M. K. (2018). CRISPR-based technologies for metabolic engineering in cyanobacteria. *Trends in Biotechnology*, 36(10). https://doi.org/10.1016/j.tibtech.2018.05.011

Bindal, G., & Rath, D. (2025). Programmable multiplex genome editing: Innovations in CRISPR effectors, crRNA engineering, and delivery strategies. *Synthetic Biology and Engineering*, 3(3), 10014–10014. https://doi.org/10.70322/sbe.2025.10014

Boch, J., Scholze, H., Schornack, S., Landgraf, A., Hahn, S., Kay, S., Lahaye, T., Nickstadt, A., & Bonas, U. (2009). Breaking the code of DNA binding specificity of TAL-type III effectors. *Science*, 326(5959). https://doi.org/10.1126/science.1178811

Cullis, P. R., & Hope, M. J. (2017). Lipid nanoparticle systems for enabling gene therapies. *Molecular Therapy*, 25(7). https://doi.org/10.1016/j.ymthe.2017.03.013

Dong, F., Xie, K., Chen, Y., Yang, Y., & Mao, Y. (2017). Polycistronic tRNA and CRISPR guide-RNA enables highly efficient multiplexed genome engineering in human cells. *Biochemical and Biophysical Research Communications*, 482(4). https://doi.org/10.1016/j.bbrc.2016.11.129

Drost, J., Van Jaarsveld, R. H., Ponsioen, B., Zimberlin, C., Van Boxtel, R., Buijs, A., Sachs, N., Overmeer, R. M., Offerhaus, G. J., Begthel, H., Korving, J., Van De Wetering, M., Schwank, G., Logtenberg, M., Cuppen, E., Snippert, H. J., Medema, J. P., Kops, G. J. P. L., & Clevers, H. (2015). Sequential cancer mutations in cultured human intestinal stem cells. *Nature*, 521(7550). https://doi.org/10.1038/nature14415

Dull, T., Zufferey, R., Kelly, M., Mandel, R. J., Nguyen, M., Trono, D., & Naldini, L. (1998). A third-generation lentivirus vector with a conditional packaging system. *Journal of Virology*, 72(11). https://doi.org/10.1128/jvi.72.11.8463-8471.1998

Friedland, A. E., Baral, R., Singhal, P., Loveluck, K., Shen, S., Sanchez, M., Marco, E., Gotta, G. M., Maeder, M. L., Kennedy, E. M., Kornepati, A. V. R., Sousa, A., Collins, M. A., Jayaram, H., Cullen, B. R., & Bumcrot, D. (2015). Characterization of *Staphylococcus aureus* Cas9: A smaller Cas9 for all-in-one adeno-associated virus delivery and paired nickase applications. *Genome Biology*, 16(1). https://doi.org/10.1186/s13059-015-0817-8

Gaj, T., Liu, J., Anderson, K. E., Sirk, S. J., & Barbas, C. F. (2014). Protein delivery using Cys2-His2 zinc-finger domains. *ACS Chemical Biology*, 9(8). https://doi.org/10.1021/cb500282g

Gao, H., Qiu, Z., Wang, X., Zhang, X., Zhang, Y., Dai, J., & Liang, Z. (2024). Recent advances in genome-scale engineering in *Escherichia coli* and their applications. *Engineering Microbiology*, 4(1). https://doi.org/10.1016/j.engmic.2023.100115

Gaudelli, N. M., Komor, A. C., Rees, H. A., Packer, M. S., Badran, A. H., Bryson, D. I., & Liu, D. R. (2017). Programmable base editing of T to G C in genomic DNA without DNA cleavage. *Nature*, 551(7681). https://doi.org/10.1038/nature24644

González, F., Zhu, Z., Shi, Z. D., Lelli, K., Verma, N., Li, Q. V., & Huangfu, D. (2014). An iCRISPR platform for rapid, multiplexable, and inducible genome editing in human pluripotent stem cells. *Cell Stem Cell*, 15(2). https://doi.org/10.1016/j.stem.2014.05.018

Han, K., Jeng, E. E., Hess, G. T., Morgens, D. W., Li, A., & Bassik, M. C. (2017). Synergistic drug combinations for cancer identified in a CRISPR screen for pairwise genetic interactions. *Nature Biotechnology*, 35(5). https://doi.org/10.1038/nbt.3834

Haurwitz, R. E., Jinek, M., Wiedenheft, B., Zhou, K., & Doudna, J. A. (2010). Sequence- and structure-specific RNA processing by a CRISPR endonuclease. *Science*, 329(5997). https://doi.org/10.1126/science.1192272

Jinek, M., Chylinski, K., Fonfara, I., Hauer, M., Doudna, J. A., & Charpentier, E. (2012). A programmable dual-RNA-guided DNA endonuclease in adaptive bacterial immunity. *Science*, 337(6096). https://doi.org/10.1126/science.1225829

Kim, S., Kim, D., Cho, S. W., Kim, J., & Kim, J. S. (2014). Highly efficient RNA-guided genome editing in human cells via delivery of purified Cas9 ribonucleoproteins. *Genome Research*, 24(6). https://doi.org/10.1101/gr.171322.113

Komor, A. C., Kim, Y. B., Packer, M. S., Zuris, J. A., & Liu, D. R. (2016). Programmable editing of a target base in genomic DNA without double-stranded DNA cleavage. *Nature*, 533. https://doi.org/10.1038/nature17946

Lee, C. S., Bishop, E. S., Zhang, R., Yu, X., Farina, E. M., Yan, S., Zhao, C., Zeng, Z., Shu, Y., Wu, X., Lei, J., Li, Y., Zhang, W., Yang, C., Wu, K., Wu, Y., Ho, S., Athiviraham, A., Lee, M. J., … He, T. C. (2017). Adenovirus-mediated gene delivery: Potential applications for gene and cell-based therapies in the new era of personalized medicine. *Genes and Diseases*, 4(2). https://doi.org/10.1016/j.gendis.2017.04.001

Li, L., Hu, S., & Chen, X. (2018). Non-viral delivery systems for CRISPR/Cas9-based genome editing: Challenges and opportunities. *Biomaterials*, 171. https://doi.org/10.1016/j.biomaterials.2018.04.031

Lv, P., Su, F., Chen, F., Yan, C., Xia, D., Sun, H., Li, S., Duan, Z., Ma, C., Zhang, H., Wang, M., Niu, X., Zhu, J. K., & Zhang, J. (2024). Genome editing in rice using CRISPR/Cas12i3. *Plant Biotechnology Journal*, 22(2). https://doi.org/10.1111/pbi.14192

Miroshnichenko, D. N., Timerbaev, V. R., Divashuk, M. G., Pushin, A. S., Alekseeva, V., Kroupin, P. Y., Bazhenov, M. S., Samarina, M. A., Ermolaev, A., Karlov, G., & Dolgov, S. V. (2023). Advancing gene editing: Multiplex mutagenesis in hexaploid triticale. *Ecological Genetics*, 21(3). https://doi.org/10.17816/ecogen568624

Naldini, L., Blömer, U., Gallay, P., Ory, D., Mulligan, R., Gage, F. H., Verma, I. M., & Trono, D. (1996). In vivo gene delivery and stable transduction of nondividing cells by a lentiviral vector. *Science*, 272(5259). https://doi.org/10.1126/science.272.5259.263

Nissim, L., Perli, S. D., Fridkin, A., Perez-Pinera, P., & Lu, T. K. (2014). Multiplexed and programmable regulation of gene networks with an integrated RNA and CRISPR/Cas toolkit in human cells. *Molecular Cell*, 54(4). https://doi.org/10.1016/j.molcel.2014.04.022

Paul, B., & Montoya, G. (2020). CRISPR-Cas12a: Functional overview and applications. *Biomedical Journal*, 43(1). https://doi.org/10.1016/j.bj.2019.10.005

Prajapati, R., & Tyagi, K. (2024). CRISPR/Cas-mediated multiplex genome editing in plants and applications. In Santosh Kumar Upadhyay (Ed.), *Applications of Genome Engineering in Plants* (pp. 20–39). John Wiley & Sons, Ltd. https://doi.org/10.1002/9781394183913.ch2

Stewart, H. J., Fong-Wong, L., Strickland, I., Chipchase, D., Kelleher, M., Stevenson, L., Thoree, V., McCarthy, J., Ralph, G. S., Mitrophanous, K. A., & Radcliffe, P. A. (2011). A stable producer cell line for the manufacture of a lentiviral vector for gene therapy of Parkinson's disease. *Human Gene Therapy*, 22(3). https://doi.org/10.1089/hum.2010.142

Urnov, F. D., Miller, J. C., Lee, Y. L., Beausejour, C. M., Rock, J. M., Augustus, S., Jamieson, A. C., Porteus, M. H., Gregory, P. D., & Holmes, M. C. (2005). Highly efficient endogenous human gene correction using designed zinc-finger nucleases. *Nature*, 435(7042). https://doi.org/10.1038/nature03556

Victoria, A. J., Toscano Selão, T., Ángel Moreno-Cabezuelo, J., Mills, L. A., Gale, G. A. R., Lea-Smith, D. J., Mccormick, A. J., & Mccormick, A. (2024). A toolbox to engineer the highly productive cyanobacterium Synechococcus sp. PCC 11901. *Plant Physiology* 196(2), 1674–90. https://doi.org/10.1093/plphys/kiae261

Wang, H., Yang, H., Shivalila, C. S., Dawlaty, M. M., Cheng, A. W., Zhang, F., & Jaenisch, R. (2013). One-step generation of mice carrying mutations in multiple genes by CRISPR/Cas-mediated genome engineering. *Cell*, 153(4). https://doi.org/10.1016/j.cell.2013.04.025

Wang, D., Zhang, F., & Gao, G. (2020). CRISPR-based therapeutic genome editing: Strategies and in vivo delivery by AAV vectors. *Cell*, 181(1). https://doi.org/10.1016/j.cell.2020.03.023

Xie, S., Shen, B., Zhang, C., Huang, X., & Zhang, Y. (2014). SgRNAcas9: A software package for designing CRISPR sgRNA and evaluating potential off-target cleavage sites. *PLoS ONE*, 9(6). https://doi.org/10.1371/journal.pone.0100448

6 Genome Editing in Stem Cells and Regenerative Medicine

Afreen Anis and Nisha Kumari

6.1 INTRODUCTION

Advances in genome editing and stem cell technologies have reshaped the landscape of modern regenerative medicine. Genome editing allows researchers to modify genetic sequences with unprecedented precision, targeting specific loci to either disrupt, correct, or replace DNA sequences. Stem cells, characterized by their self-renewal capacity and differentiation potential, provide an ideal platform for these interventions. Together, these technologies have enabled the creation of physiologically relevant disease models, improved understanding of human development, and development of novel therapies for genetic and acquired disorders (Jaenisch & Young, 2008; Doudna & Charpentier, 2014).

Traditional genetic engineering approaches, such as random transgene integration or homologous recombination, were limited by inefficiency, low reproducibility, and unpredictable genomic effects. The advent of programmable nucleases transformed the field, enabling targeted DNA cleavage and controlled genetic modification. Early systems included zinc finger nucleases (ZFNs) and transcription activator-like effector nucleases (TALENs), which rely on engineered proteins to recognize specific DNA sequences. Despite their effectiveness, their design complexity and high cost limited their widespread adoption (Gaj et al., 2013; Urnov et al., 2010). The introduction of RNA-guided CRISPR/Cas systems dramatically simplified genome editing. CRISPR/Cas9 employs a guide RNA to direct the Cas9 nuclease to a specific genomic sequence, allowing precise double-strand breaks.

Subsequent repair by the cell's endogenous mechanisms can result in gene disruption or correction. Variants such as Cas12 and Cas13 expand the scope to alternative DNA and RNA targets, while base editing and prime editing enable single-nucleotide changes without inducing double-strand breaks, reducing genomic instability (Anzalone et al., 2020; Liu et al., 2023). Stem cells are particularly well suited for genome editing due to their proliferative capacity and ability to differentiate into multiple cell types. Pluripotent stem cells, such as embryonic stem cells (ESCs) and induced pluripotent stem cells (iPSCs), allow detailed mechanistic studies, disease modeling, and creation of patient-specific therapeutic cells. Adult stem cells, including hematopoietic stem cells (HSCs), mesenchymal stem cells (MSCs), and neural stem cells (NSCs), provide translational opportunities for regenerative therapies due to their accessibility, lineage specificity, and established clinical use (Rowe & Daley, 2019). This chapter provides a comprehensive overview of genome editing in stem cells for regenerative medicine, covering stem cell types, genome editing technologies, applications in pluripotent and adult stem cells, delivery methods, and ethical and regulatory considerations.

6.2 STEM CELL PLATFORMS FOR GENOME EDITING

Stem cells serve as essential substrates for genome editing due to their self-renewal ability, developmental plasticity, and clinical relevance. They are broadly categorized into pluripotent stem cells and adult (somatic) stem cells, each offering unique advantages and limitations (Jinek et al., 2012).

DOI: 10.1201/9781003735113-6

TABLE 6.1
Comparative Features of Major Stem Cell Types

Stem Cell Type	Source	Differentiation Potential	Ethical Concerns	Clinically Utility	References
ESCs	Blastocyst	Pluripotent	High	Limited	Li et al., 2022
iPSCs	Reprogrammed somatic cells	Pluripotent	Low	High	
Adult stem cells	Tissue-specific niches	Multipotent	Minimal	Established	

6.2.1 Embryonic Stem Cells

ESCs are derived from the inner cell mass of preimplantation blastocysts. They exhibit pluripotency, the ability to differentiate into all three germ layers (endoderm, mesoderm, ectoderm), and indefinite self-renewal under defined culture conditions (Evans & Kaufman, 1981). Genome editing in ESCs has been instrumental in functional genomics, providing insights into developmental biology and gene regulatory networks. However, clinical translation is constrained by ethical considerations, immunogenicity, and risk of tumorigenicity (Hanna et al., 2010).

6.2.2 Induced Pluripotent Stem Cells

iPSCs are generated by reprogramming adult somatic cells using defined transcription factors, thereby restoring pluripotency (Takahashi & Yamanaka, 2006). iPSCs offer patient-specific genetic backgrounds, circumventing immune rejection and ethical concerns associated with ESCs. Genome editing in iPSCs enables precise correction of pathogenic mutations, creation of isogenic control lines, and modeling of genetic disorders. These features make iPSCs highly valuable for drug screening, disease modeling, and personalized regenerative therapies (Hockemeyer & Jaenisch, 2016).

6.2.3 Adult Stem Cells

Adult stem cells are tissue-resident progenitors responsible for homeostasis and repair. Examples include HSCs, MSCs, and NSCs. Table 6.1 represents comparative features of major stem cell types. Although their differentiation potential is restricted compared to pluripotent stem cells, adult stem cells have established clinical applications, lower tumorigenic risk, and ethical acceptability (Rowe & Daley, 2019). They are increasingly targeted for gene correction strategies aimed at treating blood disorders, musculoskeletal injuries, and neurological diseases. The comparative features of major stem cell types are indexed in Table 6.1.

6.3 GENOME EDITING TECHNOLOGIES

Genome editing technologies have evolved from protein-based recognition systems to RNA-guided nucleases, enabling precise genetic modifications. The evolution of programmable nucleases for genome editing demonstrates a trajectory of increasing precision and ease of use. The first-generation ZFNs, which fuse zinc finger domains to a FokI endonuclease, enable site-specific cleavage but require complex protein engineering for each new target sequence (Urnov et al., 2010). Subsequently, TALENs, also employing FokI, offered a more predictable design framework for high-specificity editing, though their construction remained labor-intensive (Gaj et al., 2013). A paradigm shift arrived with the RNA-guided CRISPR/Cas9 system, which facilitates highly efficient, scalable, and multiplexed genome editing by inducing double-strand breaks at loci specified by a guide RNA (Cong et al., 2013; Doudna & Charpentier, 2014). Most recently, advanced editing technologies have been developed to further minimize unintended alterations. Base editors achieve

TABLE 6.2
Genome Editing Platforms in Stem Cells

Technology	Mechanism	Advantages	Limitations	References
ZFNs	Protein–DNA recognition	High specificity	Complex design expensive	Boettcher & McManus, 2015; Barrangou & Doudna, 2016
TALENs	TALE–DNA recognition	Improved specificity	Labor-intensive	
CRISPR/Cas	RNA-guided DNA cleavage	High efficiency, scalable	Off-target effects possible	
Base/prime editors	Nucleotide substitution/ precise editing	High precision, fewer double-strand breaks	Emerging, technical complexity	

direct nucleotide conversions without double-strand breaks, while prime editors, combining a Cas9 nickase with a reverse transcriptase, enable precise insertions, deletions, and point mutations; both systems enhance editing fidelity and genomic integrity in sensitive applications like stem cell engineering (Anzalone et al., 2020; Liu et al., 2023). Table 6.2 gives a summarized overview for the current genome editing platforms in stem cells.

6.4 GENOME EDITING IN PLURIPOTENT STEM CELLS

Pluripotent stem cells provide a versatile platform for functional genomics, disease modeling, and regenerative applications. The integration of CRISPR/Cas9 genome editing with pluripotent stem cells has profoundly accelerated biomedical research. In ESCs and iPSCs, CRISPR/Cas9 facilitates targeted gene knockout studies, enabling the elucidation of fundamental developmental pathways and disease mechanisms (Hsu et al., 2014). A key therapeutic application is the precise correction of pathogenic mutations in patient-derived iPSCs; this process generates isogenic control lines that are indispensable for accurate disease modeling and the development of personalized therapies (Lattanzi et al., 2021). Furthermore, these genome-edited iPSCs can be differentiated into specialized cell types or complex organoids, thereby recapitulating human disease phenotypes for robust drug testing and therapeutic evaluation (Karagiannis et al., 2019; Li et al., 2022).

6.5 GENOME EDITING IN ADULT STEM CELLS FOR REGENERATIVE THERAPIES

Adult stem cells are essential for translational regenerative medicine because they reside in tissues responsible for maintenance and repair (Kim et al., 1996). They include HSCs, MSCs, and NSCs. Genome editing in these cells enables correction of disease-causing mutations, enhancement of therapeutic properties, and generation of models for preclinical studies.

6.5.1 Hematopoietic Stem Cells

HSCs are clinically relevant due to their ability to reconstitute the entire blood system. CRISPR/Cas9-mediated gene correction in HSCs has shown remarkable success in preclinical and early clinical trials for sickle cell disease and β-thalassemia (Frangoul et al., 2021; Dever et al., 2023). Edited HSCs can be transplanted autologously, reducing risks of immune rejection and graft-versus-host disease. Challenges include ensuring efficient engraftment, minimizing off-target edits, and preserving long-term self-renewal.

6.5.2 Mesenchymal Stem Cells

MSCs are multipotent cells with immunomodulatory and regenerative properties, making them attractive for musculoskeletal, cardiovascular, and autoimmune disorders. Genome editing can enhance MSC survival, differentiation potential, and secretion of therapeutic factors (Li et al., 2022).

TABLE 6.3
Therapeutic Applications of Genome-Edited Adult Stem Cells

Stem Cell Type	Target Disease	Editing Strategy	Clinical/ Preclinical Status	References
HSCs	Sickle cell disease and β-thalassemia	CRISPR/Cas9 correction	Early clinical trials	Knott & Doudna, 2018; Menzies et al., 2022
MSCs	Musculoskeletal cardiovascular disorders	Gene enhancement, overexpression	Preclinical	
NSCs	Neurodegenerative diseases	Gene correction/ modulation	Preclinical	

6.5.3 Neural Stem Cells

NSCs reside in specific brain regions and contribute to neurogenesis and tissue repair. Genome editing strategies in NSCs include correction of mutations in neurodegenerative diseases and modulation of signaling pathways to enhance neuronal differentiation and survival. Preclinical studies demonstrate that CRISPR-modified NSCs can ameliorate disease phenotypes in models of Parkinson's disease, spinal cord injury, and other neurological disorders (Hockemeyer & Jaenisch, 2016). Table 6.3 elaborates few of the applications of adult stem cells.

6.6 REGENERATIVE MEDICINE APPLICATIONS

Genome-edited stem cells have a profound impact on tissue engineering, organoid modeling, and combined gene-cell therapies.

- **Tissue Engineering**: Edited stem cells can enhance extracellular matrix production, angiogenesis, and functional integration within engineered constructs. Genetic modifications, such as overexpression of VEGF (Vascular Endothelial Growth Factor) or anti-apoptotic genes, improve survival and integration of transplanted tissues (Wu & Hochedlinger, 2011).
- **Organoid-Based Models**: Organoids derived from genome-edited iPSCs provide physiologically relevant 3D models for organ development, disease modeling, and pharmacological testing. These systems enable high-throughput drug screening while preserving patient-specific genotypes (Li et al., 2022; Karagiannis et al., 2019).
- **Cell Replacement and Gene Therapy Hybrids**: Combining gene correction with cell replacement therapy addresses both the underlying genetic defect and tissue damage. For instance, autologous iPSCs corrected for monogenic mutations can be differentiated into functional tissue-specific cells for transplantation, offering personalized regenerative solutions (Cox et al., 2015).

6.7 DELIVERY STRATEGIES FOR GENOME-EDITED CELLS

Efficient and safe delivery of genome editing components is critical for therapeutic success. Strategies can be divided into viral and nonviral approaches.

6.7.1 Viral Vectors

Viral vectors, including lentivirus, adenovirus, and adeno-associated virus, provide high transduction efficiency. They are widely used for both in vitro and in vivo delivery but present risks of immunogenicity, insertional mutagenesis, and prolonged expression of nucleases, which may increase off-target effects (Feng et al., 2023).

6.7.2 Nonviral Methods

Nonviral approaches, such as electroporation, nucleofection, and delivery of ribonucleoprotein (RNP) complexes, reduce immunogenic risks and enable transient expression. These methods are particularly advantageous for ex vivo genome editing of stem cells, where edited cells can be selected and expanded before transplantation (Pickar-Oliver & Gersbach, 2019).

6.7.3 In Vivo Versus Ex Vivo Editing

Ex vivo editing involves isolating stem cells, performing genome editing in controlled conditions, and transplanting the modified cells back into the patient. This method allows rigorous quality control and minimizes systemic exposure to genome editing tools. In vivo editing delivers genome editing components directly into the patient, bypassing cell isolation. While promising for certain organs (e.g., liver, muscle), it carries higher risks of off-target effects and immune reactions (Gillmore et al., 2021).

6.8 SAFETY, ETHICAL, AND REGULATORY CONSIDERATIONS

Safety is paramount for clinical translation of genome-edited stem cells. Key concerns include off-target mutations, chromosomal instability, immunogenicity, and tumorigenicity. Advances in high-fidelity Cas variants, base editing, and prime editing help mitigate these risks (Wilkinson et al., 2023). Ethical considerations include the sourcing of ESCs, informed consent for donor cells, equitable access to therapies, and strict prohibition of germline genome editing. Public engagement, transparency, and adherence to international guidelines are essential to maintain trust in regenerative medicine (Menzies et al., 2022). Regulatory frameworks emphasize rigorous preclinical testing, standardized manufacturing protocols, and long-term patient monitoring. Agencies such as the Food and Drug Administration, enhanced mechanical qualification, and national biotechnology authorities require clear documentation of genome editing efficiency, specificity, and safety before clinical approval (Frangoul et al., 2021).

6.9 CURRENT CLINICAL TRIALS AND CASE STUDIES

Genome-edited stem cells are transitioning from preclinical research to clinical application. The most advanced examples involve HSCs for monogenic blood disorders:

- **Sickle Cell Disease and β-Thalassemia**: CRISPR/Cas9-mediated correction of the β-globin gene in autologous HSCs has demonstrated long-term therapeutic efficacy. Patients receiving edited HSC transplants show restoration of normal hemoglobin levels and resolution of disease symptoms (Frangoul et al., 2021; Dever et al., 2023).
- **Immunodeficiency Disorders**: Edited HSCs targeting the IL2RG gene for X-linked severe combined immunodeficiency (SCID-X1) have shown proof-of-concept for functional immune reconstitution. Preclinical studies highlight the importance of off-target analysis and vector optimization for clinical safety (Charlesworth et al., 2019).
- **Mesenchymal and Neural Stem Cells**: MSCs and NSCs are under investigation for regenerative therapies in cardiac repair, neurodegenerative diseases, and musculoskeletal injuries. Early-phase clinical trials focus on feasibility, safety, and engraftment, rather than efficacy, but demonstrate encouraging regenerative potential (Wu & Hochedlinger, 2011; Li et al., 2022).

6.10 CHALLENGES AND FUTURE DIRECTIONS

Despite significant progress, several challenges remain in the clinical translation of genome-edited stem cells. The clinical translation of genome-edited stem cells faces several interconnected technical and ethical challenges. Technically, achieving high delivery efficiency without compromising cell viability necessitates methods like electroporation of RNP complexes, yet advancements in tissue-specific targeting are still required (Pickar-Oliver & Gersbach, 2019). Concurrently, minimizing off-target effects remains a priority, driving the development and validation of high-fidelity Cas variants, base editors, and prime editors to enhance genomic precision (Anzalone et al., 2020; Liu et al., 2023). Furthermore, ensuring the genomic stability and robust differentiation potential of edited cells during ex vivo expansion is critical, as epigenetic drift and culture-induced senescence can undermine therapeutic efficacy (Wilkinson et al., 2023). Ethically, the field must navigate ongoing scrutiny of ESC-based applications, issues of informed consent, and the societal implications of heritable edits, while regulatory pathways demand rigorous standardization and safety data to mitigate risks like tumorigenicity (Menzies et al., 2022; Frangoul et al., 2021). Future progress will likely be directed by AI-guided design to optimize editing parameters, the creation of hypoimmunogenic universal donor cells for off-the-shelf therapies, and the combination of genome editing with tissue engineering and organoid technologies to enable complex regenerative strategies (Zhang et al., 2022; Li et al., 2022). Collectively, overcoming these hurdles promises to realize safer, more efficient, and widely applicable genome-edited stem cell therapies, shaping the next generation of regenerative medicine.

6.11 CONCLUSION

Genome editing in stem cells represents a transformative paradigm in regenerative medicine, merging the developmental potential of stem cells with increasingly precise molecular tools. This synergy has unlocked powerful applications in disease modeling, drug discovery, and the creation of patient-specific or universal therapeutic cells. The clinical translation of these technologies, exemplified by early successes in HSC therapies, demonstrates tangible progress while underscoring the critical need to address delivery efficiency, genomic safety, and ethical oversight. Looking ahead, innovations such as AI-guided design, prime editing, and integration with tissue engineering promise to expand the scope of these therapies from rare genetic disorders to complex degenerative conditions. Ultimately, the responsible advancement of this field, guided by rigorous science and ethical governance, holds unprecedented potential for personalized and regenerative medicine (Capecchi, 2005).

REFERENCES

Anzalone, A. V., Koblan, L. W., & Liu, D. R. (2020).Genome editing with CRISPR–Cas nucleases, base editors, transposases, and prime editors. *Nature Biotechnology,* 38(7), 824–844.

Barrangou, R., & Doudna, J. A. (2016). Applications of CRISPR technologies in research and beyond. *Nature Biotechnology,* 34(9), 933–941.

Boettcher, M., & McManus, M. T. (2015).Choosing the right tool for the job: RNAi, TALEN, or CRISPR. *Molecular Cell,* 58(4), 575–585.

Capecchi, M. R. (2005). Gene targeting in mice: Functional analysis of the mammalian genome. *Nature Reviews Genetics,* 6(6), 507–512.

Charlesworth, C. T., Deshpande, P. S., Dever, D. P., Camarena, J., Lemgart, V. T., Cromer, M. K., … Corn, J. E. (2019).Identification of preexisting adaptive immunity to Cas9 proteins in humans. *Nature Medicine,* 25(2), 249–254.

Cong, L., Ran, F. A., Cox, D., Lin, S., Barretto, R., Habib, N., … Zhang, F. (2013). Multiplex genome engineering using CRISPR/Cas systems. *Science,* 339(6121), 819–823.

Cox, D. B. T., Platt, R. J., & Zhang, F. (2015). Therapeutic genome editing: Prospects and challenges. *Nature Medicine,* 21(2), 121–131.

Dever, D. P., et al. (2023). CRISPR-Cas9 gene editing in hematopoietic stem cells for inherited blood disorders. *Nature Medicine,* 29(1), 37–46.

Doudna, J. A., & Charpentier, E. (2014). The new frontier of genome engineering with CRISPR–Cas9. *Science,* 346(6213), 1258096.

Evans, M. J., & Kaufman, M. H. (1981). Establishment in culture of pluripotential cells from mouse embryos. *Nature,* 292(5819), 154–156.

Feng, R., Patil, S., Zhao, X., & Li, S. (2023). Advances in nonviral delivery systems for genome editing in stem cells. *Advanced Drug Delivery Reviews,* 195, 114731.

Frangoul, H., Altshuler, D., Cappellini, M. D., Chen, Y. S., Domm, J., Eustace, B. K., … Corbacioglu, S. (2021). CRISPR-Cas9 gene editing for sickle cell disease and β-thalassemia. *New England Journal of Medicine,* 384(3), 252–260.

Gaj, T., Gersbach, C. A., & Barbas, C. F. (2013). ZFN, TALEN, and CRISPR/Cas-based methods for genome engineering. *Trends in Biotechnology,* 31(7), 397–405.

Gillmore, J. D., Gane, E., Taubel, J., Kao, J., Fontana, M., Maitland, M. L., … Lebwohl, D. (2021). CRISPR-Cas9 in vivo gene editing for transthyretin amyloidosis. *New England Journal of Medicine,* 385(6), 493–502.

Hanna, J. H., Saha, K., & Jaenisch, R. (2010). Pluripotency and cellular reprogramming: Facts, hypotheses, unresolved issues. *Cell,* 143(4), 508–525.

Hockemeyer, D., & Jaenisch, R. (2016). Induced pluripotent stem cells meet genome editing. *Cell Stem Cell,* 18(5), 573–586.

Hsu, P. D., Lander, E. S., & Zhang, F. (2014). Development and applications of CRISPR-Cas9 for genome engineering. *Cell,* 157(6), 1262–1278.

Jaenisch, R., & Young, R. A. (2008). Stem cells, the molecular circuitry of pluripotency and nuclear reprogramming. *Cell,* 132(4), 567–582.

Jinek, M., Chylinski, K., Fonfara, I., Hauer, M., Doudna, J. A., & Charpentier, E. (2012). A programmable dual-RNA–guided DNA endonuclease in adaptive bacterial immunity. *Science,* 337(6096), 816–821.

Karagiannis, P., et al. (2019). Induced pluripotent stem cells and their use in human models of disease and development. *Physiological Reviews,* 99(1), 79–114.

Kim, Y. G., Cha, J., & Chandrasegaran, S. (1996). Hybrid restriction enzymes: Zinc finger fusions to FokI cleavage domain. *PNAS,* 93(3), 1156–1160.

Knott, G. J., & Doudna, J. A. (2018). CRISPR-Cas guides the future of genetic engineering. *Science,* 361(6405), 866–869.

Lattanzi, A., et al. (2021). Gene editing in iPSCs: Clinical translation challenges. *Molecular Therapy,* 29(5), 1664–1678.

Li, X., et al. (2022). Organoids and genome editing: Modeling human disease and regeneration. *Cell Stem Cell,* 29(7), 1003–1020.

Liu, M., et al. (2023). Prime editing in human pluripotent stem cells. *Stem Cell Reports,* 18(2), 345–358.

Menzies, F. M., et al. (2022). Regulatory perspectives on genome-edited cell therapies. *Nature Reviews Drug Discovery,* 21(5), 337–338.

Pickar-Oliver, A., & Gersbach, C. A. (2019). The next generation of CRISPR–Cas technologies. *Nature Reviews Molecular Cell Biology,* 20(8), 490–507.

Rowe, R. G., & Daley, G. Q. (2019). Induced pluripotent stem cells in disease modelling and drug discovery. *Nature Reviews Genetics,* 20(7), 377–388.

Takahashi, K., & Yamanaka, S. (2006). Induction of pluripotent stem cells from mouse fibroblasts. *Cell,* 126(4), 663–676.

Urnov, F. D., Rebar, E. J., Holmes, M. C., Zhang, H. S., & Gregory, P. D. (2010). Genome editing with engineered zinc finger nucleases. *Nature Reviews Genetics,* 11(9), 636–646.

Wilkinson, R. N., et al. (2023). Long-term genomic stability of CRISPR-edited stem cells. *Nature Communications,* 14, 1123.

Wu, S. M., & Hochedlinger, K. (2011). Harnessing the potential of induced pluripotent stem cells for regenerative medicine. *Nature Cell Biology,* 13(5), 497–505.

Zhang, Y., et al. (2022). Artificial intelligence–guided genome editing in stem cells. *Nature Machine Intelligence,* 4(6), 472–482.

7 Precision Breeding 2.0
Accelerating Crop Improvement with Genome Editing

Devendra Kumar Choudhary, Sunil Kumar Sunani, Harvinder Kumar Singh, and Sandeep Kumar

7.1 INTRODUCTION

Phenotypic selection was the initial type of breeding in which farmers selected and reproduced plants exhibiting subjects of interest like a greater yield, resistance to disease or higher quality. Although more basic, such a method is slow and inaccurate due to the high dependency of observable characteristics. With the emergence of molecular breeding, the concept of marker-assisted selection (MAS) emerged, where the breeder can now monitor genes or QTLs with the help of molecular markers and not on the phenotype by itself, which improves selection accuracy and reduces breeding cycles. MAS is effective in major-effect genes but is weak in complex polygenic traits and can be impeded by linkage drag, as well as limited diagnostic markers. Direct modification of plant genomes by transgenic technique was also made possible by sporadic breeding followed by allowing specific traits to be bred such as pest resistance traits or herbicide tolerance – traits that were previously not possible via conventional crossing. Such transgenic methods, though, posed regulatory, biosafety and public acceptance problems and were burdened with technical problems, including position effects and unwanted genomic changes. These steps combined give an indication of a progression toward faster and more accurate crop improvement (Raghavan et al., 2025).

The traditional and early molecular techniques have obvious drawbacks, although they possess advantages. The environmental variation (phenotype-based selection) and the slowness of responses in perennials or complex-genome crops to new pests or climatic changes confound phenotype-based selection. MAS is more effective at breeding single major genes than polygenic traits with many small-effect loci interacting; it also requires a large amount of prior knowledge and genotyping infrastructure, which might be unavailable in resource-constrained situations. Transgenic techniques widen the trait space, but are subject to severe regulatory systems, market failure, and technical challenges, including transgene silencing, imprecise expression due to insertion-site effects, and off-target effects. These restrictions have prompted efforts to find genome editing strategies that maintain precision but have the potential to lessen regulatory restrictions (Georges and Ray, 2017).

The genome editing, and most notably, CRISPR/Cas, characterizes a transition to what may be described as Precision Breeding 2.0. In contrast to the early transgenic methods, contemporary editors can implement specific insertions and deletions or point-based mutations without the permanent integration of foreign DNA which allows further regulation pathways and can enhance social acceptance. Programmable nucleases (zinc-finger nucleases [ZFNs], transcription activator-like effector nucleases [TALENs], CRISPR/Cas) induce site-specific breaks, which are repaired by endogenous mechanisms to induce precise edits; more recent tools like base and prime editors induce nucleotide-level edits without inducing double-strand breaks (DSBs), reducing the possibility of unintended outcomes.

The emphasis of this paradigm is on predictable and site-specific modifications of elite germplasm in order to provide targeted agronomic benefits without incurring the drag of genetic linkage and the

DOI: 10.1201/9781003735113-7

protracted requirements of backcrossing. Editing in commercial versions reduces the breeding periods and maintains desirable genetic lineages. Cutting-edge methods DNA-free delivery techniques, ribonucleoprotein (RNP) complexes, transient expression system or particle delivery, can result in edits indistinguishable to those occurring in nature, eliminating the use of transgenic constructs. RNA-targeting systems (e.g., Cas13) provide an additional layer of functionality by allowing transiently modifying the transcriptome, or resisting viruses, without making long-term changes to the genome.

The Precision Breeding 2.0 is supported by the integration with the complementary technologies. Fixation of edits can be increased by speed breeding and approaches based on doubled-haploid or haploid induction, whereas high-throughput phenotyping and omics (GWAS, transcriptomics, whole-genome sequencing [WGS]) can be used to inform target selection and enable data-driven design of multi-trait edits. The ability of multiplex editing to allow parallel editing of multiple alleles enables this, and is especially useful in polyploid crops. Such combinations can be quickly translated into better lines on yield, quality, nutrient content and stress resilience. Practical uses already exemplify the method: specific edits to genes that determine the grain size, the factors that determine susceptibility, the target of the herbicides and nutritional pathways have resulted in lines with more yield or that are more resistant to disease, or altered metabolic profiles. Numerous of these edits recapitulate alleles which may be present in nature, but are achieved more quickly and with high precision.

Significant problems remain even after the apparent promise. Primary transformants that contain off-target edits and chimerism need to undergo rigorous molecular characterization; effective, genotype-neutral delivery is a bottleneck to most crops; and effective, genome-wide validation pipelines are required to verify the presence of intended change. The regulatory frameworks and strict guidelines on products of various delivery and editing strategies are in the process of development, and will affect deployment. Overall, Precision Breeding 2.0 is a combination of genome editing, breeding accelerator methods and data science to transform crop improvement toward less stochastic selection and more directed genotype engineering and reduce the footprint of foreign DNA.

The chapter presents the history of breeding plants beyond the traditional selection process using the molecular technique to the modern technology of genome editing, emphasizing the shortcomings of the previous technologies and the need of Precision Breeding 2.0. It will outline technology foundations, uses and recent developments in crop genome editing, critically evaluate the threats and regulatory concerns, and comment on how to combine it with complementary approaches (speed breeding, phenomics, multi-omics). It is aimed at offering researchers, breeders and policymakers a succinct, practical precis of the potential of genome editing and the gaps in the research to be filled in to enable it to make its contribution to sustainable food security (Raghavan et al., 2025).

7.2 GENOME EDITING PLATFORMS UNDERPINNING PRECISION BREEDING 2.0

Genome editing has re-established plant breeding, as Precision Breeding 2.0 begins. This stage is based on platforms that can allow focused and effective engineering of plant genomes to enhance the characteristics of disease resistance, abiotic stress resistance, yield and quality. The key toolsets include early protein-based nucleases, including ZFNs and TALENs, followed by the family of RNA-guided CRISPR/Cas, and then more recently base editors and prime editors. Explaining structure, mechanism, applications and constraints explains how each platform is made to improve crop and where further improvement needs to be made.

7.2.1 Zinc-Finger Nucleases: The First-Generation Genome Editing Tools

The ZFNs are the chimeric proteins that fuse zinc-finger DNA-binding domains with FokI nuclease catalytic domain. A triplet of DNA is recognized by each zinc finger. Three to six tandem arrays of zinc fingers allow attachment of locus selectivity based on 9–18 base pairs. The FokI nuclease acts as a dimer and thus, two ZFN monomers need to bind the opposing strands with a suitable spacer, usually 5–7 base pairs, to break a strand in two. Those resulting breaks are patched by cell pathways like nonhomologous end joining (NHEJ) or homology-directed repair (HDR), which cause either disruption of genes or accurate sequence modification. The initial research on plants demonstrated that ZFNs were capable of causing directed breaks and the following edits, enabling the possibility of programmable alteration in crops (Lee et al., 2016).

ZFNs were utilized to disrupt ZFNs to disrupt susceptibility genes and modify metabolic pathways to improve resistance and quality in plants. Others can be mutagenesis of disease-disposing locus of interest and genetic manipulation of secondary metabolites to alter nutritional contents. ZFNs proved that site-specific and precise edits could be achieved in agricultural species, giving information to subsequent approaches of more friendly platforms (Sharma et al., 2021). Zinc-finger arrays are designed with the help of much protein design and empirically tested. Fingers have context-dependent binding; hence, modular assembly is not predictable as it was at first thought. Two monomers have to be coordinated in order to cleave into dimers, which increases the complexity of the design. When a binding specificity is faulty, off-target cleavage may take place. Practical constraints also existed with production and delivery of large protein constructs into plant cells. These technical overheads limited its adoption to various crop systems (Rasheed et al., 2021).

7.2.2 Transcription Activator-Like Effector Nucleases: Modular and Targeted

TALENs combine domains of FokI nuclease and domains of DNA binding based on transcription activator-like effectors. TALE repeats measure approximately 34 amino acids and specificity is identified by repeat variable diresidue at positions 12 and 13. The repeats identify one base each, allowing nearly modular design and one base resolution in targeting. Similar to ZFNs, TALENs are FokI dimers that create DSBs upon assembling two flanking arrays of binding. Other techniques like Golden Gate cloning and FLASH assembly simplified the construction of array repeats of TALEN and allowed the generation of TALENs rapidly. Functional genomics and crop improvement TALENs have been used to target gene disruption in rice, make specific edits to tomato and maize, and have been used to edit polyploid genomes, where allele-specific targeting is necessary. Multiplexing where several pairs of TALEN allowed simultaneous targeting of gene families, which became useful when it was necessary to alter pathways that regulate plant architecture or plant responses to stress (Corrigan-Curay et al., 2015).

7.2.3 CRISPR/Cas Systems: Revolution in Crop Breeding

CRISPR/Cas systems are based on the adaptive immune system of archaea and bacteria. Genome editing is a technique in genomics that uses a guide RNA (gRNA) to complement matching DNA target, and guides a Cas endonuclease to cleave that target (Corrigan-Curay et al., 2015). A protospacer adjacent motif next to the target sequence is also normally required in target recognition. crRNA and tracrRNA are combined into single programmable synthetic single-guide RNAs (sgRNA). The RNA-guided recognition system simplifies the tasks of designing targets compared to protein-based systems and allows quick retargeting between loci (Sharma et al., 2021).

7.2.4 CRISPR/Cas9

SpCas9 is directed by a 20-nucleotide sequence in the sgRNA and identifies a 5′-NGG-3′ PAM next to the target. The introduction of a blunt DSB three bases above the PAM by Cas9 is introduced. The break causes NHEJ or HDR-mediated repair. Predictable cut site and simple guide design are the basis of the widespread use of Cas9 in crop engineering (Corrigan-Curay et al., 2015).

There is the issue of off-target mutagenesis. Truncated guides are to be used to mitigate partial mismatches, paired nickases are to be used to demand two independent recognition events, and high-fidelity Cas9 variants with low nonspecific DNA affinity are engineered. Direct delivery of RNP complex into plant cells decreases off-target activity as well as transgene integration. These refinements in the engineering and delivery of products can be the idea of enhancing the safety and effectiveness of agriculture (Naeem et al., 2020).

7.2.5 Cas12a/Cpf1, Cas13, Cas14

Cas12a identifies T-rich PAMs like 5′-TTTV-3′, and is less bound by the NGG limit of targeting. It produces staggered cuts with five overhangs and excises its own crRNA using precursor transcripts, and enables multiplex editing. Smaller size can facilitate delivery in systems that have restrictions on the capacity of vectors in these systems. These properties expand the range of opportunities with editing the loci of plants that do not contain SpCas9 PAMs, and multiplexing trait editing (Gautam et al., 2025).

Cas13 is directed to RNA and not DNA and allows transient regulation of transcripts, interference with viruses and conditional regulation of gene expression in plants. Cas14 breaks single-stranded DNA, and due to its small size, has a possible application in special editing or diagnostic applications. RNA targeting finds application especially in the injection of viral plant pathogens and reversible control of gene networks in the absence of changes to genomic DNA (Gautam et al., 2025).

7.2.6 Prime Editing and Base Editing

Base editors are catalytically allosteric or nickel catalytic variants of a deaminase fused with catalytically impaired proteins to catalyze the direct conversion of one base to another without the use of donor templates or DSBs. C–T mutations are produced by cytosine base editors and A–G mutations are produced by adenine base editors. This method can be used to accurately correct or insert point mutations that are the basis of agronomically significant alleles with minimal collateral damage (Yang et al., 2024). Prime editing involves a Cas9 nickase that contains reverse transcriptase and a gRNA that is prime editing and contains an encoded code of the desired edit. It is capable of installing specific deletions, insertions and any base conversion with minimal use of cellular HDR. Compared to single base and small indel edits which can be made with ease in plants, prime editing upscales the scale of edits possible at a single base and small scale, and eliminates the use of donor templates, which is beneficial in tissues with low HDR activity (Manghwar et al., 2020).

7.2.7 Multiplex Genome Editing

This is done in CRISPR systems through expression of sgRNA arrays using different promoters or as polycistronic transcripts cleaved by endogenous RNases. TALEN arrays are also modularly assembled in order to target locus concurrently. Multiplex technologies allow a coordinated mechanism of changing pathways of complex traits, reduce breeding times and allow redundant copies of a gene to be simultaneously edited in polyploid crops.

7.3 GENOME EDITING IN CROP IMPROVEMENT

CRISPR/Cas systems and their fast offshoots are transforming the philosophy of crop improvement by replacing large-scale phenotypic selection with hypothesis-based manipulation of genes and networks. Instead of applying lengthy backcrossing, recent editors can use targeted knockouts, specific nucleotide replacements, promoter rewiring and multiplexed edits that control physiological responses. This has been used in the primary breeding focus, which is abiotic and biotic stressfulness, yield and quality, herbicide resistance and resource-use productivity, as well as providing an opening to minimize agrochemical inputs and speed up varietal release.

These objectives are already being implemented at large scale by a wide variety of crop-level achievements and methodological improvements that have been documented in the literature (Inam et al., 2024; Kumar et al., 2023). In the case of abiotic stresses, such as drought, salinity, heat and cold, CRISPR strategies are aimed at nodes that mediate perception and downstream acclimation: transcription factors, stomatal regulators, ion transporters, osmoprotectants biosynthetic enzymes and ABA signaling components. Examples of practical approaches are loss-of-function of negative regulators, promoter editing to regulate the timing of expression, and targeted base edits to modify the activity of proteins without leaving transgenic footprints. In cereals, legumes and horticultural crops, improved water-use efficiency, root architecture and ionic homeostasis have been reported following reprogramming of key loci, and the recent use of prime editing and multiplexed guides have hastened the production of complex, quantitative tolerance phenotypes (Salava et al., 2021; Karunarathne et al., 2023).

Under the biotic stress, genome editing has been used to decrease host vulnerability, enhance recognition and reconfigure metabolite defenses against infection. Coverage of the full range of susceptibility genes, or editing of receptor and signaling elements, offers pan, long-term resistance without dependency on specific large R-gene introgressions. Other significant applications include interference of the genes of susceptibility in banana and citrus, alteration of recognition pathways in cereals and alteration of secondary-metabolite pathways that repel insect herbivores. These host-targeted strategies are complementary to pathogen-targeted approaches and are already demonstrating effectiveness in fungal, bacterial and viral diseases in a wide range of crops (Vinay and Manivannan, 2025; Tyagi et al., 2020). Direct edits of metabolic and developmental regulators are beneficial to yield and product quality.

The modification of genes that regulate grain size, fruit morphology, and fruit ripening has resulted in crops with better market value or processing attributes without affecting or reducing their yield. Equally, biofortification via the precise edits in vitamin and mineral pathways can be achieved without transgenic insertion, as do manipulations of ethylene biosynthesis/signaling, which provide tangible benefits in shelf-life and postharvest stability – examples are edits that reduce ethylene synthesis in banana and ethylene biosynthesis-transduction pathways in cereals (Ku and Ha, 2020; Rogo et al., 2024).

Lastly, herbicide tolerance and nutrient-use efficiency editing solve the input reduction and sustainability. Accurate modification of herbicide target sites eliminates linkage drag, even in polyploid genomes, and modifications in transporters and regulatory factors can enhance nitrogen and phosphorus capture in deplorable soils. Combined, such as on abiotic tolerance, pest resistance and nutrient efficiency, edited varieties can reduce pesticide and fertilizer dependency and lead to more resource-efficient systems of production (Roychowdhury et al., 2023; Yadav et al., 2023).

Collectively, these applications indicate that CRISPR is not a one-size-fits-all solution but a flexible toolset: knockouts, base editors, prime editors and multiplexing all have answers to a variety of biological questions. Omics-based target discovery with careful phenotyping and stacking edits of polygenic effects have now become the focus of transforming molecular edits into socially acceptable growers (Table 7.1).

7.4 CHALLENGES AND LIMITATIONS IN GENOME EDITING IN CROP IMPROVEMENT

Genome editing has provided potent pathways to the enhancement of crops but a number of biological, technical and regulatory obstacles constrain its full utilization. The key pitfalls are the lack of specificity (off-target effects), polygenic and polyploid genomes, and delivery and regenerative bottlenecks. These limitations are summarized below together with their causes and solutions to curb them currently with the examples and references being specific to plants.

TABLE 7.1
Applications of Genome Editing in Crop Improvement

Crop Species	Target Gene(s)/Locus	Editing Tool/Strategy	Trait Improvement/Resulting Phenotype	References
Rice (*Oryza sativa*)	*GS3* (grain size 3)	CRISPR/Cas9 (editing of coding region)	Superior yield and grain quality (longer grains, increased grain yield per plant, improved translucency, reduced chalkiness in hybrid rice)	Huang et al. (2024)
Rice (*O. sativa*)	*OsSNB* (novel grain size gene)	CRISPR/Cas9 (knockout)	Increased grain length, width, and weight (negative regulator knockout)	Ma et al. (2019)
Rice (*O. sativa*)	*OsALS* (acetolactate synthase)	CRISPR/Cas9 (gene editing)	Novel herbicide tolerance allele (G628W mutation)	Wang et al. (2021)
Rice (*O. sativa*)	*Pi21* and *OsSULTR3;6* (EBE: Effector-Binding Element)	CRISPR/Cas9 (multiplex mutagenesis)	Improved resistance to rice blast and bacterial leaf streak simultaneously	Yang et al. (2023)
Rice (*O. sativa*)	*OsER1* and *OsER2* (ERECTA family)	CRISPR/Cas9 (loss-of-function mutants)	Shortened plant stature and reduced panicle size	Zhang et al. (2018)
Rice (*O. sativa*)	*OsPIN5b*, *GS3*, *OsMYB30*	Multiplex CRISPR/Cas9	Improved yield and cold tolerance (simultaneously editing three genes for combined traits)	Zeng et al. (2020)
Wheat (*Triticum aestivum*)	*TaGW2*, *TaLpx-1*, *TaMLO*	Multiplex CRISPR/Cas9	Increased seed size and thousand grain weight (knockout of all three homoeologous copies of *TaGW2*)	Wang et al. (2018)
Wheat (*T. aestivum*)	*TaARE1* orthologs (nitrogen use efficiency)	CRISPR/Cas9 (targeted mutagenesis)	Enhanced tolerance to N starvation, delayed senescence and increased grain yield	Karunarathne et al. (2022)
Wheat (*T. aestivum*)	*ALS* and *ACC* genes	Base editing	Herbicide tolerance (tolerance to sulfonylurea-, imidazolinone- and aryloxyphenoxy propionate-type herbicides)	Fu et al. (2025)
Wheat (*T. aestivum*)	*TaRPK1* (receptor-like protein kinase 1)	CRISPR/Cas9	Enhanced yield by modifying root architecture system (increased tillers and total grain weight)	Rahim et al. (2024)
Maize (*Zea mays*)	*CLE* genes (CLAVATA-WUSCHEL pathway)	CRISPR/Cas9 (promoter editing)	Increased multiple grain yield–related traits by making weak promoter alleles	Liu et al. (2021)
Soybean (*Glycine max*)	*GmIPK1* (inositol polyphosphate kinase 1)	CRISPR/Cas9 (knockout)	Low-phytate soybean (6.6–7.05-fold reduction in phytate content)	Krishnan et al. (2023)
Tomato (*Solanum lycopersicum*)	*LeMADS-RIN* (ripening gene)	CRISPR/Cas9	Reduced ethylene production (resulting in incompletely ripening fruits)	Jung et al. (2018)
Tomato (*S. lycopersicum*)	*SlHyPRP1* (hybrid proline-rich protein 1)	CRISPR/Cas9 (precise domain editing)	Multi-stress tolerance (high salinity tolerance, moderate heat tolerance and increased resistance to *Pseudomonas syringae*)	Tran et al. (2023)

(Continued)

TABLE 7.1 (*Continued*)
Applications of Genome Editing in Crop Improvement

Crop Species	Target Gene(s)/Locus	Editing Tool/Strategy	Trait Improvement/Resulting Phenotype	References
Tomato (*S. lycopersicum*)	*eIF4E1* (eukaryotic translation initiation factor)	CRISPR/Cas9 (targeted mutagenesis)	Enhanced resistance to Pepper Mottle Virus	Yoon et al. (2020)
Tomato (*S. lycopersicum*)	*GF* (GREENFLESH) locus	CRISPR/Cas9	Improved fruit quality (reduced chlorophyll degradation, increased carotenoid/tocopherol levels and reduced susceptibility to *Botrytis cinerea*)	Gianoglio et al. (2022)
Tomato (*S. lycopersicum*)	*KLUH* (promoter region)	CRISPR/Cas9 (editing *cis*-regulatory element)	Increased fruit weight (gain-of-function allele created by promoter editing)	Li et al. (2022)
Apple (*Malus x domestica*)	*PDS* (phytoene desaturase)	CRISPR/Cas9	Targeted gene knockout (albino phenotype, used for validation)	Charrier et al. (2019); Li et al. (2023)
Potato (*Solanum tuberosum*)	*StCBP80* (cap-binding protein)	CRISPR/Cas9	Enhanced drought tolerance (reduced yield penalties under water restriction)	Oneto et al. (2025)
Barley (*Hordeum vulgare*)	*HvPM19* (multicopy gene)	RNA-guided Cas9 nuclease	Induction of heritable mutations	Lawrenson et al. (2015)
Alfalfa (*Medicago sativa*)	*MsC3H* (coumarate 3-hydroxylase)	Multiplex CRISPR/Cas9	Reduced lignin deposition and improved forage quality	Wolabu et al. (2024)
Cotton (*Gossypium hirsutum*)	*TAC1* homoeologs	CRISPR/Cas9	Reduced branch angle and altered petiole angles (improved plant architecture)	Kangben et al. (2024)
Tobacco (*Nicotiana tabacum*)	*NtHL1* (hybrid lethality 1)	Gene editing	Reverse interspecific seedling lethality (expanding the gene pool for crop improvement)	Ma et al. (2020)
Sweet orange (*Citrus sinensis*)	*CsPDS* (phytoene desaturase)	Cas9/sgRNA	Targeted genome modification (albino phenotype, used for functional study)	Jia and Wang (2014)
Sugarcane	Lignin biosynthesis genes	Genome editing (TALENs and CRISPR/Cas9)	Modified lignin composition (for improved bioenergy potential)	Mohan et al. (2022)
Cassava (*Manihot esculenta*)	*EPSPS* (5-enolpyruvylshikimate-3-phosphate synthase)	Gene editing (HRD/NHEJ strategies)	Robust glyphosate tolerance	Hummel et al. (2018)
Cabbage (*Brassica oleracea*)	*BoPDS* (phytoene desaturase)	CRISPR/Cas9	Targeted gene editing (albino phenotype, used for variety improvement)	Ma et al. (2019)
Field pea (*Pisum sativum*)	*PsPDS* (phytoene desaturase)	CRISPR/Cas9	Targeted mutagenesis (albino phenotype, first report of successful gene-edited pea)	Li et al. (2023)

7.4.1 Off-Target Effects

Off-target effects are unwanted edits at loci that are not the target site and are due to the fact that nucleases like CRISPR/Cas9 are able to tolerate partial mismatches between gRNA and genomic DNA. The cleavage propensity is highly sensitive to the gRNA seed sequence next to the PAM: any mismatch at the distal end will allow cleavage, which creates off-target DSBs (Zhang et al., 2015). Exposure of potential off-target sites is context-accessible by chromatin and modulates by nucleoside positioning and DNA methylation, and therefore open chromatin enhances off-target risk. Other intrinsic gRNA characteristics (secondary structure, nucleotide composition), as well as nuclease properties, also play a role, and off-target activity is a multifactorial issue (Ahmadikhah et al., 2025).

Strong evaluation of off-targets is a combination of objective and focused methods. WGS is a survey without any bias, but is expensive and can underreport low-frequency edits. Specific assays, like T7E1, are cheap and fast but they need a priori candidate loci (Zhang et al., 2015). GUIDE-seq and analogous DSB-capture technologies map off-targets at high resolution, but have a tendency to be used in cultured cells. In silico predictors (CRISPRScan, CCTop, Cas-OFFinder) are used to assist in designing gRNAs, as they rank off-targets (ranking per their sequence homology) and, more recently, chromatin properties (Corrigan-Curay et al., 2015). Finally, the combination of computational prediction and experimental confirmation provides the best reliable off-target profile, but sensitivity and scalability are still disadvantages (Zhang et al., 2015).

A number of technical advances minimize off-target mutagenesis. Improved the specificity of the interaction and reduced off-target cleavage without sacrificing the activity of Cas9. High-fidelity Cas9 (SpCas9-HF1, eSpCas9) weakens the nonspecific interactions of Cas9 with DNA (Zhang et al., 2020). A strategy of truncated gRNAs or paired nickases is more specific, as it allows less tolerance to mismatches or requires two adjacent binding events to make a productive edit. The use of editing reagents as RNP complexes minimizes the duration of exposure to nuclease in contrast to the case of DNA vectors and off-target accumulation, and the use of RNPs also does not lead to the stable integration of foreign DNA (Zhang et al., 2018; Erdogan et al., 2023). The risk in a variety of crop germplasm is further reduced by careful gRNA design based on pan-genome variation and chromatin maps.

7.4.2 Polygenic Crop Complexity Traits

Polyploid many of the staple crops are polyploid (e.g., wheat, cotton) and have many homoeologous copies that can obscure the phenotypic impacts of edits. To obtain a clear phenotype, it is often necessary to simultaneously edit all functional copies, and it is necessary to multiplex and design gRNA efficiently to prevent cross-reactivity (Luo et al., 2021). Polyploid genomes are more repetitive and larger, making it more difficult to predict and downstream genotype off-target. The possible solutions to these challenges are multiplex editing systems and doubled-haploid strategies (Luo et al., 2021). Agronomic traits – yield, drought tolerance, disease resistance – are often polygenic, and are the result of small additive effects and complex gene networks. The edits of single genes can hence lead to a minimal returns or undesirable trade-offs associated with pleiotropy. Detailed functional dissection will be required to pinpoint regulatory nodes that to be modulated will produce desirable results with no adverse side effects (Ahmadikhah et al., 2025). The inducible or tissue-specific edits will be able to minimize pleiotropic penalties by limiting the alterations to the relevant developmental stages or organs.

To handle polygenic architectures, multiplex CRISPR arrays (e.g., tRNA-gRNAs tandems) can achieve simultaneous targeting of multiple loci or homoeologs in a single step, which is required to pyramid desirable alleles (Li et al., 2022). Base and prime editors allow accurate allele-specific nucleotide modifications with no DSBs, which are useful in refining alleles in polygenic networks. Combining genomics-aided breeding with editing enhances their selection of candidates and reduces unwanted effects.

7.4.3 Technical Production Blocks in Delivery and Regeneration

One of the fundamental constraints is delivery: *Agrobacterium* and biolistics are efficient, yet work on a genotype-specific basis and in many elite cultivars is inefficient (Sandhya et al., 2020). Viral vectors are eventually transient and systemic, but have host-range limitations and regulation issues. Carrers based on nanotechnology (MSNs, AuNPs, CNTs, LDHs) are promising in species-independent delivery of nucleic acids and proteins with no DNA integration; however, field-relevant use requires solving practical challenges – stability, targeting, scale and cost-related (Zhi et al., 2022). New methods such as pollen magnetofection are under investigation to circumvent the use of tissue culture, but they are still in the experimental phase (Sandhya et al., 2020).

Regardless of the success of the delivery, the production of entire plants with altered cells is frequently inefficient, genotype-dependent, and slow, especially in perennials and clonally propagated crops (Goralogia et al., 2021). Somaclonal variation that is introduced in the course of tissue culture makes it complicated to measure phenotypes (Luo et al., 2021). There are solutions based on developmental regulator genes to enhance regeneration, embryo rescue, and combining with speed breeding to reduce the duration, but crop-specific optimization should be done (Varshney et al., 2021).

The central focus to regulatory and public acceptance is the production of transgene-free edited plants. It is possible to generate edits without stable transgenes, allowing RNP delivery and transient expression to ease regulatory classification in certain jurisdictions (Zhang et al., 2016). New methods of grafting incorporate editing parts of the transgenic rootstock into the wild-type scion, producing heritable, transgene-free edits (Simoni et al., 2025), but with scale-up problems. The regulatory landscapes are still diverse across the globe; stringent, scientifically grounded policies and stringent off-target testing will be needed to transform genome editing into a widespread and safe crop modification (Waghmode et al., 2024).

7.5 FUTURE PERSPECTIVES OF GENOME EDITING IN CROP IMPROVEMENT

Genome editing has evolved to be a flexible system of enhancing crops with precision, speed and minimal perturbation of genetic backgrounds. The future direction of agriculture is to stop attempting single-gene fixes and start using comprehensive approaches that involve next-generation nucleases, enhanced delivery, epigenetic regulation, and breeding pipelines that can create resilient, productive plant lines. The growing list of Cas nucleases enables greater targetability and fidelity at plant genomes. Relaxed PAM variants of SpCas9 (SpCas9-NRRH, SpG, SpCas9-NG) and high-fidelity variants (SpCas9-HF1, HypaCas9, HiFi Cas9) expand the editable sequence space and minimize off-target activity, a favorable trait of polyploid crops and regulatory acceptability (Hassan et al., 2021). Other systems like Cas12a/Cpf1, Cas12b and small effectors like Cas14 offer other PAMs and cleavage mechanisms that may be used to target specific plant loci or to accommodate temperature and developmental requirements (Bandyopadhyay et al., 2020). Guide-RNA design and multiplexing advances are also equally crucial. Prime editing and optimized pegRNAs make it possible to accurately mutate bases without DSBs, which could be useful in fine-tuning the engineering of traits in plants (Hassan et al., 2021).

Simultaneously editing of numerous loci can be undertaken using multiplex approaches-tRNA arrays, Csy4 processing and self-cleaving ribozymes, enabling the rapid pyramid of alleles underlying complex traits. The further enhancement of expression and transformation in recalcitrant species relies on construct engineering (i.e., intron inclusion) (Vats et al., 2019). The implementation depends on the delivery methods that do not contact stable foreign DNA. Direct delivery of Cas protein-gRNA RNPs into the cell allows the removal of transgene footprints and off-target persistence without lysis of DNA (Razzaq et al., 2019). Viral vectors (e.g., TRV-based sgRNA delivery) can be used to deliver genome-wide, transiently modulable (e.g., targeted demethylation or transcriptional activation) but nanoparticles and carbon-nanotube carriers are species-agnostic, non-integrative, and can be used in fields (Ghoshal et al., 2020; Ojha et al., 2023). All these modalities increase editing of otherwise intractable crops.

In addition to sequence editing, epigenome editing provides an alternative non-sequence-editing approach to controlling traits, which uses dCas9 complexes with epigenetic effectors to interact with DNA methylation or histone marks to activate or silence genes; an example of endogenous epigenomic tooling is epigenetic demethylases (green ROS1). Epigenetic therapies have an ability to enhance stress reactions, nutrition and characteristics, and – crucially – meaningful states are partially transgenerational, which offers an alternative to irreversible genomic modifications. Issues persist in the specificity of delivery, intergenerational consistency of marks, and targeting of regulatory loci, and combining methylome and transcriptome data will be crucial to focus on actionable loci (Kumar and Mohapatra, 2021). The de novo domestication takes advantage of editing to domesticate wild or underused species by editing key domestication loci (seed shattering, architecture, seed size) without removing adaptive alleles lost during classical domestication.

Multiplex editing, pan-genomics and fast breeding shorten the schedules to agronomically proficient lines and draw on the wild gene pools to stay resilient to stress (Rogo et al., 2024). Agronomic customization, commercial and cultural acculturation, and ecological assimilation problems may be best met with participatory breeding and socioeconomic evaluation to implement successful deployments (Hebelstrup et al., 2023). Precision Breeding 2.0 is a combination of pan-genome discovery, targeted editing, speed breeding and genomic selection that stacks major and minor alleles of drought, salinity and heat tolerance at an increased rate than classical breeding methods (Varshney et al., 2021). Integration of omics and high-throughput phenotyping-paired with AI/ML-enhances predicting genotype×environment interactions at the systems-level and improves selection criteria to increase genetic gain faster (Varshney et al., 2021).

The alignment of policies, involvement of the population and integrated agronomic practice are needed in order to scale these innovations into sustainable food systems. Genome editing has the potential to contribute to SDGs, such as improving yields, nutrition and resource efficiency, and the heterogeneity of regulations and concerns in society is subject to open dialogue and governance with proportional risks (Purnhagen et al., 2021). Genome-edited crops can help to sustain biodiversity, decrease the use of chemicals and produce resistance to climate change (when combined with agroecological practices, low-input adapted varieties and precision agriculture) (Fiaz et al., 2024).

7.6 CONCLUSION

The evolution of Precision Breeding 2.0 represents a paradigm shift in agricultural science, transitioning from the random insertions of traditional transgenics to the surgical accuracy of genome editing. By leveraging advanced tools like CRISPR/Cas9, base editing, and prime editing, researchers can now bypass the lengthy backcrossing cycles of conventional breeding to introduce complex traits, such as enhanced nutritional profiles, climate resilience, and disease resistance with unprecedented speed and specificity. Ultimately, these technologies do more than just refine crop genetics; they provide a scalable solution to the dual pressures of a burgeoning global population and a volatile climate. As regulatory frameworks begin to align with the scientific reality of these "nature-equivalent" modifications, Precision Breeding 2.0 stands as the cornerstone of a more sustainable and secure global food system.

REFERENCES

Ahmadikhah, A., Zarabizadeh, H., Nayeri, S., & Abbasi, M. S. (2025). Advancements in genome editing tools for genetic studies and crop improvement. *Frontiers in Plant Science,* 15, 1370675.

Bandyopadhyay, A., Kancharla, N., & Sinha, N. (2020). CRISPR-Cas12a (Cpf1): A versatile tool in the plant genome editing toolbox for agricultural advancement. *Frontiers in Plant Science,* 11, 584151.

Charrier, A., Vergne, E., Dousset, N., Richer, A., Petiteau, A., & Chevreau, E. (2019). Efficient targeted mutagenesis in apple and first time edition of pear using the CRISPR-Cas9 system. *Frontiers in Plant Science,* 10, 40. https://doi.org/10.3389/fpls.2019.00040

Corrigan-Curay, J., O'Reilly, M., Kohn, D. B., et al. (2015). Genome editing technologies: Defining a path to clinic. *Molecular Therapy,* 23(5), 796–806.

Erdogan, İ., Cevher-Keskin, B., Bilir, O., et al. (2023). Recent developments in CRISPR/Cas9 genome-editing technology related to plant disease resistance and abiotic stress tolerance. *Biology,* 12(7), 1037.

Fiaz, S., Uzair, M., & Zhang, B. (2024). Modern plant breeding for achieving global food security. *Physiologia Plantarum,* 176(6), Article e70014. https://doi.org/10.1111/ppl.70014

Fu, X., Wang, N., Li, L. N., Qiao, D. X., Qi, X. T., Liu, C. L., Gao, Z. X., Xie, C. X., & Zhu, J. J. (2025). Development of cytosine and adenine base editors for maize precision breeding. *Journal of Integrative Plant Biology*. https://doi.org/10.1111/jipb.13964

Gautam, K. K. G., & B. V., S. B. V. (2025). Contributions of CRISPR and gene-editing technologies to plant breeding and biotechnology: Advances, integrative approaches, and future directions. *Journal of Research in Agricultural and Applied Sciences,* 11(7), 23–34. https://doi.org/10.35629/9459-11072228

Georges, F., & Ray, H. (2017). Genome editing of crops: A renewed opportunity for food security. *GM Crops & Food,* 8(1), 1–12.

Ghoshal, B., Vong, B., Picard, C. L., et al. (2020). A viral guide RNA delivery system for CRISPR-based transcriptional activation and heritable targeted DNA demethylation in *Arabidopsis thaliana*. *PLOS Genetics,* 16(12), e1008983.

Gianoglio, S., Comino, C., Moglia, A., Acquadro, A., Garcia-Carpintero, V., Diretto, G., Sevi, F., Rambla, J. L., Dono, G., Valentino, D., Moreno-Gimenez, E., Fullana-Pericas, M., Conesa, M. A., Galmes, J., Lanteri, S., Mazzucato, A., Orzaez, D., & Granell, A. (2022). In-depth characterization of greenflesh tomato mutants obtained by CRISPR/Cas9 editing: A case study with implications for breeding and regulation. *Frontiers in Plant Science,* 13, 936089. https://doi.org/10.3389/fpls.2022.936089

Goralogia, G. S., Strauss, S. H., & Redick, T. P. (2021). Gene editing in tree and clonal crops: Progress and challenges. *In Vitro Cellular & Developmental Biology – Plant,* 57, 683–699.

Hassan, M. M., Zhang, Y., Yuan, G., De, K., Chen, J.-G., Muchero, W., Tuskan, G. A., Qi, Y., & Yang, X. (2021). Construct design for CRISPR/Cas-based genome editing in plants. *Trends in Plant Science,* 26(11), 1133–1152. https://doi.org/10.1016/j.tplants.2021.06.015

Hebelstrup, K. H., Azariadis, A., Cordes, A., Henriksen, P. S., & Brinch-Pedersen, H. (2023). Prehistoric plant exploitation and domestication: An inspiration for the science of de novo domestication in present times. *Plants,* 12(12), 2310. https://doi.org/10.3390/plants12122310

Huang, J., Chen, W. W., Gao, L. J., Qing, D. J., Pan, Y. H., Zhou, W. Y., Wu, H., Li, J. C., Ma, C. L., Zhu, C. L., Dai, G. X., & Deng, G. F. (2024). Rapid improvement of grain appearance in three-line hybrid rice via CRISPR/Cas9 editing of grain size genes. *Theoretical and Applied Genetics,* 137(7). https://doi.org/10.1007/s00122-024-04627-8

Hummel, A. W., Chauhan, R. D., Cermak, T., Mutka, A. M., Vijayaraghavan, A., Boyher, A., Starker, C. G., Bart, R., Voytas, D. F., & Taylor, N. J. (2018). Allele exchange at the EPSPS locus confers glyphosate tolerance in cassava. *Plant Biotechnology Journal,* 16(7), 1275–1282. https://doi.org/10.1111/pbi.12868

Inam, S., Muhammad, A., Irum, S., Rehman, N., Riaz, A., Uzair, M., & Khan, M. R. (2024). Genome editing for improvement of biotic and abiotic stress tolerance in cereals. *Functional Plant Biology, 51*. https://doi.org/10.1071/FP24092

Jia, H., & Wang, N. (2014). Targeted genome editing of sweet orange using Cas9/sgRNA. *PloS One,* 9(4), e93806.

Jung, Y. J., Lee, G. J., Bae, S., & Kang, K. K. (2018). Reduced ethylene production in tomato fruits upon CRSPR/Cas9-mediated LeMADS-RIN mutagenesis. *Horticultural Science & Technology,* 36(3), 396–405. https://doi.org/10.12972/kjhst.20180039

Kangben, F., Kumar, S., Li, Z. G., Sreedasyam, A., Dardick, C., Jones, D., & Saski, C. A. (2024). Phylogenetic and functional analysis of tiller angle control homeologs in allotetraploid cotton. *Frontiers in Plant Science,* 14, 1320638. https://doi.org/10.3389/fpls.2023.1320638

Karunarathne, S. D., Han, Y., Zhang, X. Q., & Li, C. D. (2022). CRISPR/Cas9 gene editing and natural variation analysis demonstrate the potential for HvARE1 in improvement of nitrogen use efficiency in barley. *Journal of Integrative Plant Biology,* 64(3), 756–770. https://doi.org/10.1111/jipb.13214

Karunarathne, S. D., Walker, E., Sharma, D., Li, C., & Han, Y. (2023). Genetic resources and precise gene editing for targeted improvement of barley abiotic stress tolerance. *Journal of Zhejiang University – Science B,* 24(12), 1069–1092. https://doi.org/10.1631/jzus.B2200552

Krishnan, V., Jolly, M., Vinutha, T., Manickavasagam, M., & Sachdev, A. (2023). RETRACTED: Low phytate soybean: next generation metabolic engineering using CRISPR-Cas 9 genome editing technology. *Journal of Plant Biochemistry and Biotechnology,* 32(4), 846–861. https://doi.org/10.1007/s13562-023-00845-1

Ku, H.-K., & Ha, S.-H. (2020). Improving nutritional and functional quality by genome editing of crops: Status and perspectives. *Frontiers in Plant Science,* 11, 577313.

Kumar, M., Prusty, M. R., Pandey, M. K., Singh, P. K., Bohra, A., Guo, B., & Varshney, R. K. (2023). Application of CRISPR/Cas9-mediated gene editing for abiotic stress management in crop plants. *Frontiers in Plant Science,* 14, Article 1157678. https://doi.org/10.3389/fpls.2023.1157678

Kumar, S., & Mohapatra, T. (2021). Dynamics of DNA methylation and its functions in plant growth and development. *Frontiers in Plant Science,* 12, Article 596236. https://doi.org/10.3389/fpls.2021.596236

Lawrenson, T., Shorinola, O., Stacey, N., Li, C. D., Ostergaard, L., Patron, N., Uauy, C., & Harwood, W. (2015). Induction of targeted, heritable mutations in barley and *Brassica oleracea* using RNA-guided Cas9 nuclease. *Genome Biology,* 16. https://doi.org/10.1186/s13059-015-0826-7

Lee, C. M., Fine, E. J., Cradick, T. J., & Bao, G. (2016). Nuclease target site selection for maximizing on-target activity and minimizing off-target effects in genome editing. *Molecular Therapy,* 24(3), 475–487.

Li, F. R., Kawato, N., Sato, H., Kawaharada, Y., Henmi, M., Shinoda, A., Hasunuma, T., Nishitani, C., Osakabe, Y., Osakabe, K., Wada, M., Tanaka, N., Watanabe, M., Zhang, C. F., Deng, S., & Komori, S. (2023). Release of chimeras and efficient selection of editing mutants by CRISPR/Cas9-mediated gene editing in apple. *Scientia Horticulturae,* 316, 112011. https://doi.org/10.1016/j.scienta.2023.112011

Li, H., Si, L., Zhu, Z., et al. (2022). Multiplex precision gene editing by a surrogate prime editor in rice. *Molecular Plant,* 15(7), 1077–1080.

Liu, L., Gallagher, J., Arevalo, E. D., Chen, R., Skopelitis, T., Wu, Q. Y., Bartlett, M., & Jackson, D. (2021). Enhancing grain-yield-related traits by CRISPR-Cas9 promoter editing of maize CLE genes. *Nature Plants,* 7(3), 287. https://doi.org/10.1038/s41477-021-00858-5

Luo, J., Li, S., Xu, J., Yan, L., Ma, Y., & Xia, L. (2021). Pyramiding favorable alleles in an elite wheat variety in one generation by CRISPR-Cas9-mediated multiplex gene editing. *Molecular Plant,* 14(6), 847–850. https://doi.org/10.1016/j.molp.2021.03.024

Ma, J., Hancock, W. G., Nifong, J. M., Kernodle, S. P., & Lewis, R. S. (2020). Identification and editing of a hybrid lethality gene expands the range of interspecific hybridization potential in Nicotiana. *Theoretical and Applied Genetics,* 133(10), 2915–2925. https://doi.org/10.1007/s00122-020-03641-w

Ma, X. S., Feng, F. J., Zhang, Y., Elesawi, I. E., Xu, K., Li, T. F., Mei, H. W., Liu, H. Y., Gao, N. N., Chen, C. L., Luo, L. J., & Yu, S. W. (2019). A novel rice grain size gene OsSNB was identified by genome-wide association study in natural population. *PLOS Genetics,* 15(5), e1008191. https://doi.org/10.1371/journal.pgen.1008191

Manghwar, H., Li, B., Ding, X., et al. (2020). CRISPR/Cas systems in genome editing: Methodologies and tools for sgRNA design, off-target evaluation, and strategies to mitigate off-target effects. *Advanced Science,* 7(6), 1902312.

Mohan, C., Easterling, M., & Yau, Y. Y. (2022). Gene editing technologies for sugarcane improvement: Opportunities and limitations. *Sugar Tech,* 24(1), 369–385. https://doi.org/10.1007/s12355-021-01045-8

Naeem, M., Majeed, S., Hoque, M. Z., & Ahmad, I. (2020). Latest developed strategies to minimize the off-target effects in CRISPR-Cas-mediated genome editing. *Cells, 9*(7), Article 1608. https://doi.org/10.3390/cells9071608

Ojha, A., Patil, G. B., & Zhang, F. (2023). Genome editing and chromosome engineering in plants. *The Plant Genome,* 16(2), e20352.

Oneto, C. A. D., Massa, G. A., Echarte, L., Burusco, M. F. R., Gonzalez, M. N., Alfonso, C. S., Laserna, M. P., Norero, N. S., Divito, S. B., & Feingold, S. E. (2025). CRISPR/Cas9 editing of CBP80 enhances drought tolerance in potato (*Solanum tuberosum*). *Frontiers in Plant Science,* 16, 1598947. https://doi.org/10.3389/fpls.2025.1598947

Purnhagen, K. P., Clemens, S., Eriksson, D., Fresco, L. O., Tosun, J., Qaim, M., Visser, R. G. F., Weber, A. P. M., Wesseler, J. H. H., & Zilberman, D. (2021). Europe's Farm to Fork Strategy and its commitment to biotechnology and organic farming: Conflicting or complementary goals? *Trends in Plant Science,* 26(6), 600–606. https://doi.org/10.1016/j.tplants.2021.03.012

Raghavan, M., Pradhan, V., Singh, S., Jain, S., Kumari, N., Kaur, R., & Kumar, G. (2025). Advances in genetic plant breeding for horticultural crop improvement. *Plant Cell Biotechnology and Molecular Biology,* 26(7–8), 432–444. https://doi.org/10.56557/pcbmb/2025/v26i7-89576

Rahim, A. A., Uzair, M., Rehman, N., Fiaz, S., Attia, K. A., Abushady, A. M., Yang, S. H., & Khan, M. R. (2024). CRISPR/Cas9 mediated TaRPK1 root architecture gene mutagenesis confers enhanced wheat yield. *Journal of King Saud University — Science,* 36(2), 103063. https://doi.org/10.1016/j.jksus.2023.103063

Rasheed, A., Gill, R. A., Hassan, M. U., Mahmood, A., Qari, S., Zaman, Q. U., … Wu, Z. (2021). A critical review: recent advancements in the use of CRISPR/Cas9 technology to enhance crops and alleviate global food crises. *Current Issues in Molecular Biology,* 43(3), 1950–1976.

Razzaq, A., Saleem, F., Kanwal, M., Mustafa, G., Yousaf, S., Imran Arshad, H. M., Hameed, M. K., Khan, M. S., & Joyia, F. A. (2019). Modern trends in plant genome editing: An inclusive review of the CRISPR/Cas9 toolbox. *International Journal of Molecular Sciences,* 20(16), 4045. https://doi.org/10.3390/ijms20164045

Rogo, U., Simoni, S., Fambrini, M., Giordani, T., Pugliesi, C., & Mascagni, F. (2024). Future-proofing agriculture: De novo domestication for sustainable and resilient crops. *International Journal of Molecular Sciences,* 25(4), Article 2374. https://doi.org/10.3390/ijms25042374

Roychowdhury, R., Das, S. P., Gupta, A., Parihar, P., Chandrasekhar, K., Sarker, U., Kumar, A., Ramrao, D. P., & Sudhakar, C. (2023). Multi-omics pipeline and omics-integration approach to decipher plant's abiotic stress tolerance responses. *Genes,* 14(6), 1281. https://doi.org/10.3390/genes14061281

Salava, H., Mohan, V., Thula, S., et al. (2021). Application of genome editing in tomato breeding: Mechanisms, advances, and prospects. *International Journal of Molecular Sciences,* 22(2), 682.

Sandhya, D., Jogam, P., Madugula, P., et al. (2020). The present and potential future methods for delivering CRISPR/Cas9 components in plants. *Journal of Genetic Engineering and Biotechnology,* 18, 25.

Sharma, G., Sharma, A. R., Bhattacharya, M., Lee, S. S., & Chakraborty, C. (2021). CRISPR-Cas9: A preclinical and clinical perspective for the treatment of human diseases. *Molecular Therapy,* 29(2), 571–586.

Simoni, S., Fambrini, M., Pugliesi, C., & Rogo, U. (2025). Genome editing by grafting. *International Journal of Molecular Sciences,* 26(19), 9294. https://doi.org/10.3390/ijms26199294

Tran, M. T., Son, G. H., Song, Y. J., Nguyen, N. T., Park, S., Thach, T. V., … Kim, J. Y. (2023). CRISPR-Cas9-based precise engineering of SlHyPRP1 protein towards multi-stress tolerance in tomato. *Frontiers in Plant Science,* 14, 1186932.

Tyagi, S., Kesiraju, K., Saakre, M., Rathinam, M., Raman, V., Pattanayak, D., & Sreevathsa, R. (2020). Genome editing for resistance to insect pests: An emerging tool for crop improvement. *ACS Omega,* 5(33), 20674 20683. https://doi.org/10.1021/acsomega.0c01435

Varshney, R. K., Bohra, A., Yu, J., Graner, A., Zhang, Q., & Sorrells, M. E. (2021). Designing future crops: Genomics-assisted breeding comes of age. *Trends in Plant Science,* 26(6), 631–649. https://doi.org/10.1016/j.tplants.2021.03.010

Vats, S., Kumawat, S., Kumar, V., Patil, G. B., Joshi, T., Sonah, H., … Deshmukh, R. (2019). Genome editing in plants: Exploration of technological advancements and challenges. *Cells,* 8(11), 1386.

Vinay, K., & Manivannan, M. I. (2025). Genome editing for biotic and abiotic stress management in banana: A comprehensive review. *Plant Science Today,* 12(sp1), 1–12. https://doi.org/10.14719/pst.8933

Waghmode, P., Pitha, C. C., & Banka, V. (2024). The role of genome editing technologies in sustainable agriculture. *Journal of Advances in Biology & Biotechnology,* 27(8), 543–558. https://doi.org/10.9734/jabb/2024/v27i81168

Wang, B. X., Liu, Y., Wang, Y. F., Li, J. F., Sun, Z. G., Chi, M., Xing, Y. G., Xu, B., Yang, B., Li, J., Liu, J. B., Chen, T. M., Fang, Z. W., Lu, B. G., Xu, D. Y., & Bello, B. K. (2021). OsbZIP72 is involved in transcriptional gene-regulation pathway of abscisic acid signal transduction by activating rice high-affinity potassium transporter OsHKT1;1. *Rice Science,* 28(3), 257–267. https://doi.org/10.1016/j.rsci.2021.04.005

Wang, W., Pan, Q. L., He, F., Akhunova, A., Chao, S. M., Trick, H., & Akhunov, E. (2018). Transgenerational CRISPR-Cas9 activity facilitates multiplex gene editing in allopolyploid wheat. *CRISPR Journal,* 1(1), 65–74. https://doi.org/10.1089/crispr.2017.0010

Wolabu, T. W., Mahmood, K., Chen, F., Torres-Jerez, I., Udvardi, M., Tadege, M., Cong, L. L., Wang, Z. Y., Wen, J. Q. (2024). Mutating alfalfa COUMARATE 3-HYDROXYLASE using multiplex CRISPR/Cas9 leads to reduced lignin deposition and improved forage quality. *Frontiers in Plant Science,* 15, 1363182. https://doi.org/10.3389/fpls.2024.1363182

Yadav, R. K., Tripathi, M. K., Tiwari, S., Tripathi, N., Asati, R., Chauhan, S., … Payasi, D. K. (2023). Genome editing and improvement of abiotic stress tolerance in crop plants. *Life,* 13(7), 1456.

Yang, J. L., Fang, Y. Y., Wu, H., Zhao, N., Guo, X. Y., Mackon, E., Peng, H. W., Huang, S., He, Y. Q., Qin, B. X., Liu, Y. G., Liu, F., & Chen, S. W. (2023). Improvement of resistance to rice blast and bacterial leaf streak by CRISPR/Cas9-mediated mutagenesis of Pi21 and OsSULTR3;6 in rice (*Oryza sativa* L.). *Frontiers in Plant Science,* 14, 1209384. https://doi.org/10.3389/fpls.2023.1209384

Yang, X., Zhu, P., & Gui, J. (2024). Advancements of CRISPR-mediated base editing in crops and potential applications in populus. *International Journal of Molecular Sciences,* 25(15), 8314.

Yoon, Y. J., Venkatesh, J., Lee, J. H., Kim, J., Lee, H. E., Kim, D., & Kang, B. C. (2020). Genome editing of eIF4E1 in tomato confers resistance to pepper mottle virus. *Frontiers in Plant Science,* 11, 1098. https://doi.org/10.3389/fpls.2020.01098

Zeng, Y. F., Wen, J. Y., Zhao, W. B., Wang, Q., & Huang, W. C. (2020). Rational improvement of rice yield and cold tolerance by editing the three genes OsPIN5b, GS3, and OsMYB30 with the CRISPR-Cas9 system. *Frontiers in Plant Science,* 10, 1663. https://doi.org/10.3389/fpls.2019.01663

Zhang, D., Hussain, A., Manghwar, H., et al. (2020). Genome editing with the CRISPR-Cas system: An art, ethics and global regulatory perspective. *Plant Biotechnology Journal,* 18(8), 1651–1669.

Zhang, X. H., Tee, L. Y., Wang, X. G., Huang, Q. S., & Yang, S. H. (2015). Off-target effects in CRISPR/Cas9-mediated genome engineering. *Molecular Therapy Nucleic Acids, 4,* e264. https://doi.org/10.1038/mtna.2015.37

Zhang, Y. C., Li, S. M., Xue, S. J., Yang, S. H., Huang, J., & Wang, L. (2018). Phylogenetic and CRISPR/Cas9 studies in deciphering the evolutionary trajectory and phenotypic impacts of rice ERECTA genes. *Frontiers in Plant Science,* 9, 473. https://doi.org/10.3389/fpls.2018.00473

Zhang, Y., Liang, Z., Zong, Y., Wang, Y., Liu, J., Chen, K., Qiu, J.-L., & Gao, C. (2016). Efficient and transgene-free genome editing in wheat through transient expression of CRISPR/Cas9 DNA or RNA. *Nature Communications,* 7, Article 12617. https://doi.org/10.1038/ncomms12617

Zhi, H., Peng, W., & Zhou, S. (2022). The promising nanovectors for gene delivery in plant genome engineering. *International Journal of Molecular Sciences,* 23(15), 8501.

8 CRISPR-Cas Genome Editing and Artificial Intelligence for Resilient Crop Disease Control

Aqleem Abbas, Yanyin Guo, Nana Ji, Huihui Xu, and Shanshan Hao

8.1 INTRODUCTION

Global food security faces growing threats from crop diseases, which account for an estimated 10% and 16% of annual yield losses worldwide, translating into billions of dollars in economic damage (Adigun et al., 2021; Savary et al., 2019). These losses impact food availability and impose substantial economic burdens, undermining agricultural economies and exacerbating food insecurity. As the world population is projected to reach approximately 9.8 billion by 2025, according to predictions, innovative strategies in plant disease management are urgently needed to ensure sustainable crop yields (Ray et al., 2013). Traditionally, farmers have relied on chemical pesticides, cultural practices, and conventional breeding to manage plant diseases (Leadbeater & Gisi, 2009). These management strategies are crucial; however, they face significant limitations. Chemical pesticides cause environmental pollution, pose health risks, and drive the emergence of resistant pathogenic strains. Conventional breeding is slow, labor-intensive, and limited by the genetic variation available within a species (Acquaah, 2015). Furthermore, climate change speed up the emergence and spread of plant pathogens by shifting their geographical distribution and creating favorable conditions for new diseases. Besides, Intensive monocultures and high-input agricultural systems have reduced crop diversity, leaving crops more vulnerable to large-scale disease epidemics (Acquaah, 2015).

The 21st century, however, has brought revolutionary advances in biological and computational sciences, that offer new ways to overcome these challenges. At the forefront of this revolution are two cutting-edge technologies: genome editing and artificial intelligence (AI). Together, they promise to shift plant disease management from traditional reactive approaches toward predictive, precise, and sustainable solutions (Manzoor et al., 2024). Genome-editing technologies, especially the clustered regularly interspaced short palindromic repeats (CRISPR)-Cas system, have redefined what is possible in crop improvement (Manzoor et al., 2024). These molecular "scissors" allow scientists to modify specific DNA sequences with exceptional precision to confer disease resistance, enhance innate immunity, or eliminate susceptibility factors, thereby facilitating the development of resilient crops (Gómez-Lama Cabanás & Mercado-Blanco, 2025). CRISPR-Cas has become a robust tool for developing resilient pathogen-resistant crops due to its high precision, minimal off-target effects, and capacity for simultaneous editing of several genes.

CRISPR-based genome editing can incorporate specific, desirable resistance traits without introducing foreign DNA (transgenes, similar to those in genetically modified organisms [GMOs]) as observed in conventional genetic modifications (Wagh et al., 2021). The field of AI is based on the basic technologies like machine learning (ML), deep learning (DL), and computer vision, which enable systems to execute functions such as the analysis of massive datasets and the identification of intricate patterns within them that typically require human intelligence. In the field of crop disease management, AI applications cover early disease detection, diagnosis, and prediction of epidemics and optimization of treatment strategies (González-Rodríguez et al., 2024). In this chapter, the use

DOI: 10.1201/9781003735113-8

of CRISPR to disrupt susceptibility genes in host plants is examined in order to enhance natural immunity and engineer durable resistance across diverse crops.

8.2 GENOME-EDITING TECHNOLOGIES FOR CROP DISEASE RESISTANCE

Genome editing encompasses a suite of molecular tools that enable targeted and precise modifications to an organism's genetic code (Wagh et al., 2021). The most dominant and widely utilized system in this field is CRISPR and its associated Cas proteins, collectively known as CRISPR-Cas (Wagh et al., 2021). Before CRISPR's rise, gene editing relied on engineered nucleases like zinc-finger nucleases (ZFNs) and transcription activator-like effector nucleases (TALENs). While these were the standard until approximately 2013, they necessitated complex and costly protein redesign for every new DNA target, making the process less efficient (Ochiai & Yamamoto, 2015). ZFNs combine zinc-finger DNA-binding domains with a nuclease to cleave specific sites, while TALENs use TALE proteins with repeat variable diresidues, both involving complex design and optimization (Ni et al., 2021). In contrast, CRISPR requires only a short guide RNA (gRNA) to direct the nuclease, making it faster, cheaper, and more efficient, with the added ability to simultaneously edit multiple genes, a key advantage for addressing multigenic traits such as disease resistance (Ochiai & Yamamoto, 2015). Originally discovered as an adaptive immunological mechanism in prokaryotes, CRISPR records viral or plasmid DNA sequences as "spacers" in clustered arrays, enabling RNA-guided Cas proteins to identify and slice invading genomes during reinfection. First described in *Escherichia coli* in 1987 as 32-nucleotide repeats near the *iap* gene and later named "CRISPR" in 2000, the system's immunological role was confirmed when spacer sequences were shown to match plasmid and phage DNA (Mitić et al., 2023).

8.2.1 How CRISPR-Cas Works

The CRISPR-Cas machinery can be likened to a molecular "search-and-cut" mechanism. These systems are categorized as Class 1 (incorporating types I, III, and IV) or Class 2 (comprising types II, V, and VI), a distinction rooted in the evolutionary relationships, architecture, and functional roles of their Cas proteins (Ochiai & Yamamoto, 2015). The Type II Cas9 nuclease from *Streptococcus pyogenes* emerged as the initial genome-editing enzyme to achieve widespread utilization (Wheatley & Yang, 2020). The CRISPR-Cas9 system functions through a programmed Cas9-sgRNA complex that induces targeted DNA breaks. The synthetic sgRNA, a fusion of crRNA and tracrRNA, guides Cas9 to a specific genomic locus adjacent to a PAM sequence (5′-NGG-3′). Cas9 then generates a double-strand break (DSB) via its HNH and RuvC nuclease domains, which cleave the complementary and non-complementary DNA strands, respectively. The cell's subsequent repair via the non-homologous end joining pathway is inherently error-prone, frequently resulting in frameshift indels (insertion or deletion) that disrupt the function of plant susceptibility genes and ultimately confer pathogen resistance. Homology-directed repair, present in bacteria and numerous fungi, facilitates accurate gene repairs, the insertion of novel genetic material (e.g., a resistance gene from a different variety or species), or targeted alterations of existing genes (Wheatley & Yang, 2020). Figure 11.1 provides a schematic summary of these phases. Recently, base editors, including cytidine base editors (CBEs) that convert base pair C–G to T–A, and adenine base editors that convert A–T to G–C, as well as prime editors (Cas9 nickase fused to M-MLV reverse transcriptase), have facilitated precise substitutions, insertions, and deletions without inducing DSBs. Furthermore, employing endogenous RNA polymerase III promoters, such as U3 and U6, to facilitate gRNA or pegRNA expression has enhanced editing efficiency in multiple plant species (Ochiai & Yamamoto, 2015).

CRISPR-Cas12 (type V class) differs from Cas9 in both evolutionary origin and structural organization. Cas12 proteins primarily rely on a single RuvC-like nuclease domain for target DNA cleavage, in contrast to Cas9, which utilizes two nuclease domains (RuvC and HNH) to create DSBs. Multiple Cas12 subtypes (Cas12a–Cas12k) have been identified, each with distinct properties and functions. Among the Class II systems, Cas12a (Cpf1) has seen the most widespread use in plants (Wagh et al., 2021).

It recognizes T-rich PAM sequences and creates staggered DSBs, producing sticky ends that support precise genome modifications. Following target recognition and cleavage, the activated RuvC domain of Cas12 also exhibits collateral single-stranded DNA nuclease activity. Unlike the above DNA-targeting systems, the type VI CRISPR-Cas13 system (Four subtype ~Cas13a–Cas13d) functions as an RNA-guided RNA-targeting ribonuclease (Wheatley & Yang, 2020). Its Cas13a subtype is particularly used in plants to target and cleave single-stranded RNA. This activity not only degrades specific viral RNAs but also exhibits collateral RNase activity, establishing a robust and programmable platform for defending crops against RNA viral diseases (Wagh et al., 2021). By precisely altering susceptibility genes, CRISPR-Cas9 has been leveraged to develop disease-resistant varieties of crops like rice, wheat, and tomato, while Cas13 systems provide new tools for controlling RNA viruses through targeted RNA degradation (Yu et al., 2024). The workflow for developing a disease-resistant plant using CRISPR-Cas9 is shown in Figure 8.1 (Rani et al., 2016).

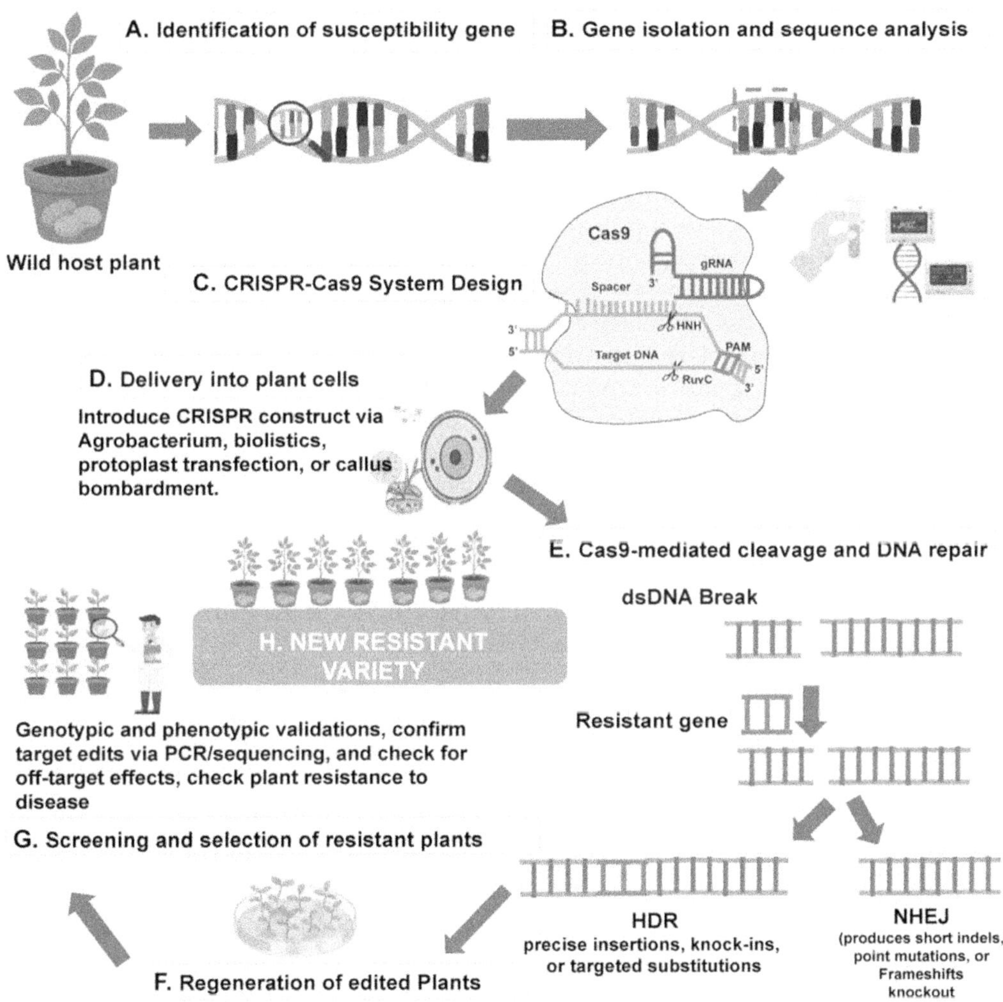

FIGURE 8.1 Workflow for developing a disease-resistant plant using CRISPR-Cas9. (a) Identification of a susceptibility gene. (b) Sequence analysis for guide RNA (gRNA) design. (c) Complex formation: the Cas9 nuclease and sequence-specific gRNA form ribonucleoprotein (RNP) complex. (d) Delivery: CRISPR–Cas9 components into plant cells. (e) DNA cleavage: Cas9 generates a targeted double-strand break (DSB) in the genomic DNA. (f) DNA repair: the cellular machinery repairs the DSB. (g) Regeneration and screening: edited cells are regenerated and screened for enhanced resistance. (h) Plant propagation: resistant lines are propagated to obtain stable, heritable disease-resistant plants.

8.2.2 Applications in Crop Disease Resistance

8.2.2.1 Advanced Pathogen Detection and Diagnostics

CRISPR-Cas systems offer high specificity and sensitivity for nucleic acid detection, which allow for precise identification of major plant pathogens such as *Magnaporthe oryzae*, grapevine red-blotch virus, apple viruses, and viroids (Farrall et al., 2024). By integrating isothermal amplification methods, including recombinase polymerase amplification (RPA) and loop-mediated isothermal amplification (LAMP), with nuclease activation upon target recognition, these technologies achieve sensitivities ranging from femtomolar to zeptomolar levels. Their portability and precision offer early detection of pathogens such as citrus greening (*Candidatus Liberibacter asiaticus*) in asymptomatic tissues. Moreover, CRISPR-based biosensors and portable devices enable point-of-care testing, offering benefits over traditional methods such as PCR by lowering equipment requirements, enhancing specificity, and minimizing false negatives.

8.2.2.2 Functional Genomics in Host–Pathogen Interactions

CRISPR technology, initially applied in *E. coli*, has been extended to several bacterial species. Precise genome editing in *Pseudomonas putida* KT2440 was achieved with near-total efficiency using either CRISPR-SpCas9 or CRISPR-FnCas12a for a range of modifications such as gene deletions, insertions, and allelic replacements (Sun et al., 2018). A base-editing tool, pnCasPA-BEC, combining APOBEC1 (a cytidine deaminase) with a Cas9 nickase, was developed to achieve highly efficient C-to-T substitutions in several species of plant-pathogenic bacteria (Chen et al., 2018). In addition, CRISPR interference using dCas9 has been utilized for targeted transcriptional repression in *Pseudomonas* spp. CRISPR-Cas9 has also been used to target multiple fungal pathogens, including *M. oryzae* and *Fusarium* spp. To deliver Cas9 machinery and sgRNA into plant cells, several approaches are available, such as polyethylene glycol–mediated transformation, *Agrobacterium*-mediated transformation, electroporation, or biolistic approaches. Additionally, ribonucleoprotein (RNP) complexes have been effectively used in *M. oryzae* and *Fusarium* spp. Furthermore, CRISPR has helped identify virulence factors in oomycetes, such as *Phytophthora*. For example, disrupting the *Avr4/6* effector gene in *Phytophthora sojae* prevented it from being recognized by resistant soybean plants. Moreover, knocking out the *PcAvh1* gene in *Phytophthora capsici* reduced its ability to cause disease. On the plant side, CRISPR has been used to study the function of immunity-related genes. For example, knockout of the *NRG1* gene in *Nicotiana benthamiana* revealed that the gene plays a key role in the TIR-NLR–mediated immune signaling, and gene editing of *Ptr* gene in rice confirmed that *Ptr* gene has the potential to provide broad-spectrum resistance to blast disease.

8.2.2.3 Precision Editing of Promoters and Alleles

CRISPR-Cas facilitates the precise editing of promoters and alleles, permitting modulation of host susceptibility genes while preserving essential gene function. For example, researchers employed CRISPR-Cas9 to disrupt the effector binding elements in the citrus *CsLOB1* promoter, an intervention that led to significantly improved resistance against citrus canker (Peng et al., 2017). In rice, editing of the *OsSWEET11* and *OsSWEET14* promoters impaired TALE effector binding and resulted in the resistance to bacterial blight disease (Ni et al., 2021). In addition to gene knockouts, the CBE rBE5 corrected a recessive *pi-d2* allele in the rice variety Kitaake via a G-to-A substitution (*M411I*), and consequently restored disease resistance (Ren et al., 2018). Hence, CRISPR-Cas promoter engineering and base editing strategy can develop resistant alleles without compromising other agronomic traits, providing durable resistance and reducing unintended effects.

8.2.2.4 Direct Targeting of Viral Genomes or Editing Host Genes against Viruses

CRISPR-Cas systems confer immunity through the direct cleavage of viral genome or the engineering of host susceptibility genes (Yu et al., 2024). In DNA viruses such as geminiviruses, CRISPR-Cas9 disrupts replication (Rep) genes, coat protein, and intergenic regions, effectively blocking viral replication and enhancing resistance to Tomato yellow leaf curl virus, Beet curly top virus, and Bean yellow dwarf virus (Yang et al., 2019). Use of multiplex gRNAs has broadened resistance to Chili leaf curl virus and Cotton leaf curl viruses, and Cas9 has also effectively targeted Cauliflower mosaic virus and endogenous Banana streak virus in plantain (Mubarik et al., 2021). For RNA viruses, Cas13 systems have demonstrated superior efficacy. LshCas13a targeting the HC-Pro gene of Turnip mosaic virus suppressed infection in *N. benthamiana* (Aman et al., 2018), and the same system conferred resistance to multiple Potato virus Y strains. Building on its success, Cas13 technology has also been deployed to shield rice crops from infections caused by the Southern rice black-streaked dwarf virus and Rice stripe mosaic virus, with *RfxCas13d* effectively suppressing mixed infections (Yu et al., 2024). Editing host factors is also critical for viral replication and has proven effective. CRISPR-Cas9-mediated modification of host-encoded eukaryotic translation initiation factors, specifically eIF4E, eIF(iso)4E, and eIF4G, has generated broad-spectrum resistance against PVY, ZYMV, PRSV-W, and CVYV (Kan et al., 2023). Other examples include editing NISP and NIG to block begomovirus movement and disruption of SlPelo in tomato to limit TYLCV spread (Pereira-Carvalho et al., 2015).

8.2.2.5 Disruption of Susceptibility (S) Genes, Enhancement of Resistance Genes and Defense Pathways

Pathogens exploit these susceptibility (S) genes to establish infection, and their targeted knockout by CRISPR-Cas has conferred durable, broad-spectrum resistance in multiple crops as shown in Table 8.1. For example, by modifying *MLO* homologs in wheat, barley, and tomato, the resistance to powdery mildew was enhanced, resulting in cultivars such as the commercially valuable tomato "Tomelo" (Nekrasov et al., 2017). Scientists have found that disrupting the DMR6 gene in tomatoes induces a broad-spectrum resistance, effectively protecting the plants from various bacteria and oomycetes (Paula de Toledo Thomazella et al., 2016). DNA-free mutagenesis using CRISPR-Cas9 RNPs in grapevine and apple conferred enhanced resistance to powdery mildew and fire blight without introducing transgenic DNA, highlighting the versatility and precision of S-gene targeting strategies (Nekrasov et al., 2017). CRISPR-Cas enables broader recognition of pathogen effectors, stacking of multiple R-genes, and accelerated deployment in elite cultivars, effectively complementing traditional breeding methods by avoiding linkage drag and shortening development timelines (Han et al., 2025). Beyond R-gene modification, CRISPR-Cas can engineer additional defense pathways, including hormone signaling, callose deposition, and production of antimicrobial compounds. Together with S-gene disruption, promoter editing, and viral genome targeting, these strategies offer a comprehensive approach to enhancing crop resilience (Nekrasov et al., 2017). The CRISPR-Cas for editing plant susceptibility genes to enhance resistance against pathogens is compiled in Table 8.1.

8.2.2.6 Challenges

a. **Off-target Effects:** The specificity of CRISPR-Cas systems is sometime affected by off-target modifications. Minimizing these off-target effects is necessary for ensuring the safety and reliability of engineered crops.
b. **Delivery Systems:** Effective, efficient, and stable transformation of genome-editing components (Cas protein, gRNA) into plant cells remains a challenge, particularly in some resistant plant species.
c. **Ploidy and Gene Redundancy:** Moreover, many crops are polyploid, having more than one set of chromosomes (ploidy); therefore, it is difficult to editing all homologous copies of a gene. In addition, gene redundancy, where multiple genes perform similar functions, may require editing several genes to achieve the desired phenotype.
d. **Regulatory and Ethical Challenges:** Globally, there is a significant lack of harmonization in the regulatory frameworks governing genome-edited crops, impacting their adoption and commercialization. The distinction between modified crops that are considered equivalent to conventionally bred varieties and those regulated as GMOs is a topic of ongoing discussion. Besides, a vigorous debate continues regarding the ethics of Cas CRISPR-Cas, and public acceptance is important for the widespread adoption of CRISPR-Cas technology.

TABLE 8.1
CRISPR-Cas for Editing Plant Susceptibility Genes to Enhance Resistance against Pathogens

Host Plant	Targeted Gene(s)	Pathogen(s)	Disease	Outcome	References
Rice	*OsSWEET11*, *OsSWEET14*	*Xanthomonas oryzae* pathovars	Bacterial blight and leaf streak	Promoter editing disrupted TALE binding, conferred resistance	Ni et al. (2021)
Wheat	*TaMLO-A1*	*Blumeria graminis* f. sp. *tritici*	Powdery mildew	Knockout improved resistance without a growth penalty	Wang et al. (2020)
Tomato	*SlDMR6-1*	*Pseudomonas syringae*, *Phytophthora capsici*, *Xanthomonas* spp.	Bacterial speck, root rot, bacterial spot	Frameshift mutation increased Salicylic acid (SA) levels; resistance without growth impact	Paula de Toledo Thomazella et al. (2016)
Tomato	*SlMLO1*	*Oidium neolycopersici*	Powdery mildew	Developed non-transgenic resistant cultivar "Tomelo"	Nekrasov et al. (2017)
Tomato	*SlPelo*, *SlMLO1*	TYLCV, *O. neolycopersici*	Viral yellow leaf curl, powdery mildew	Trait introgression in elite line BN-86; strong resistance in mutant lines	Pramanik et al. (2021)
Grapevine	*MLO-7*	*Erysiphe necator*	Powdery mildew	RNP-based editing; DNA-free targeted mutagenesis	Malnoy et al. (2016)
Apple	*DIPM-1*,2, and *4*	*Erwinia amylovora*	Fire blight	RNP-mediated knockout of susceptibility genes	Malnoy et al. (2016)
Citrus	*CsLOB1* (TI LOBP, TII LOBP), *CsLOB1G*	*Xanthomonas citri*	Citrus canker	Allelic editing required in grapefruit (TI+TII); single allele (*CsLOB1G*) sufficient in orange	Jia et al. (2022)
Arabidopsis	*eIF4E1*	Clover yellow vein virus	Viral disease	Editing conferred virus resistance through translation initiation factor disruption	Bastet et al. (2019)
Tomato	*SlJAZ2*	*P. syringae*	Bacterial speck	Prevented coronatine-induced stomatal opening; enhanced resistance without compromising other defenses	Ortigosa et al. (2019)
Rice	*OsHPP04*	*Root-knot nematodes*	Root knot	Enhanced resistance	Huang et al. (2023)
Potato	*StPM1*	*Phytophthora infestans*	Late blight	Enhanced resistance	Bi et al. (2023)
Tomato	*XSP10, SlSAMT*	*Fusarium* spp.	Wilt	Tolerance to wilt	Debbarma et al. (2023)

e. **Economic and Practical Challenges:** CRISPR-Cas system provides a cost-effective alternative to traditional breeding methods; however, the initial setup and technical expertise required can be prohibitive for some regions. Moreover, most CRISPR-Cas systems have been used in controlled environments; however, applying them in field conditions is crucial for practical agricultural applications.

8.3 ARTIFICIAL INTELLIGENCE FOR CROP DISEASE MANAGEMENT

8.3.1 How AI Works

As a dynamic field of computer science, AI develops systems capable of simulating intelligent human behaviors to perform complex tasks autonomously. These tasks consist of experiential learning, pattern recognition, addressing complex challenges, and facilitating strategic decision-making (Jafar et al., 2024). While its intellectual origins date back to the early theorizing of Ada Lovelace and Alan Turing, followed by the development of symbolic AI in the 1950s, the field of AI faced significant challenges for decades. A transformative shift occurred in the 21st century, driven by the trifecta of enhanced computational resources, the availability of big data, and the rise of DL. These factors enabled groundbreaking progress in applied areas such as computer vision (CV), speech recognition, and natural language processing (NLP). The wider disciplines of AI now include key subdisciplines like ML, DL, CV, NLP, and expert systems (ES) (Jafar et al., 2024).

ML uses data to predict disease outbreaks or assess disease severity based on weather, historical, and crop datasets (Sasikumar et al., 2024). Moreover, based on learning paradigm, the field of ML includes supervised, unsupervised, and reinforcement learning. In the supervised approach, models learn from a predefined set of corrected answers, which allows them to map inputs to known outputs (Jafar et al., 2024). This category includes the use of powerful algorithms models such as Support Vector Machines (SVMs), Random Forests, and Artificial Neural Networks (ANNs). These models are often used for the categorization of plant leaves and/or plant as healthy or diseased for the prediction of disease. Unsupervised learning encompasses K-means clustering and principal component analysis, which assist in uncovering the underlying structures and pattern of unlabeled data to group similar disease symptoms, and detect novel pathogen strains (Sasikumar et al., 2024).

Reinforcement learning learns through trial and feedback, which is not common in plant diagnosis, however, could be optimized in the field operations such scheduling of autonomous fungicides or fertilizers spraying or irrigation based on real-time disease risk. DL is another specialized subfield of ML, that uses a multilayered neural network architecture, to automatically extract complex features from the raw data. DL is a robust and powerful tool in plant disease management due to its versatile ability to analyze images and videos with high accuracy. Convolutional Neural Networks (CNNs) models are widely used for detecting visual disease symptoms on leaves, stems, and fruits, as well as analyzing UAVs, satellites, drone imagery, or microscopic images for early pathogen identification. At the same time, Recurrent Neural Networks and Long Short-Term Memory networks excel at time-series data, and helpful in studying disease progression, environmental influences, and pathogen population dynamics. CV, which frequently uses DL models like CNNs, allows computers to "see" and interpret plant images or videos, offering automated disease detection, phenotyping, and crop health monitoring (Sasikumar et al., 2024). NLP extracts valuable information from scientific literature, field reports, and farmer feedback to support disease surveillance and data-driven decision-making. ES utilizes structured rule-based knowledge (e.g., "if–then" logic) to assist in diagnosing plant diseases and suggesting management strategies. AI is also increasingly integrated with robotics, hyperspectral, and multispectral and/or other sensors and cameras, ML, DL, and CV models to enable automated monitoring, precision spraying, and targeted disease interventions in crop fields (Pajjuri et al., 2022). Large language models such as ChatGPT (OpenAI), Bard (Google), and Claude (Anthropic) has further advanced generative AI, enhancing NLP's ability to process and reason with complex agricultural data. Current research emphasizes the safety, reliability, and ethical use of AI, forming a strong foundation for sustainable, intelligent plant disease management (Jafar et al., 2024).

8.3.2 Application of AI in Plant Disease Management

8.3.2.1 AI in Disease Detection and Diagnosis

Traditional CNN models such as LeNet, AlexNet, and VGG16 have been widely used to classify diseased and healthy crops from publicly available images (e.g., PlantVillage images) (Pajjuri et al., 2022). These models were effective in the controlled laboratory conditions, however, could not applicable in the real-world field environments due to variations in lighting, background, crop variety, and overlapping leaves, resulting in reduced accuracy and poor generalizability. In contrast, improved CNN models, including ResNet, MobileNet, Inception, EfficientNet-b0, YOLO, and TensorFlow-based models, with ensemble models or reinforcement learning, are capable of real-time disease detection and classification, enabling precise infection-risk evaluation and treatment guidance (Sasikumar et al., 2024).

The widespread availability of smartphones with high-resolution cameras have made easy to develop mobile applications that utilize pre-trained CNNs to analyze leaf images uploaded by farmers (Pajjuri et al., 2022). These tools provide real-time diagnoses and advise plant health management, democratizing access to expert-level plant health insights even in remote farming. In addition to smartphones, UAVs equipped with RGB, multispectral, and hyperspectral cameras have extended farm disease surveillance (Pajjuri et al., 2022). AI algorithms process high-throughput drone imagery to detect subtle stress signatures and early signs of infection invisible to the human eye, allowing for targeted management strategies that minimizes pesticide usage and environmental impact. While drone imagery offers detailed coverage at the farm level, satellite imagery offers a wider regional surveillance, allowing for extensive monitoring of disease outbreaks. In controlled environments such as greenhouses, a camera system with AI continuously monitors plants to identify initial disease lesions symptoms, allowing immediate intervention and preventing outbreaks in these controlled spaces (Akshalin Jefita et al., 2025; González-Rodríguez et al., 2024). Recent advancements have improved the accuracy and robustness of AI-driven detection systems. CNNs reinforced by the attention mechanisms have reduced computational costs while maintaining high accuracy in detecting diseases such as late blight. Lightweight models such as EfficientNet-b0 have achieved almost 98% accuracy in diagnosing several rice diseases, including bacterial blight, downy mildew, hispa, and tungro virus (Mishra et al., 2024). Similarly, hyperspectral imaging integrated with deep neural networks provided remarkable precision, reaching over 94% accuracy in wheat rust detection and 97% accuracy in distinguishing virus-infected from healthy grapevines (Guo et al., 2020; Mickey Wang et al., 2024). Other innovations, such as a spectral feature pseudo-graph-based residual network, referred to as SFPGRN, have achieved 93% accuracy in classifying apple leaf diseases, while Fourier-based CNNs have improved both detection speed and model interpretability, contributing to better stress diagnostics (Yi et al., 2023). Besides image-based detection, AI is being applied to other forms of data for disease detection. For example, multispectral and hyperspectral sensors capture reflected light across various wavelengths, showing alterations in plant physiological processes (e.g., chlorophyll content, water stress) before the appearance of symptoms (Wei et al., 2021). AI algorithms are trained to interpret these spectral signatures to identify specific pathogens or disease states. Besides, plants under disease attack release volatile organic compounds. Electric noses integrated with ML detect and distinguish these complex odor profiles, offering a noninvasive method for early disease detection. Integration AI with the data from weather stations, soil moisture sensors, and other environmental monitors offers insights into disease-favorable conditions, aiding in early warning systems (González-Rodríguez et al., 2024). The workflow of AI-based disease detection and diagnosis in crops is shown in Figure 8.2.

8.3.3 AI in Disease Surveillance and Prediction

ML models such as SVMs, Random Forests, and ANNs have been employed to classify plant diseases based on visual symptoms (e.g., healthy vs. diseased leaves) or to predict disease severity and epidemic trends (Akshalin Jefita et al., 2025; González-Rodríguez et al., 2024). These models have

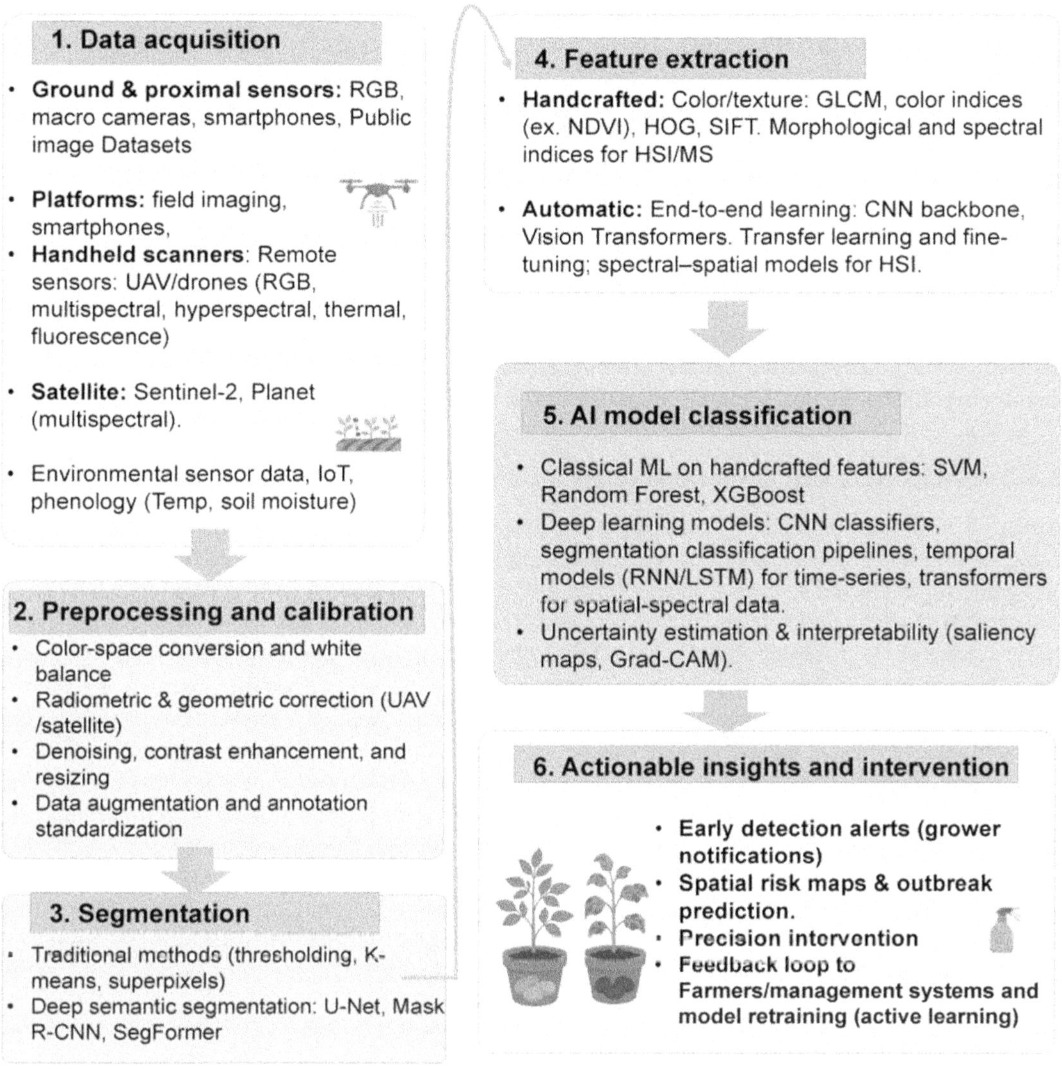

FIGURE 8.2 Workflow of AI-based disease detection and diagnosis in crops. AI models, especially deep learning, analyze various data inputs, including visual images, hyperspectral data, and environmental sensor data, to identify disease symptoms, diagnose pathogens, and provide actionable insights for farmers.

achieved infection-risk prediction accuracies above 90% and over 80% accuracy in the treatment recommendation (Wójcik-Gront et al., 2024). AI-driven surveillance systems are increasingly integrating remote sensing technologies with advanced analytics to monitor crop health at a large scale. Satellite imagery, drones, and ground-based IoT sensors produce huge datasets that are processed by AI models to detect presymptomatic infections. Multisource integration, such as hyperspectral (400–2,500 nm) and thermal imaging, in combination with DL, facilitates the early identification of presymptomatic infections, especially caused by vascular pathogens (Yi et al., 2023). Optimized CNNs, vision transformers, and ensemble models routinely reach a classification accuracy over 90%, often detecting infections 7–10 days before visual symptoms appear. Lightweight architectures such as EfficientNet-b0 enable on-device AI-based diagnosis of rice and potato diseases, while the integration of hyperspectral imaging with AI models can identify virus-infected grapevines with near-perfect accuracy. The reliability of these systems is enhanced by AI-enhanced preprocessing, class imbalance correction, and sensor fusion techniques. Drone-based monitoring offers scalable data acquisition, while cloud-based AI platforms combine real-time disease mapping, forecasting, and diagnostics into integrated decision-support tools (Yi et al., 2023). Together, these advances create proactive frameworks, supported by AI for timely intervention, informed resource allocation, and resilient crop disease management. ML models can assist in identifying the best timing, dosage, and site of pesticides application by assessing the historical disease incidence, weather patterns (temperature, humidity, rainfall), host susceptibility, and inoculum pressure to model and forecast the potential occurrence and temporal dynamics of future disease outbreaks. Such predictive models allow for proactive rather than reactive disease management, facilitating preventative treatments only when necessary, reducing the use of pesticides. This precision agriculture consequently reduces the use of fungicides, lowers costs, and minimizes environmental impact while maximizing efficacy (Yi et al., 2023). Besides, through the analysis of large-scale genomic and phenotypic datasets, AI can facilitate the identification of genetic markers associated with disease resistance, thereby expediting the development of resilient crop varieties (Gupta et al., 2024).

8.3.4 Challenges and Limitations of AI

Although AI holds immense promise for plant disease management, its practical adoption is hindered by multiple challenges. Robust AI models require large, diverse, and well-annotated datasets, and gathering collected such hug dataset especially for rare diseases or early symptoms is laborious and costly (Akshalin Jefita et al., 2025; González-Rodríguez et al., 2024). Moreover, data heterogeneity, arising from variations in image quality, lighting conditions, and crop types further complicates the training of AI models. Another concern is the model's generalizability, because algorithms trained on one region or crop variety may perform poorly when applied elsewhere due to differences in symptoms, environments, or pathogen strains. The development and implementation of advanced AI systems demand substantial computational and specialized expertise in AI, data science, and plant pathology, which are not always available to farmers or local institutions (Akshalin Jefita et al., 2025; González-Rodríguez et al., 2024). In addition, the inherent, "black-box" nature, of DL models restricts their interpretability and weakens trust among farmers seeking transparent, causative reasoning. Finally, successful implementation depends on the smooth integration of AI tools with existing disease management practices, precision agriculture technologies, and extension services to ensure they are practical and widely applicable (Akshalin Jefita et al., 2025; González-Rodríguez et al., 2024).

8.4 INTEGRATION OF CRISPR AND AI

The integration of AI with CRISPR-Cas systems creates a predictive and precise framework for plant disease management (Zhao et al., 2024). By mining multi-omics data, AI algorithms effectively recognize key susceptibility and resistance genes as optimal targets for CRISPR editing. AI is further optimizing the process by designing highly efficient gRNAs with minimized off-target effects. After editing, AI-driven phenotyping platforms immediately screen for resistant traits, while predictive models speed up the breeding by stacking these edits into elite varieties. This synergistic loop between computational prediction through AI and genome editing is helpful for the development of crop varieties with enhanced resilience to environmental changes and emerging diseases with unprecedented speed and precision, creating a scalable defense against evolving pathogens (Zhao et al., 2024).

8.5 CONCLUSION

In conclusion, CRISPR has proved its versatility, offering a multifaceted toolkit for engineering resistance to disease by knocking out host susceptibility genes, fine-tuning resistance genes via prime and base editing, and directly disabling viral genomes. At the same time, AI has revolutionized disease management by enabling rapid, high-throughput diagnosis from field imagery, predicting outbreaks using data from weather station, historical records, field, and controlled environment images, and guiding resource-efficient management strategies. Individually, each technology represents a monumental leap forward; however, their true potential in management of plant diseases can be unlocked through synergy. While challenges in regulation, ethics, and practical deployment remain, the concerted application of CRISPR and AI paves the way for a future of predictive, precise, and proactive plant disease management, which is critical for developing robust agricultural systems to withstand climate change and emerging pathogen challenges.

REFERENCES

Acquaah, G. (2015). Conventional plant breeding principles and techniques. In J. M. Al-Khayri, S. M. Jain, & D. V. Johnson (Eds.), *Advances in plant breeding strategies: Breeding, biotechnology and molecular tools* (pp. 115–158). Springer International Publishing. https://doi.org/10.1007/978-3-319-22521-0_5

Adigun, O. A., Nadeem, M., Pham, T. H., Jewell, L. E., Cheema, M., & Thomas, R. (2021). Recent advances in bio-chemical, molecular and physiological aspects of membrane lipid derivatives in plant pathology. *Plant, Cell & Environment,* 44(1), 1–16. https://doi.org/10.1111/pce.13904

Akshalin Jefita, R. J., Deepti, S., Indhumathi, M., Muthu Petchi, G., & Magesh, D. (2025). Plant disease prediction using convolutional neural network. *Journal of Computer Science Engineering,* 11(1), 20–26. https://doi.org/10.53555/cse.v11i1.6324

Aman, R., Mahas, A., Butt, H., Ali, Z., Aljedaani, F., & Mahfouz, M. (2018). Engineering RNA virus interference via the CRISPR/Cas13 machinery in Arabidopsis. *Viruses,* 10(12), 732. https://doi.org/10.3390/v10120732

Bastet, A., Zafirov, D., Giovinazzo, N., Guyon-Debast, A., Nogué, F., Robaglia, C., & Gallois, J.-L. (2019). Mimicking natural polymorphism in eIF4E by CRISPR-Cas9 base editing is associated with resistance to potyviruses. *Plant Biotechnology Journal,* 17(9), 1736–1750. https://doi.org/10.1111/pbi.13106

Bi, W., Liu, J., Li, Y., He, Z., Chen, Y., Zhao, T., & Dou, D. (2023). CRISPR/Cas9-guided editing of a novel susceptibility gene in potato improves Phytophthora resistance without growth penalty. *Plant Biotechnology Journal,* 22(1), 4–6. https://doi.org/10.1111/pbi.14188

Chen, W., Zhang, Y., Zhang, Y., Pi, Y., Gu, T., Song, L., Wang, Y., & Ji, Q. (2018). CRISPR/Cas9-based genome editing in *Pseudomonas aeruginosa* and cytidine deaminase-mediated base editing in Pseudomonas species. *iScience,* 6, 222–231. https://doi.org/10.1016/j.isci.2018.07.024

Debbarma, J., Saikia, B., Singha, D. L., Das, D., Keot, A. K., Maharana, J., Hazarika, D. J., Dutta, P., Modi, M. K., & Chikkaputtaiah, C. (2023). CRISPR/Cas9-mediated mutation in XSP10 and SlSAMT genes impart genetic tolerance to Fusarium wilt disease of tomato (*Solanum lycopersicum* L.). *Genes,* 14(2), 488. https://doi.org/10.3390/genes14020488

Farrall, T., Abeynayake, S. W., Webster, W., Fiorito, S., Dinsdale, A., Whattam, M., Campbell, P. R., Chi, Y., Pagay, V., Powell, K. S., Wang, Y. M., & Gambley, C. (2024). Development of a rapid, accurate, and field-deployable LAMP-CRISPR-Cas12a integrated assay for *Xylella fastidiosa* detection and surveillance. *Australasian Plant Pathology,* 53(1), 115–120. https://doi.org/10.1007/s13313-023-00954-4

Gómez-Lama Cabanás, C., & Mercado-Blanco, J. (2025). Groundbreaking technologies and the biocontrol of fungal vascular plant pathogens. *Agronomy,* 11(1), 77. https://doi.org/10.3390/agronomy11010077

González-Rodríguez, V. E., Izquierdo-Bueno, I., Cantoral, J. M., Carbú, M., & Garrido, C. (2024). Artificial intelligence: A promising tool for application in phytopathology. *Horticulturae,* 10(3), 291. https://doi.org/10.3390/horticulturae10030291

Guo, A., Huang, W., Ye, H., Dong, Y., Ma, H., Ren, Y., & Ruan, C. (2020). Identification of wheat yellow rust using spectral and texture features of hyperspectral images. *Remote Sensing,* 12(9), 1419. https://doi.org/10.3390/rs12091419

Gupta, D. K., Pagani, A., Zamboni, P., & Singh, A. K. (2024). AI-powered revolution in plant sciences: Advancements, applications, and challenges for sustainable agriculture and food security. *Exploration of Foods and Foodomics,* 2(5), 443–459. https://doi.org/10.37349/eff.2024.00045

Han, X., Li, S., Zeng, Q., Sun, P., Wu, D., Wu, J., Guo, X., Wang, H., Chen, R., Wu, X., Liu, J., Yi, X., Li, C., Song, B., & Li, G. (2025). Genetic engineering, including genome editing, for enhancing broad-spectrum disease resistance in crops. *Plant Communications,* 6(2), 101195. https://doi.org/10.1016/j.xplc.2024.101195

Huang, Q., Lin, B., Cao, Y., Zhang, Y., Song, H., Huang, C., Liao, J., & Zhuo, K. (2023). CRISPR/Cas9-mediated mutagenesis of the susceptibility gene OsHPP04 in rice confers enhanced resistance to rice root-knot nematode. *Frontiers in Plant Science,* 14, 1134653. https://doi.org/10.3389/fpls.2023.1134653

Jafar, A., Bibi, N., Naqvi, R. A., Sadeghi-Niaraki, A., & Jeong, D. (2024). Revolutionizing agriculture with artificial intelligence: Plant disease detection methods, applications, and their limitations. *Frontiers in Plant Science,* 15, 1356260. https://doi.org/10.3389/fpls.2024.1356260

Jia, H., Omar, A. A., Orbović, V., & Wang, N. (2022). Biallelic editing of the LOB1 promoter via CRISPR/Cas9 creates canker-resistant 'Duncan' grapefruit. *Phytopathology,* 112(2), 308–314. https://doi.org/10.1094/PHYTO-04-21-0144-R

Kan, J., Cai, Y., Cheng, C., Chen, S., Jiang, C., He, Z., & Yang, P. (2023). CRISPR/Cas9-guided knockout of eIF4E improves Wheat yellow mosaic virus resistance without yield penalty. *Plant Biotechnology Journal,* 21(5), 893–895. https://doi.org/10.1111/pbi.14002

Leadbeater, A., & Gisi, U. (2009). The challenges of chemical control of plant diseases. In U. Gisi, I. Chet, & M. L. Gullino (Eds.), *Recent developments in management of plant diseases* (pp. 3–17). Springer Netherlands. https://doi.org/10.1007/978-1-4020-8804-9_1

Malnoy, M., Viola, R., Jung, M.-H., Koo, O.-J., Kim, S., Kim, J.-S., Velasco, R., & Kanchiswamy, C. N. (2016). DNA-free genetically edited grapevine and apple protoplast using CRISPR/Cas9 ribonucleoproteins. *Frontiers in Plant Science,* 7, 1904. https://doi.org/10.3389/fpls.2016.01904

Manzoor, S., Nabi, S. U., Rather, T. R., Gani, G., Mir, Z. A., Wani, A. W., Ganai, S. A., Mansoor, S., Sharma, V., & Manzar, N. (2024). Advancing crop disease resistance through genome editing: A promising approach for enhancing agricultural production. *Frontiers in Genome Editing,* 6, 1399051. https://doi.org/10.3389/fgeed.2024.1399051

Mickey Wang, Y., Ostendorf, B., & Pagay, V. (2024). Evaluating the potential of high-resolution hyperspectral UAV imagery for grapevine viral disease detection in Australian vineyards. *International Journal of Applied Earth Observation and Geoinformation,* 130, 103876. https://doi.org/10.1016/j.jag.2024.103876

Mishra, S., Yadav, C. S. S., Raju, A., Balasundaram, A., & Shaik, A. (2024, September 18–20). EfficientNetB0-based rice disease classification: A strategic fine-tuning approach. *2024 5th International Conference on Smart Electronics and Communication (ICOSEC),* Trichy, India. (pp. 1682–1687). IEEE. https://doi.org/10.1109/ICOSEC61587.2024.10722610

Mitić, D., Bolt, Edward L., & Ivančić-Baće, I. (2023). CRISPR-Cas adaptation in *Escherichia coli. Bioscience Reports, 43*(3), BSR20221198. https://doi.org/10.1042/BSR20221198

Mubarik, M. S., Wang, X., Khan, S. H., Ahmad, A., Khan, Z., Amjid, M. W., Razzaq, M. K., & Azhar, M. T. (2021). Engineering broad-spectrum resistance to cotton leaf curl disease by CRISPR-Cas9 based multiplex editing in plants. *GM Crops & Food,* 12(2), 647–658. https://doi.org/10.1080/21645698.2021.1938488

Nekrasov, V., Wang, C., Win, J., Lanz, C., Weigel, D., & Kamoun, S. (2017). Rapid generation of a transgene-free powdery mildew resistant tomato by genome deletion. *Scientific Reports,* 7(1), 482. https://doi.org/10.1038/s41598-017-00578-x

Ni, Z., Cao, Y., Jin, X., Fu, Z., Li, J., Mo, X., Xu, Z., Li, J., & Huang, S. (2021). Engineering resistance to bacterial blight and bacterial leaf streak in rice. *Rice,* 14(1), 38. https://doi.org/10.1186/s12284-021-00482-z

Ochiai, H., & Yamamoto, T. (2015). Genome editing using zinc-finger nucleases (ZFNs) and transcription activator–like effector nucleases (TALENs). In T. Yamamoto (Ed.), *Targeted genome editing using site-specific nucleases: ZFNs, TALENs, and the CRISPR/Cas9 system* (pp. 3–24). Springer Japan. https://doi.org/10.1007/978-4-431-55227-7_1

Ortigosa, A., Gimenez-Ibanez, S., Leonhardt, N., & Solano, R. (2019). Design of a bacterial speck resistant tomato by CRISPR/Cas9-mediated editing of SlJAZ2. *Plant Biotechnology Journal,* 17(3), 665–673. https://doi.org/10.1111/pbi.13006

Pajjuri, N., Kumar, U., & Thottolil, R. (2022, July 8–10). Comparative evaluation of the convolutional neural network based transfer learning models for classification of plant disease. *2022 IEEE International Conference on Electronics, Computing and Communication Technologies (CONECCT)* (pp. 1–6). IEEE, India. https://doi.org/10.1109/CONECCT55679.2022.9865836

Paula de Toledo Thomazella, D., Brail, Q., Dahlbeck, D., & Staskawicz, B. (2016). CRISPR-Cas9 mediated mutagenesis of a DMR6 ortholog in tomato confers broad-spectrum disease resistance. *bioRxiv*, 064824. https://doi.org/10.1101/064824

Peng, A., Chen, S., Lei, T., Xu, L., He, Y., Wu, L., Yao, L., & Zou, X. (2017). Engineering canker-resistant plants through CRISPR/Cas9-targeted editing of the susceptibility gene CsLOB1 promoter in citrus. *Plant Biotechnology Journal,* 15(12), 1509–1519. https://doi.org/10.1111/pbi.12733

Pereira-Carvalho, R. C., Díaz-Pendón, J. A., Fonseca, M. E. N., Boiteux, L. S., Fernández-Muñoz, R., Moriones, E., & Resende, R. O. (2015). Recessive resistance derived from tomato cv. Tyking limits drastically the spread of Tomato yellow leaf curl virus. *Viruses,* 7(5), 2518–2533. https://doi.org/10.3390/v7052518

Pramanik, D., Shelake, R. M., Park, J., Kim, M. J., Hwang, I., Park, Y., & Kim, J.-Y. (2021). CRISPR/Cas9-mediated generation of pathogen-resistant tomato against Tomato yellow leaf curl virus and powdery mildew. *International Journal of Molecular Sciences,* 22(4), 1878. https://doi.org/10.3390/ijms22041878

Rani, R., Yadav, P., Barbadikar, K. M., Baliyan, N., Malhotra, E. V., Singh, B. K., Kumar, A., & Singh, D. (2016). CRISPR/Cas9: A promising way to exploit genetic variation in plants. *Biotechnology Letters,* 38(12), 1991–2006. https://doi.org/10.1007/s10529-016-2195-z

Ray, D. K., Mueller, N. D., West, P. C., & Foley, J. A. (2013). Yield trends are insufficient to double global crop production by 2050. *PLOS ONE,* 8(6), e66428. https://doi.org/10.1371/journal.pone.0066428

Ren, B., Yan, F., Kuang, Y., Li, N., Zhang, D., Zhou, X., Lin, H., & Zhou, H. (2018). Improved base editor for efficiently inducing genetic variations in rice with CRISPR/Cas9-guided hyperactive hAID mutant. *Molecular Plant,* 11(4), 623–626. https://doi.org/10.1016/j.molp.2018.01.005

Sasikumar, A., Ravi, L., Devarajan, M., Selvalakshmi, A., & Subramaniyaswamy, V. (2024). A lightweight CNN architecture for prediction of plant diseases. In G. Kaushik, D. Gupta, M. D. Khari, & V. Shrivastava (Eds.), *Cognitive analytics and reinforcement learning: Theories, techniques and applications* (pp. 203–222). Wiley. https://doi.org/10.1002/9781394214068.ch11

Savary, S., Willocquet, L., Pethybridge, S. J., Esker, P., McRoberts, N., & Nelson, A. (2019). The global burden of pathogens and pests on major food crops. *Nature Ecology & Evolution,* 3(3), 430–439. https://doi.org/10.1038/s41559-018-0793-y

Sun, J., Wang, Q., Jiang, Y., Wen, Z., Yang, L., Wu, J., & Yang, S. (2018). Genome editing and transcriptional repression in *Pseudomonas putida* KT2440 via the type II CRISPR system. *Microbial Cell Factories,* 17(1), 41. https://doi.org/10.1186/s12934-018-0893-z

Wagh, S. G., Pohare, M. B., & Kale, R. R. (2021). CRISPR/Cas in food security and plant disease management. In A. Kumar & S. Droby (Eds.), *Food security and plant disease management* (pp. 171–191). Woodhead Publishing. https://doi.org/10.1016/B978-0-12-821843-3.00020-9

Wang, X., Whalley, W. R., Miller, A. J., White, P. J., Zhang, F., & Shen, J. (2020). Sustainable cropping requires adaptation to a heterogeneous rhizosphere. *Trends in Plant Science,* 25(12), 1194–1202. https://doi.org/10.1016/j.tplants.2020.07.006

Wei, X., Johnson, M. A., Langston, D. B., Mehl, H. L., & Li, S. (2021). Identifying optimal wavelengths as disease signatures using hyperspectral sensor and machine learning. *Remote Sensing,* 13(14), 2833. https://doi.org/10.3390/rs13142833

Wheatley, M. S., & Yang, Y. (2020). Versatile applications of the CRISPR/Cas toolkit in plant pathology and disease management. *Phytopathology,* 111(7), 1080–1090. https://doi.org/10.1094/PHYTO-08-20-0322-IA

Wójcik-Gront, E., Zieniuk, B., & Pawełkowicz, M. (2024). Harnessing AI-powered genomic research for sustainable crop improvement. *Agriculture,* 14(12), 2299. https://doi.org/10.3390/agriculture14122299

Yang, X., Zhou, H., & Zhou, X. (2019). Rock paper scissors: CRISPR/Cas9-mediated interference with geminiviruses in plants. *Science China Life Sciences,* 62(10), 1389–1391. https://doi.org/10.1007/s11427-019-9825-4

Yi, Z., Zeng, P., Zuhao, L., Tingwei, H., & Jingwen, Y. (2023). SFPGRN: Spectral detection method of plant diseases based on deep learning. *Proceedings of the 3rd International Conference on Artificial Intelligence, Automation, and High-Performance Computing (AIAHPC 2023)* (Vol. 12941, pp. 415–423). SPIE, Wuhan, China. https://doi.org/10.1117/12.3008342

Yu, L., Zou, J., Hussain, A., Jia, R., Fan, Y., Liu, J., Wang, Y., Liu, Y., Liu, W., Liu, Y., Li, J., Yan, Z., Wang, J., & Jin, S. (2024). Systemic evaluation of various CRISPR/Cas13 orthologs for knockdown of targeted transcripts in plants. *Genome Biology,* 25(1), 307. https://doi.org/10.1186/s13059-024-03448-8

Zhao, X., Chen, B., Ji, M., Wang, X., Yan, Y., Zhang, J., Sun, H., & Lv, C. (2024). Implementation of large language models and agricultural knowledge graphs for efficient plant disease detection. *Agriculture,* 14(8), 1359. https://doi.org/10.3390/agriculture14081359

9 CRISPR and Beyond

Next-Generation Tools for Environmental and Industrial Biotechnology Applications

Subhashree Pattnaik, Sitanshu Sekhar Satapathy, and Kailash Chandra Samal

9.1 INTRODUCTION

Genome editing allows scientists to precisely add, delete, or alter DNA in an organism's genetic material, with clustered regularly interspaced short palindromic repeats (CRISPR)-Cas9 being a leading technology adapted from a bacterial immune system that uses guide RNA (gRNA) and a Cas9 enzyme to cut DNA at specific sites. Next-generation genome-editing tools build on this by enhancing accuracy through features like double-strand breaks and developing methods to directly edit individual bases or alter gene expression without cutting DNA, such as with base editors and epigenome editors.

The CRISPR and CRISPR-associated (Cas) system constitute an adaptive immune mechanism in prokaryotes, conferring resistance to foreign genetic elements. This innate biological system has been harnessed and repurposed into a revolutionary toolkit for precise genome manipulation, with the CRISPR-Cas9 system emerging as the most widely adopted platform. Genome editing refers to a collection of scientific techniques that allow for precise, intentional alterations to the DNA sequence within a living organism. This capability enables researchers to: Insert: introduce new genetic code into the genome; Delete: remove specific sections of existing DNA; and Modify: change the genetic letters at precise genomic locations.

These changes are facilitated by engineered molecular tools, primarily nucleases, which function as precise scissors to create breaks in the DNA double helix. The cell's innate repair systems subsequently fix this break. One method, non-homologous end joining (NHEJ), often results in gene disruption. Alternatively, if a donor DNA template is supplied, the cell can use the homology-directed repair (HDR) pathway to insert new genetic material or correct an error (Yang et al. 2020).

CRISPR-Cas9 is a transformative gene-editing technology derived from a natural defense mechanism in bacteria. CRISPR-Cas9 originated as a form of adaptive immunity in bacteria, where it functions as a defense mechanism against viral infections. In its natural role, bacteria incorporate fragments of viral DNA into their own genome within regions known as CRISPR arrays. When the same virus attacks again, these stored sequences are transcribed into RNA molecules that guide the Cas9 enzyme to the corresponding viral DNA. The Cas9 enzyme then cleaves and neutralizes the invading genetic material. Scientists have adapted this system for research purposes by designing synthetic gRNAs that can be programmed to match virtually any target DNA sequence. These gRNAs direct the Cas9 enzyme to a specific genomic location, where it creates a precise cut. Compared to earlier gene-editing technologies, CRISPR-Cas9 is more efficient, cost-effective, accurate, and easier to use (Hossain 2021).

 DOI: 10.1201/9781003735113-9

9.2 NEXT-GENERATION GENOME EDITING TECHNOLOGIES

Building on CRISPR, newer editing tools are being developed, that provide enhanced accuracy and expanded functionality beyond simple DNA cutting. Several advanced CRISPR-based tools have been developed to improve precision and expand functionality. Base editors enable the direct conversion of one DNA base pair into another, such as changing a C–G pair to a T–A pair, without cutting both strands of the DNA helix, thereby reducing errors and unintended mutations. Epigenome editors employ a catalytically deactivated Cas9 (dCas9) protein fused to effector domains that modify epigenetic markers chemical modifications on DNA that regulate gene expression without altering the underlying DNA sequence, allowing genes to be switched on or off (Khalil 2023). High-fidelity Cas variants are engineered forms of the Cas9 enzyme designed to exhibit stronger and more specific binding to their intended target sites, significantly minimizing off-target cleavage. Additionally, CRISPR nickases are created by mutating one of Cas9's two nuclease domains so that only a single DNA strand is cut; when two gRNAs are used to generate adjacent single-strand nicks, the effect mimics a double-strand break while greatly enhancing specificity, as two independent off-target events would need to occur in close proximity (Losito 2022).

9.3 CRISPR-CAS SYSTEM: DISCOVERY AND MECHANISM

The CRISPR-Cas system is a sophisticated adaptive immune system naturally present in bacteria and archaea (Barrangou et al. 2007). Its fundamental function is to recognize and cleave foreign nucleic acids, such as those from bacteriophages and plasmids. The ancient origins of this system into a programmable genome-editing technology represents a paradigm shift in molecular biology, with broad applications across biomedical research, biotechnology, and therapeutic development (Ishino et al. 2018). The most extensively characterized and utilized variant is the Type II CRISPR-Cas9 system.

9.3.1 Discovery of the CRISPR-Cas System

The discovery of the CRISPR-Cas system involved a series of independent findings over several decades. The discovery of CRISPR began in 1987 when Yoshizumi Ishino and colleagues at Osaka University identified unusual repeated DNA sequences in *Escherichia coli* while analyzing the *iap* gene, though their function was unknown at the time. In the early 2000s, Francisco Mojica characterized these structures in archaea, coined the term CRISPR, and proposed that the spacer sequences matching viral and plasmid DNA formed part of an adaptive immune system in prokaryotes. Experimental confirmation came in 2007 when Philippe Horvath and his team at Danisco demonstrated that CRISPR provides acquired viral immunity in *Streptococcus thermophilus* by incorporating phage DNA fragments as spacers to confer resistance to later infections. The field was transformed in 2012 when Jennifer Doudna and Emmanuelle Charpentier showed that the Type II CRISPR-Cas9 system could be engineered as a programmable genome-editing tool using a synthetic gRNA, enabling targeted DNA cleavage. Soon after, research groups led by Feng Zhang and George Church validated its effectiveness in eukaryotic cells, including mouse and human models, establishing CRISPR-Cas9 as a revolutionary technology for genome engineering.

9.3.2 Molecular Mechanism of the Type II CRISPR-Cas9 System

In bacteria, the native CRISPR-Cas immune system functions through three coordinated stages that have been simplified for modern biotechnology. First, during adaptation (spacer acquisition), when a virus infects a bacterium, the Cas1–Cas2 complex captures short fragments of foreign DNA (protospacers) and integrates them as new spacers into the CRISPR locus, creating a molecular memory of past infections. Second, in the expression and maturation stage, this locus is transcribed into a

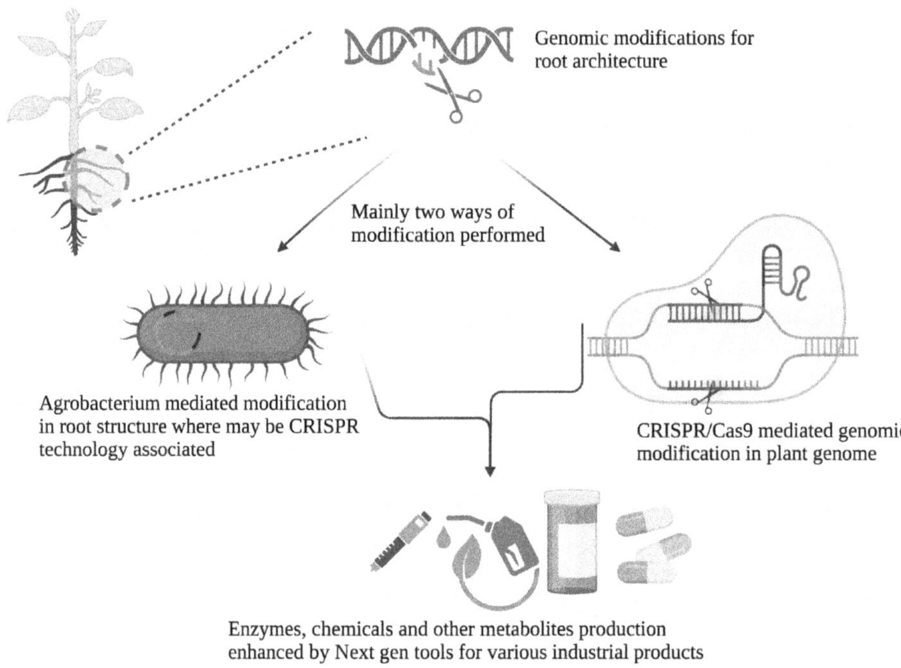

FIGURE 9.1 Diagram showing how CRISPR and *Agrobacterium* tools modify plant root architecture. This modification helps increase the production of industrial enzymes and metabolites.

long precursor CRISPR RNA (crRNA), which is then processed into shorter, mature crRNAs, each carrying a single spacer sequence; in Type II systems, this processing requires a trans-activating crRNA (tracrRNA). Third, during interference, the mature crRNA, tracrRNA, and Cas9 nuclease assemble into an effector complex that scans the cell for DNA sequences matching the crRNA spacer. Recognition depends on the presence of a specific protospacer adjacent motif (PAM), and once the correct sequence is identified, Cas9 generates a double-strand break in the target DNA, neutralizing the invading genetic material.

For genome editing applications, this natural defense mechanism has been streamlined into a two-component system consisting of the Cas9 endonuclease and a synthetic single-guide RNA (sgRNA), which fuses crRNA and tracrRNA functions. The sgRNA directs Cas9 to a chosen DNA sequence adjacent to a PAM, where Cas9 induces a double-strand break. Cellular DNA repair pathways then determine the genetic outcome: error-prone NHEJ typically creates small insertions or deletions that disrupt gene function, while HDR, when provided with a donor template, enables precise sequence modification or targeted insertion of new genetic material. Figure 9.1 displays genomic alteration in plants for abundant synthesis of chemicals, resulting in more industrial goods.

9.4 CASE STUDIES ON NEXT-GENERATION TOOLS FOR ENVIRONMENTAL BIOTECHNOLOGY

Many researchers work to enhance technologies in the field of environmental biotechnology. Some recent studies have used consortia, CRISPR-based changes in algae, synthetic microorganisms, and other approaches. Excessive plastic manufacture has become a severe environmental and health issue as a result of inefficient waste management. Hence, researchers directed their focus toward sustainable and complete degradation of polyethylene terephthalate (PET) pollution. Mutations and protein engineering of the PETase enzyme (IsPETase), as well as pathway optimization in organisms or enzyme combinations, have been shown to boost PET hydrolysis to MHET/TPA. A previous study found that modified PETases depolymerize PET far quicker than wild-type enzymes, and they were tested on PET film and bottle material in lab/bench reactors (Sevilla et al. 2023). This study supports the use of in situ or wastewater-based methods for reducing PET microplastic burden.

This research resulted to sustainable degradation of PET waste. CRISPR-enabled metabolic gene modifications (introducing or upregulating dioxygenases, dehalogenases, or efflux/transforming systems) to broaden substrate range or improve action against PFAS, PCBs, organochlorines, or persistent pesticides. Experimental reports demonstrate improved degradation kinetics in lab strains and the identification of target pathways for converting recalcitrant pollutants to less toxic products. CRISPR enables precise pathway rewiring (knock-ins and promoter tuning). Most work stays at lab scale; environmental deployment poses containment, monitoring, and governance difficulties. Efficacy in complicated soils/sediments is still being established (Minhas et al. 2024).

As photosynthetic organisms, microalgae can create biomolecules for use in food, feed, cosmetics, nutraceuticals, fuel, and other purposes. Researchers using CRISPR-Cas technology to increase lipid synthesis and reroute carbon flux toward triacylglycerols for biofuels, engineer carbon concentrating mechanisms, Rubisco, and CO_2 uptake/transport pathways to increase CO_2 fixation and use multiplexed CRISPR (CRISPRi/CRISPRa and multiplex editing) to coordinate multiple pathway changes at once. Promising improvements in lipid yields and quicker CO_2 absorption are seen in lab studies, but there are still significant obstacles to overcome, including scaling from lab photobioreactors to open/industrial cultivation, off-target and regulatory concerns, and transformation efficiency across species (Dhokane et al. 2023). The ability to create reliable, highly productive microalgal cell factories is made possible by the thoughtful integration of this CRISPR toolset, ultimately fulfilling their potential as sustainable platforms for next-generation biomanufacturing. More recent toolkits like CRISPRi/CRISPRa go "beyond cutting" by employing multiplexed guides and CRISPR interference/activation to simultaneously upregulate targeted pathways and downregulate competing sinks, enhancing flux control without permanently disrupting vital genes. This allows for more precise management of trade-offs (growth vs. product) (Yang and Lu 2025).

Heavy metals ions like cadmium, mercury, lead, arsenic, and chromium are considered to be harmful chemicals for the ecosystem and also carcinogenic in nature. Graphene, metal/metal-oxide nanoparticles, and composites are being used extensively by researchers to alter electrodes, resulting in significantly increased surface area, quicker electron transfer, and sub-ppb sensitivity for Pb^{2+}, Cd^{2+}, Hg^{2+}, and organophosphate pesticide residues. These methods have dominated current research on inexpensive, on-site heavy-metal monitoring (Dweh et al., 2023; Sulthana et al. 2024). Enzymes, bacteria, fungi, and plant and animal cells are examples of biological heavy-metal detectors. This research was to assess the potential of biosensors in environmental analysis, taking into account the ease and environmental friendliness of analyzing individual compounds (e.g., heavy metals) or their mixtures in various samples directly in the environment, as well as continuous monitoring of such compounds. Additionally, biosensors are used to monitor the decomposition of plastics in the aquatic environment, with recombinant whole-cell microbial biosensors serving as detectors for plastic monomers such as acrylic acid. New optical and electrochemical biosensing technologies utilize tailored recognition elements (protein/aptamer probes, microbial proteins) and electrochemical signatures to allow faster, lower-cost detection of microplastics in water samples (Popenda et al. 2024).

9.5 CASE STUDIES ON NEXT-GENERATION TOOLS FOR INDUSTRIAL BIOTECHNOLOGY

Numerous technologies of the next generation are being applied in order to either accelerate or advance the production of industrial goods. Enzymes and microbes are used in industrial biotechnology to create valuable compounds from renewable resources. Industrial biotechnology is a fast-expanding sector because of its relationship with lower energy usage, greenhouse gas emissions, and waste creation (Tang and Zhao 2009).

The generation of bioethanol from food waste might lead to a sustainable solution that would address the issue of food waste disposal as well as the expanding population's energy demands. This research focused on order to increase yields and reduce costs, the focus is on converting lignocellulosic and waste feedstocks utilizing modified yeasts (*Saccharomyces cerevisiae*) and bacteria, consolidated bioprocessing, thermotolerant/nonconventional strains, and AI-assisted process optimization. Researchers examine the production of bioethanol from different feedstocks and analyze existing technologies, such as fermentation processes, gives industrial manufacturers information about the resources and technology that are already available as well as the potential for future innovation (Kazmi et al. 2025).

An important organic acid, lactic acid has many industrial uses, especially in the food, pharmaceutical, cosmetic, and biomedical industries. Because the traditional method of producing lactic acid from refined sugars is expensive and has major negative effects on the environment, more sustainable methods and alternative raw materials are being investigated. Recent research uses nontraditional microorganisms and lactic acid bacteria to create high-purity L- or D-LA from lignocellulose, agricultural waste, and even methane (methanotroph engineering). Engineering addresses downstream purity, stereospecificity, and acid tolerance. Recent studies have produced high-purity L- or D-LA from lignocellulose, agricultural waste, and even methane (methanotroph engineering) using unconventional microbes and lactic acid bacteria. Stereospecificity, acid tolerance, and downstream purity are all addressed via engineering (Díaz-Orozco et al. 2025).

The convergence of platforms for genome editing and enzymatic modification in industrial biotechnology led to advancement in the field of industrial biotechnology. Terpenoids are the widely used for flavors medicines, cosmetics, and natural products. Engineered microbes for biosynthesis of terpenoids are developed by using various genome editing technologies such as CRISPR-Cas9 system. Researchers selected microbial strains such as *Streptococcus pyogenes* and *E. coli*. Many research studies were conducted to obtain stable bacterial strains for yielding various terpenoids ranging from sesquiterpenoids to tetraterpenoids (Sun et al. 2024).

CRISPR activation and interference (CRISPRi/a) are programmable, nuclease-dead or catalytically disabled CRISPR systems that are utilized for multigene regulation in microorganisms and cell factories, metabolic-pathway balance, and genome-scale screening without permanently altering DNA (Enright et al. 2024). Researchers provide an effective dual-mode CRISPRa/i system that unites an evolving PAM-flexible dxCas9 with an engineered *E. coli* cAMP receptor protein. The possibility for rewiring bacterial metabolic pathways is advanced by this dual-mode CRISPRa/i technology, which offers accurate and adaptable control for a variety of biotechnological applications (Moon et al. 2025). Recent research and innovation of next-gen tools for industrial and environmental biotechnology applications are summarized in Table 9.1. Figure 9.2 shows that, directly or indirectly, CRISPR system-based changes contribute to industrial and environmental biotechnology.

9.6 FUTURE ASPECTS BASED ON NEXT-GEN TOOLS FOR ENVIRONMENT AND INDUSTRIAL BIOTECHNOLOGY

In recent times, researchers and scientists have been investing their skills and knowledge for the integration and advancement of technologies that led to sustainable and accurate products. Next-generation genome editing (CRISPR-Cas systems, base editors, prime editors, RNA-targeting Cas enzymes, and engineered gene-drive architectures) is progressing from laboratory demonstrations toward real-world environmental tools, such as programmable biosensors for water/soil monitoring, engineered microbes/algae for pollutant and plastic degradation, and local-confined gene drives to control invasive pests or disease vectors. These advancements offer more accuracy and

TABLE 9.1
Overview of Recent Research and Innovation of Next-Gen Tools for Industrial and Environmental Biotechnology Applications

S. No.	Research Study	Next-Gen Tools or Platforms	Year	Purpose	References
1	Precision genome writing/ therapeutic and environmental strain design	Prime editing for microbes and plants	2023–2025	For strain optimization and safer environmental releases, precise, DSB-free editing can be used to install complicated alterations (point changes, tiny insertions).	Lee et al. (2025)
2	Precise single-base changes in industrial strains	Base editors (dCas-derived dual-function editors, cytidine/ adenine base editors)	2022–2025	Expanding host range (e.g., cyanobacteria); rapid reprogramming of metabolic genes in industrial microorganisms (no DSBs).	Zheng et al. (2025)
3	Site-specific DNA insertions without DSB repair	RNA-guided transposases and CRISPR-associated transposons	2019–2024	Large DNA cargos can be programmed and scarlessly inserted into bacterial genomes, which is helpful for creating metabolic pathways in industrial and environmental strains.	Strecker et al. (2019)
4	Environmental nucleic acid biosensing	Electrochemical and fluorescent CRISPR-Cas13 (collateral RNA cleavage) biosensors	2020–2025	Point-of-use monitoring and quick, sensitive identification of pollutants, RNA markers, and microbiological hazards in soil and water.	Zhang et al. (2024)
5	Robust bioremediation/ complex pollutant degradation	Biofilm engineering, DIET (Direct Interspecies Electron Transfer) stimulation, and synthetic consortia are examples of engineered microbial consortia.	2022–2025	In wastewater and soil remediation, combine orthogonal metabolisms to break down resistant xenobiotics and increase resistance.	Lü et al. (2024)
6	Data-driven enzyme discovery for pollutant breakdown	Platforms for chemo-informatics and machine learning (enzyme prediction tools, such as XenoBug)	2024–2025	Quick in silico prediction of potential enzymes and pathways from metagenomes to expedite the design of biodegradation and bioremediation.	Malwe et al. (2025)
7	CRISPR-enabled environmental monitoring and rapid QA (Quality Assurance)	Nucleic acid sensors based on CRISPR and integrated water quality monitoring systems	2023–2025	Integrate CRISPR sensors into regular environmental quality assurance processes and monitor infections and pollutants in environmental samples in real time.	Sun et al. (2024)

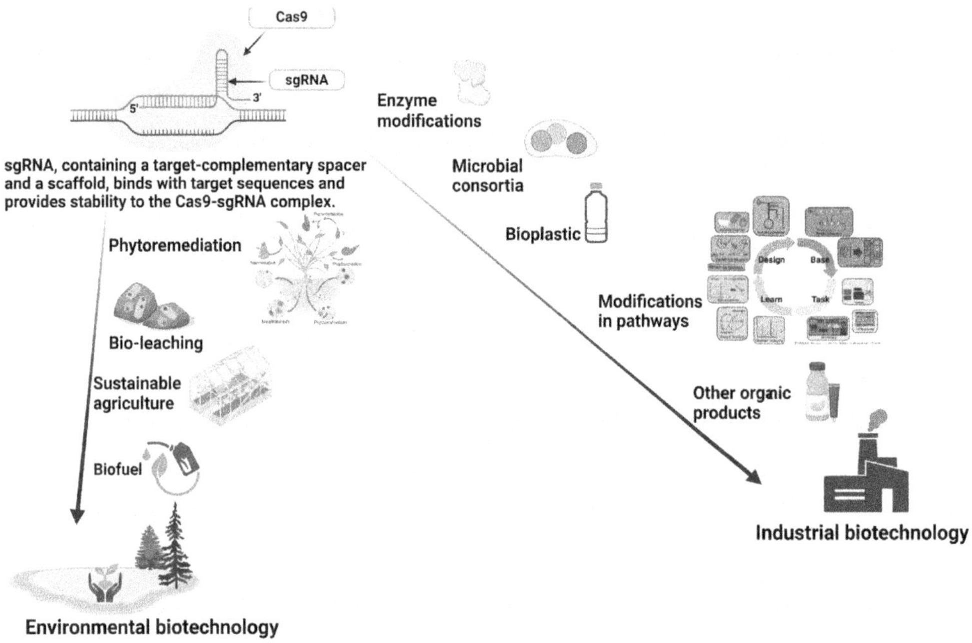

FIGURE 9.2 Schematic illustration of the applications of CRISPR in environmental and industrial biotechnology, ranging from biofuel production to the creation of bioplastics.

novel functionality (single-base alterations, RNA editing, self-limiting drives), but they also present ecological, regulatory, and ethical concerns that necessitate multi-layered risk assessment, monitoring, and community governance (Miklau et al. 2024). Wastewater treatment plants are one of the key avenues for microplastics to enter the environment. In general, microplastics are contaminants of global concern that pose dangers to ecosystem and human health. By delivering recombinant DNA to bacteria in wastewater to facilitate the breakdown of PET, researchers demonstrated a proof-of-concept for reducing microplastic pollution released from wastewater treatment facilities. Researchers also made it possible for several bacterial species from a municipal wastewater sample to express FAST-PETase, which was then released into the extracellular environment, using a broad-host-range conjugative plasmid (Yip et al. 2024).

Many aquatic plant species may naturally absorb and break down harmful substances and metabolites in aquatic environments. *Eichhornia crassipes* metabolizes PCP (Pentachlorophenol), while *Lemna gibba* detoxifies phenolic chemicals. Algal turf scrubbers use filamentous algae to remove pollutants and nutrients, whereas built wetlands mimic natural wetlands with aquatic plant species such as *Typha* spp. and *Phragmites* spp. Genomic modifications to these plants' metabolic pathways can improve phytoremediation (Satapathy et al. 2025). Because of its significance to human life and health, agricultural pollution has emerged as one of the world's most pressing environmental challenges. Microbial remediation is an efficient way for reducing heavy-metal pollution in agriculture, but assessing its efficacy has proven problematic. ML (Machine Learning) has shown promise in determining the effectiveness of microbial remediation of heavy-metal pollution in agriculture, and it is expected to be an effective pollution management method (Wu and Zhao 2023).

CRISPR 3.0 is the most recent evolution in genome engineering, combining compact nucleases, programmable transcriptional control, RNA-targeting methods, and multi-layered cellular reprogramming. These advancements in industrial and environmental biotechnology provide dramatic opportunities for sustainable manufacturing, precision biocatalysis, and better cleanup (Onyeyili et al. 2025). CRISPR 3.0 introduces new Cas families (Cas12f, CasΦ, Cas7–11) that are effective in many environmental microorganisms (Bagi et al. 2025). Its future impacts like engineered consortia of bacteria, fungi, and algae engineered to break down plastics (PET, PVC), insecticides, hydrocarbons, or heavy metals by route upgrades, CRISPR-guided metagenome mining to uncover new enzymes from harsh environments for pollutant breakdown, etc. and many more applications were performed for advancement in industrial and environmental biotechnology (Chathangad et al. 2025; Johnson et al. 2025).

9.7 CONCLUSION

Current developments in genome engineering show that CRISPR (Clustered Regularly Interspaced Short Palindromic Repeats), TALEN (Transcription Activator-Like Effector Nuclease), and ZFN (Zinc Finger Nuclease) technologies are not only transforming plant and microbial systems but also offering accurate, dependable, and scalable solutions for industrial and environmental problems. Although TALEN and ZFN laid the groundwork for targeted genome modification, CRISPR has become the most accessible and versatile tool, allowing for exceptional specificity at lower costs and with more efficiency. Beyond gene cutting, advanced CRISPR platforms including CRISPR 3.0, Cas 12, and Cas 13 broaden the technological landscape to include gene regulation, biosensing, RNA targeting, and precise epigenetic control. The creation of engineered microorganisms that degrade pollutants, upcycle waste, monitor environmental toxins, and maximize bioprocessing is made possible by these next-generation tools. At the same time, microbial consortia engineering is a very promising approach, since community-based systems more accurately reflect natural ecology and allow for multistep bioremediation, waste degradation, and the formation of value-added products. These consortia may be logically designed to carry out complex tasks that single-species systems are unable to accomplish when combined with CRISPR and synthetic biology. Integrated platforms that integrate systems biology, community engineering, and next-generation genome editing hold the key to the future of industrial and environmental biotechnology. These developments will guarantee a cleaner, safer, and more resilient environment in the future by enabling sustainable manufacturing, strong bioremediation techniques, and sophisticated biosensing.

REFERENCES

Bagi, M., Jamalzadegan, S., Steksova, A., & Wei, Q. (2025). CRISPR–Cas based platforms for RNA detection: Fundamentals and applications. *Chemical Communications*, 61(72), 13571–13600. https://doi.org/10.1039/D5CC03257A

Barrangou, R., Fremaux, C., Deveau, H., Richards, M., Boyaval, P., Moineau, S., … Horvath, P. (2007). CRISPR provides acquired resistance against viruses in prokaryotes. *Science*, 315(5819), 1709–1712. https://doi.org/10.1126/science.1138140

Chathangad, S. N., Bovas, N. A., Kothari, M., & Sadhukhan, S. (2025). Gene modification and omics technology for pollution management. In Pradeep Verma (Ed.), *Biotechnology for Environmental Sustainability* (pp. 469–506). Springer Nature Singapore. https://doi.org/10.1007/978-981-97-7221-6_17

Dhokane, D., Shaikh, A., Yadav, A., Giri, N., Bandyopadhyay, A., Dasgupta, S., & Bhadra, B. (2023). CRISPR-based bioengineering in microalgae for production of industrially important biomolecules. *Frontiers in Bioengineering and Biotechnology*, 11, 1267826. https://doi.org/10.3389/fbioe.2023.1267826

Díaz-Orozco, L., Moscosa Santillán, M., Delgado Portales, R. E., Rosales-Colunga, L. M., Leyva-Porras, C., & Saavedra-Leos, Z. (2025). Advances in L-lactic acid production from lignocellulose using genetically modified microbial systems. *Polymers*, 17(3), 322. https://doi.org/10.3390/polym17030322

Dweh, T. J., Pattnaik, S., & Sahoo, J. P. (2023). Assessing the impact of meta-genomic tools on current cutting-edge genome engineering and technology. *International Journal of Biochemistry and Molecular Biology*, 14(4), 62–75.

Enright, A. L., Heelan, W. J., Ward, R. D., & Peters, J. M. (2024). CRISPRi functional genomics in bacteria and its application to medical and industrial research. *Microbiology and Molecular Biology Reviews*, 88(2), e00170–22. https://doi.org/10.1128/mmbr.00170-22

Hossain, M. A. (2021). CRISPR-Cas9: A fascinating journey from bacterial immune system to human gene editing. *Progress in Molecular Biology and Translational Science*, 178, 63–83. https://doi.org/10.1016/bs.pmbts.2021.01.001

Ishino, Y., Krupovic, M., & Forterre, P. (2018). History of CRISPR-Cas from encounter with a mysterious repeated sequence to genome editing technology. *Journal of Bacteriology*, 200(7), 10–1128. https://doi.org/10.1128/jb.00580-17

Johnson, K. A., Cooper, C., Philippe, C., Catchpole, R. J., Mitchell, S., & Terns, M. P. (2025). A phage variable region encodes anti-CRISPR proteins inhibiting all *Streptococcus thermophilus* CRISPR immune systems. *The CRISPR Journal*, 8(5), 333–352. https://doi.org/10.1177/25731599251369720

Kazmi, A., Sultana, T., Ali, A., Nijabat, A., Li, G., & Hou, H. (2025). Innovations in bioethanol production: A comprehensive review of feedstock generations and technology advances. *Energy Strategy Reviews*, 57, 101634. https://doi.org/10.1016/j.esr.2024.101634

Khalil, A. M. (2023). Epigenome engineering: Understanding, managing, and improving technical aspects. *Pharmacophore*, *14*(3–2023), 119–130. https://doi.org/10.51847/wv4pyDAtfP

Lee, J., Kweon, J., & Kim, Y. (2025). Emerging trends in prime editing for precision genome editing. *Experimental & Molecular Medicine*, 57(7), 1381–1391. https://doi.org/10.1038/s12276-025-01463-8

Losito, M. (2022). *Functional characterization of CRISPR-Cas interactions with DNA* (Doctoral dissertation, Durham University). Available at: https://etheses.dur.ac.uk/14700/

Lü, H., Wei, J. L., Tang, G. X., Chen, Y. S., Huang, Y. H., Hu, R., … Li, Q. X. (2024). Microbial consortium degrading of organic pollutants: Source, degradation efficiency, pathway, mechanism and application. *Journal of Cleaner Production*, 451, 141913. https://doi.org/10.1016/j.jclepro.2024.141913

Malwe, A. S., Longwani, U., & Sharma, V. K. (2025). XenoBug: Machine learning-based tool to predict pollutant-degrading enzymes from environmental metagenomes. *NAR Genomics and Bioinformatics*, 7(2), lqaf037. https://doi.org/10.1093/nargab/lqaf037

Miklau, M., Burn, S. J., Eckerstorfer, M., Dolezel, M., Greiter, A., Heissenberger, A., … Hagen, K. (2024). Horizon scanning of potential environmental applications of terrestrial animals, fish, algae and microorganisms produced by genetic modification, including the use of new genomic techniques. *Frontiers in Genome Editing*, 6, 1376927. https://doi.org/10.3389/fgeed.2024.1376927

Minhas, W. R., Saleemi, H., Raheem, M. F., Khalid, S. S., Aslam, H., & Ahmad, W. (2024). Crispr-engineered microbes for enhanced biodegradation of recalcitrant pollutants a genomic approach to environmental remediation. *Frontiers in Health Informatics*, *13*(3). https://doi.org/10.63682/fhi787

Moon, S. Y., Kim, M. R., An, N. Y., Noh, M. H., & Lee, J. Y. (2025). Dual-mode CRISPRa/i for genome-scale metabolic rewiring in *Escherichia coli*. *Nucleic Acids Research*, 53(15), gkaf818. https://doi.org/10.1093/nar/gkaf818

Onyeyili, I. N., Ogbozor, F. I., Kehinde, A. B., Ohaekwe, C. C., Ovada, E. O., Ali, V. E., & Hussaini, M. (2025). Phage-based CRISPR delivery systems for targeted eradication of drug-resistant bacterial infections. *International Journal of Life Sciences Research*, 9, 6–34. https://doi.org/10.53771/ijlsra.2025.9.1.0042

Popenda, A., Wiśniowska, E., & Manuel, C. (2024). Biosensors in environmental analysis of microplastics and heavy metal compounds–A review on current status and challenges. *Desalination and Water Treatment*, 319, 100456. https://doi.org/10.1016/j.dwt.2024.100456

Satapathy, S. S., Nayak, J., Choudhury, N., Dash, S., Anis, A., Sahu, S., … Naik, B. (2025). Mitigating aquatic toxicity through the use of aquatic plants. *Journal of Advances in Biology & Biotechnology*, 28(5), 637–652. https://doi.org/10.9734/jabb/2025/v28i52326

Sevilla, M. E., Garcia, M. D., Perez-Castillo, Y., Armijos-Jaramillo, V., Casado, S., Vizuete, K., … Cerda-Mejía, L. (2023). Degradation of PET bottles by an engineered *Ideonella sakaiensis* PETase. *Polymers*, *15*(7), 1779. https://doi.org/10.3390/polym15071779

Strecker, J., Ladha, A., Gardner, Z., Schmid-Burgk, J. L., Makarova, K. S., Koonin, E. V., & Zhang, F. (2019). RNA-guided DNA insertion with CRISPR-associated transposases. *Science*, 365(6448), 48–53. https://doi.org/10.1126/science.aax9181

Sulthana, S. F., Iqbal, U. M., Suseela, S. B., Anbazhagan, R., Chinthaginjala, R., Chitathuru, D., … Kim, T. H. (2024). Electrochemical sensors for heavy metal ion detection in aqueous medium: A systematic review. *ACS Omega*, 9(24), 25493–25512. https://doi.org/10.1021/acsomega.4c00933

Sun, X., Zhang, H., Jia, Y., Li, J., & Jia, M. (2024). CRISPR-Cas9-based genome-editing technologies in engineering bacteria for the production of plant-derived terpenoids. *Engineering Microbiology*, 4(3), 100154. https://doi.org/10.1016/j.engmic.2024.100154

Tang, W. L., & Zhao, H. (2009). Industrial biotechnology: Tools and applications. *Biotechnology Journal: Healthcare Nutrition Technology*, 4(12), 1725–1739. https://doi.org/10.1002/biot.200900127

Wu, J., & Zhao, F. (2023). Machine learning: An effective technical method for future use in assessing the effectiveness of phosphorus-dissolving microbial agroremediation. *Frontiers in Bioengineering and Biotechnology*, 11, 1189166. https://doi.org/10.3389/fbioe.2023.1189166

Yang, L., & Lu, Q. (2025). Beyond cutting: CRISPR-driven synthetic biology toolkit for next-generation microalgal metabolic engineering. *International Journal of Molecular Sciences*, 26(15), 7470. https://doi.org/10.3390/ijms26157470

Yang, H., Ren, S., Yu, S., Pan, H., Li, T., Ge, S., … Xia, N. (2020). Methods favoring homology-directed repair choice in response to CRISPR/Cas9 induced-double strand breaks. *International Journal of Molecular Sciences*, 21(18), 6461. https://doi.org/10.3390/ijms21186461

Yip, A., McArthur, O. D., Ho, K. C., Aucoin, M. G., & Ingalls, B. P. (2024). Degradation of polyethylene terephthalate (PET) plastics by wastewater bacteria engineered via conjugation. *Microbial Biotechnology*, 17(9), e70015. https://doi.org/10.1111/1751-7915.70015

Zhang, Y., Li, S., Li, R., Qiu, X., Fan, T., Wang, B., … Zhang, L. (2024). Advances in application of CRISPR-Cas13a system. *Frontiers in Cellular and Infection Microbiology*, 14, 1291557. https://doi.org/10.3389/fcimb.2024.1291557

Zheng, Y., Du, F., Hang, Y., Ma, W., Yang, G., Wu, N., & Sun, X. (2025). Recent advances in genome base editing technology and its applications in industrial microorganism. *Biotechnology Journal*, 20(6), e70052. https://doi.org/10.1002/biot.70052

10 Next-Generation Guide RNAs
Design, Delivery, and Specificity for Advanced Editing

Adeeb Rahman and Neeti Sanan Mishra

10.1 INTRODUCTION

CRISPR (clustered regularly interspaced short palindromic repeats) has revolutionized genome editing by enabling targeted, precise, and programmable modification of nucleic acids. CRISPR-Cas systems defend against invading genetic elements by recognizing and cleaving foreign DNA or RNA sequences through a sequence-specific RNA guide. From their discovery till now, two major classes of CRISPR-Cas system have been established, which include six different subtypes (Makarova et al., 2020).

Among them, Type II CRISPR-Cas9 is the most common and well established (Doudna & Charpentier, 2014). Here, the single-guide RNA (sgRNA) consists of fused CRISPR RNA (crRNA) and trans-activating crRNA (tracrRNA). This chimeric sgRNA guides Cas9 to make site-specific cut in the DNA, adjacent to protospacer adjacent motif (PAM) and generates double-strand breaks (DSBs) (Jinek et al., 2012). DSBs are further repaired through non-homologous end joining (NHEJ) or homology-directed repair. Despite the potential, this technique suffers from insufficient editing and off-target activity (Guo et al., 2023). Therefore, the accuracy and efficiency of genome editing are fundamentally dependent on the performance of the sgRNA. Modifications in guide RNA (gRNA) engineering have expanded both the functional diversity and reliability of CRISPR systems (Sakovina et al., 2022). For designing efficient sgRNAs, there are computational tools available which not only design sgRNAs but search for potential off-targets in the genome (Guo et al., 2023). Also, base and prime editors are introduced to enhance precise gene editing and lower the risk of genome instability (Anzalone et al., 2020). Delivery continues to be a major challenge in both research and applied genome editing. Even the most precisely engineered sgRNA remains ineffective without robust delivery into plant or animal cells (Kocsisova & Coneva, 2023; Yin et al., 2017). *In silico* evaluations, combined with experimental techniques, provide complementary insights into off-target landscapes and the sgRNA specificity (Guo et al., 2023). To better control CRISPR-Cas-mediated editing, strategies to optimize sgRNA are crucial. These RNA molecules if well-designed, can improve the editing outcomes.

10.2 GENERAL FEATURES OF A GUIDE RNA

CRISPR-Cas pathways differ not only in their effector proteins but also in the structure and composition of the guiding RNAs. In class I (Type I) systems, the effector is a multi-subunit surveillance complex called Cascade (CRISPR-associated complex for antiviral defence). It binds a mature crRNA, also known as Cascade-crRNA, and typically carries a short 5′ repeat-derived handle, a spacer of ~32–35 nucleotides, and a 3′ repeat-derived handle. The handles are recognized and anchored by different Cascade subunits (e.g., Cas5, Cas7, Cas6) (Beloglazova et al., 2015).

The DNA-editing system is based on Cas9, which uses a sgRNA composed of a crRNA and a tracrRNA. The crRNA consists of a variable guide sequence, known as the spacer which binds with target DNA, along with conserved direct repeat regions that base pair with the tracrRNA. The

 DOI: 10.1201/9781003735113-10

tracrRNA-derived scaffold comprising three hairpins necessary for Cas9 binding and stabilization (Allen et al., 2021). Studies have linked crRNA with tracrRNA through artificial tetraloop (like GAAA) which provides better folding and stability (Kelley et al., 2016). Guide sequence is generally 16–20-nucleotide long and a sequence less than 16 nucleotides is not efficient to cut the DNA (Dahlman et al., 2015). The 20-nucleotide-long spacer region present at 5′ end is crucial for determining the target specificity. Adjacent to this lies a 76-nucleotide-long scaffold consisting of repeat-anti-repeat region (RAR), lower stem having tetraloop (also called proximal duplex), a bulge, and an upper duplex (also known as distal duplex). Next to the RAR duplex, the stem loop 1 (SL-1) is present, which is followed by stem loop 2 (SL-2) and stem loop 3 (SL-3) (Figure 15.1). At 3′ end, the poly U-tail is present, which is formed due to termination at thymidine stretch of DNA template (De Saeger, 2025). Importantly, 10–12-nucleotide region proximal to 3′ end is most crucial for DNA cleavage and this is known as seed sequence. Remaining non-seed region tolerates 4–5 mismatches (Cong et al., 2013). For efficient Cas9-sgRNA assembly, base pairing occurs between the repeat/anti-repeat regions of crRNA and the tracrRNA (Nishimasu et al., 2014). This region can tolerate substitutions and sequence change till the basic duplex is maintained. The RAR, SL-1, and linker between SL-1 and SL-2 of sgRNA makes strong interaction with Cas9 and have major impact on DNA cleavage, while the tetraloop and SL-2 are remaining outwards of the Cas9 and have lesser impact on Cas9 activity (Jinek et al., 2012; Jiang et al., 2015). The schematic representation of common gRNAs is presented in (Figure 10.1).

Studies have shown that disruption of RAR duplex or SL-1 compromises the DNA cleavage activity of Cas9, whereas the disruption is minimal at the SL-2/3 site (Jinek et al., 2012). The gRNA

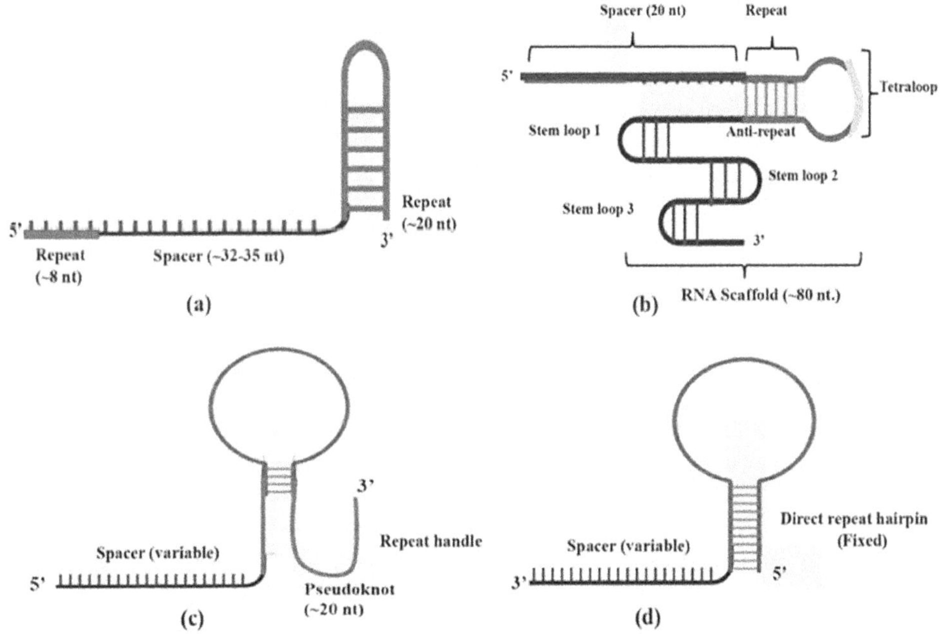

FIGURE 10.1 Diagrams displaying the structural components of different guide RNA scaffolds. They highlight essential elements like spacers, repeats, and stem-loop structures. (a) **Cascade–crRNA** used in prokaryotic Type I systems, consisting of a short 5′ repeat-derived handle, a long spacer (~32–35 nucleotide), and a 3′ repeat-derived handle (b) Synthetic **single-guide RNA (sgRNA)** used in CRISPR/Cas9, generated by fusing crRNA and tracrRNA; includes spacer region and three tracrRNA hairpins required for Cas9 binding (c) **Cas12a (Cpf1) guide RNA**, consisting only of a crRNA (no tracrRNA) with a 20–24 nucleotide spacer and a structured direct repeat forming a pseudoknot for Cas12a recognition (d) **Cas13 crRNA** used for RNA targeting, composed of a spacer region (28–30 nucleotide) and a direct repeat forming a conserved hairpin essential for Cas13 loading.

sequence plays an important role in conformational changes in Cas9, which leads to exposure of catalytically active site (Sternberg et al., 2015). On the other hand, Cas12a uses a single crRNA (no tracrRNA or sgRNA fusion) consisting of a 20–24-nucleotide spacer and a structured direct-repeat handle often forming a pseudoknot required for Cas12a binding. Cas12a-crRNA complexes recognize T-rich PAMs, unwind target DNA, and create staggered DSB with 5′ overhangs (Paul & Montoya, 2020). A new type is the RNA-targeting Cas13 systems that use crRNA guides to direct Cas13 to complementary single-stranded RNA. The crRNA typically features a direct-RAR that folds into a conserved hairpin recognized by Cas13, and a spacer (commonly 28–30-nucleotide long) complementary to the target RNA (Kavuri et al., 2022).

10.3 STRATEGIES FOR DESIGNING GUIDE RNA

The editing efficiency depends upon the gRNA composition, its stability, ability to form secondary structure, accessibility of chromatin, expression and function of Cas9 and gRNA. The RNA polymerase III–dependent U6 promoter drives the transcription of gRNA (Hsu et al., 2013). But the presence of thymidine stretch interferes in transcription, which may generate abortive sgRNAs. Modification of TTTT with TTTA/TTTG/C enhances accumulation of Cas9-sgRNA on target sites (Dang et al., 2015). A very common challenge in gene editing is the off-target activity. Due to large size of eukaryotic genome, it is difficult to completely get rid of nonspecific binding. To minimize the chances of off-targets, researchers have designed modified gRNAs.

10.3.1 Structure-Based sgRNAs

Structural engineering modifies the non-spacer part of the sgRNA comprises the scaffold having tetraloop, stem loops 1–3, and linker regions. Modifications in this region enables sgRNA to fold more stably, avoid premature termination, and improve Cas9 binding. These changes often increase on-target activity, reduce failure from misfolding, and enable additional functions like aptamer recruitment (De Saeger, 2025). It was reported that the scaffold of sgRNA can tolerate mutations and changes without any compromise of sgRNA activity that led to the generation of scaffold variants (Briner et al., 2014).

One study used truncated sgRNA (tru-sgRNA), in which the guide sequence is shortened to 17–18 nucleotides to improve the specificity of the Cas9 assembly and reduce off-targets (Fu et al., 2014). The off-targets further decrease when the tru-sgRNAs are coupled with paired Cas9 nickases (Figure 10.2a) (Cho et al., 2014). In an alternative study, engineered hairpin structure at 5′ end of spacer (hp-sgRNA) was found to be effective for on-target editing (Kocak et al., 2019). Yin et al. replaced RNA nucleotides with DNA in sgRNA, as DNA/RNA can form more stable duplex. At 5′ end of sgRNA, 8–10 nucleotides are replaced with DNA bases (Figure 10.2b) (Yin et al., 2018). Variants like F+E design combines two changes: first extending the upper stem of the scaffold to mimic the native crRNA:tracrRNA duplex and second flipping an A/U pair near the tetraloop to remove a short poly-T motif that can act as a Pol-III termination signal (De Saeger, 2025). The stem extension strengthens Cas9–sgRNA interactions, and the flip reduces premature transcription termination when sgRNAs are expressed from Pol-III promoters (U6/U3). In later publications, F+E was renamed as HEAT (Hybridization-Extended A–T inversion) (Riesenberg et al., 2022). Many spacers (the 20-nucleotides guide) form unwanted base pairs with scaffold sequences causing misfolding and loss of activity. To overcome this, studies increased the length or repaired SL-2 or sometimes the tetraloop to make folding more reliable. As we know that, SL-2 is an exposed loop in the sgRNA that usually sits outside the Cas9 and is capable of tolerating changes. Strengthening SL-2 with thermodynamically favourable extension provides a folding or nucleation site so that scaffold folds correctly, independent of the length of spacer. This type of additional hairpin is known as T-lock scaffold (Riesenberg et al., 2022). Another example of structure-based modification is prime editing-guide RNA (pegRNA) that recruits nickase to introduce only single-strand break so the risk of large insertions/deletions (indels) is much lower than classical CRISPR-Cas9. The pegRNA

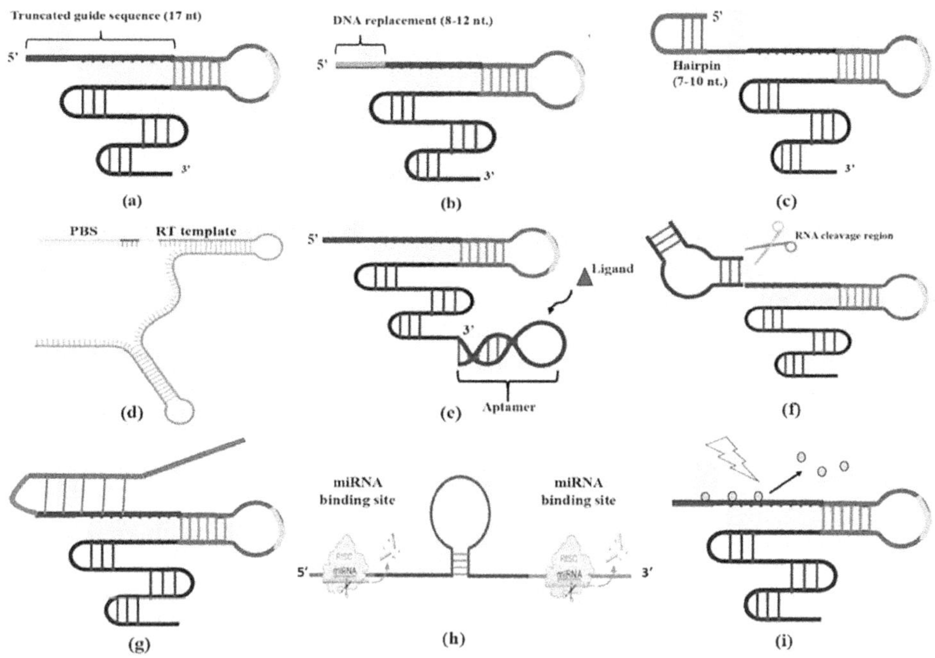

FIGURE 10.2 Various chemical modifications made to gRNA to improve precision. These include (a) truncated guide RNA: sgRNA containing a shortened guide sequence to enhance target specificity and reduce off-target effects (b) RNA-DNA hybrid: sgRNA in which selected RNA nucleotides are replaced with DNA bases to improve specificity and stability (c) hp-sgRNA: sgRNA containing an additional hairpin structure to modulate Cas9 interaction and editing efficiency (d) pegRNA: prime editing guide RNA containing a primer binding site and reverse transcription template for precise genome editing (e) Aptamer-based riboswitch: sgRNA containing an aptamer domain that regulates Cas9 activity in response to ligand binding (f) Aptamer-based ribozyme: sgRNA engineered with a ligand-responsive ribozyme for conditional activation of CRISPR activity (g) Tochold-based sgRNA: sgRNA engineered with a toehold switch for programmable and conditional CRISPR activation (h) miRNA-based sgRNA: sgRNA regulated by microRNA-responsive elements for cell- or tissue-specific genome editing (i) Photocaged sgRNA: light-responsive sgRNA containing photocleavable groups for spatiotemporal control of CRISPR activity.

consists a spacer for target, a sgRNA scaffold, and an extended 3′ end which has primer binding site (PBS) complementary to the protospacer and the RT template (RTT). There is no need for an additional donor template, as the RTT for the desired edit is provided internally in the pegRNA (Nelson et al., 2022). The dCas9-RT-pegRNA complex binds to the target site and nick the DNA strand near PAM sequence. Following nicking, the exposed DNA strand hybridizes with the primer binding site (PBS) of the pegRNA, enabling reverse transcription of the RT template into the target DNA (Figure 10.2d). The extra 3′ extensions are essential for initiating reverse transcription but also prone to misfolding and exonuclease degradation which sometimes generates truncated pegRNAs and hampers prime editing efficiency by making RNPs incompetent. To avoid degradation, researchers have engineered pegRNAs having structured RNA motifs (tevopreQ1 and mpknot) at the 3′ end of the PBS (Nelson et al., 2022). The PBS and RTT are the major sources of variation that decide editing outcome.

In a study, a dual-pegRNA approach is utilized, where one pegRNA edits using the forward strand, while the other edits using the reverse strand. Here, each pegRNA has its own PAM site, nick position, PBS, and RTT, but both are engineered so their reverse-transcribed DNA flaps encode the identical final sequence. Because some genomic loci are more responsive from one strand than the other, using two pegRNAs provides two independent editing opportunities, enhances flap conversion, and compensates for strand-specific limitations (Lin et al., 2021).

10.3.2 Conditional sgRNAs

10.3.2.1 Ligand-Based sgRNAs

These are engineered gRNAs whose activity turns ON or OFF in the presence of a specific small-molecule ligand. These are built by inserting an RNA aptamer into a noncritical region of the sgRNA scaffold. The aptamer is a 20-nucleotide spacer-blocking hairpin structure added to the 5′ end. For instance, a sgRNA precursor is made that is temporarily activated in presence of antisense oligos/Csy4-mediated RNase H cleavage (Figure 10.2e) (Ferry et al., 2017). After cleavage, the sgRNA is released and guides CRISPR machinery. The common integration sites for aptamers are tetraloop and SL-2/3 that can tolerate insertions without compromising the Cas9 binding. Another type includes aptazyme-based sgRNAs; here aptazyme is a ligand-responsive self-cleaving ribozyme that is inserted in tetraloop, SL-2, or spacer region. The ligand binding triggers the cleavage and release of sgRNA (Figure 10.2f) (Hu et al., 2023). sgRNA can also be modified by insertion of a riboswitch module at their 3′ end. A riboswitch is a *cis*-acting structured RNA element that directly binds with small-molecule ligand and undergoes a conformational change to regulate gene expression. Here, riboswitch module is an aptamer that blocks sgRNA spacer by antisense sequences (Zhan et al., 2020). When a ligand such as MS2 coat protein (MCP)/tetracycline/theophylline is present, then the conformation of aptamer changes, exposing spacer sequence for their target (Hu et al., 2023).

Toehold-based sgRNAs include an inhibitory hairpin sequester in the part of spacer known as toehold that undergoes trigger-mediated displacement. The toehold strand gets displaced only in the presence exo- or endogenous RNA triggers (Figure 10.2g) (Hu et al., 2023). Instead of RNA toehold, one study used DNA toehold strand to block repeat/SL-1 which is activated by the complementary DNA trigger (Hao et al., 2020). But there are chances of misfolding, unwanted base pairing with the hairpin, and also if the trigger and the target (part of spacer) are from endogenous origin, this can break the OFF-state or prevent proper unfolding by the trigger. To overcome this sequence constraints, Lin et al. gave two hairpin solution where one is blocking the spacer and the another absorbs structural variation caused by any spacer sequence. The hairpin 2 is placed between toehold and hairpin 1. Addition of hairpin 2 ensures hairpin 1 forms correctly regardless of spacer composition and prevents misfolding (Lin et al., 2020).

Integration of gRNAs with regulatory RNA, such as miRNA elements either by embedding sgRNAs within pri-miRNA scaffolds or by adding miRNA target sites to the sgRNA transcript, allows the sgRNA to be processed through the endogenous miRNA pathway or to become responsive to cell-specific miRNA expression, resulting in tissue-restricted or condition-specific CRISPR activity. This is also termed as miRNA-inducible CRISPR regulation (MICR). Examples include MICR-ON for miRNA-inducible transcriptional activation (via dCas9-VPR), MICR-i for repression (via dCas9-KRAB), and MICR-BE used for base editing through induction of specific miRNAs (Figure 10.2h) (De Saeger, 2025).

10.3.2.2 Light-Based sgRNAs

Light-responsive chemical moieties are used to modify sgRNA for spatiotemporal control and decrease invasiveness (Ruble et al., 2015). This approach comes under structure-guided chemical modification of sgRNAs, where chemicals such as 2-nitrobenzyl (NB) and 6-nitropiperonyloxymethylene (NPOM) act as photocleavable agents, thereby enabling light-activated CRISPR system. Photolinker, NB attaches to the specific nucleotides of spacer region at the nucleobase reactive sites to block sgRNA activity until the light exposure. When NB-caged sgRNAs are exposed to ultraviolet (UV) or blue light (typically 365–405 nm), the nitrobenzyl group absorbs photons and undergoes photochemical cleavage, and thus, restores normal base-pairing ability of sgRNA (Jain et al., 2016). The schematic representation of modified gRNAs is shown in (Figure 10.2i).

10.3.3 Functional/Multiplexed sgRNAs

Multiplex genome editing allows to target multiple genes at the same time. Earlier, this was done through delivering multiple plasmids encoding sgRNAs under the regulation of U6 pol-III promoter. However, this has limitations like poor efficiency and toxicity (Kurata et al., 2018). Studies have designed efficient CRISPR multiplexing by expressing multiple sgRNAs as a single Pol II–driven gRNA array, each flanked by Csy4 recognition sites. Co-expression of the Csy4 endo-ribonuclease precisely processed the long transcript into individual mature sgRNAs, enabling simultaneous editing of different genomic loci (Kurata et al., 2018). A study developed tRNA-gRNA array–based CRISPR-Cas9 platform to engineer cellobiose production pathway in *Trichoderma reesei*. Here, multiple sgRNAs were placed in a single polycistronic transcript, separated by endogenous tRNA spacers where the host RNase machinery precisely released individual guide (Dong et al., 2025).

10.3.4 Reporter-Based sgRNAs

The CRISPR machinery can also be harnessed for visualization of genomic loci by simply localizing the nuclease-deactivated dCas9 tagged with fluorescent proteins like Green fluorescent protein (GFP) or mCherry (Figure 10.3a) (Chaudhary et al., 2021). But these protein fusion systems suffer from high background fluorescence as the signal was always "on," regardless of whether dCas9 was bound to DNA or not. To improve signal, fluorogenic sgRNA scaffolds were developed using RNA aptamers like MS2 or PP7 hairpins (Chen et al., 2016). The MS2 and PP7 are the RNA-binding proteins utilized for sgRNA labelling. The fluorogenic sgRNA contains RNA hairpins to recruit MCP-fused fluorescent proteins, thereby concentrating fluorescence at dCas9-bound genomic loci (Ma et al., 2016).

10.3.5 Chemical-Based sgRNAs

The sgRNAs are prone to degradation by nucleases in cellular environment. To protect them, chemical-based modifications are utilized. In RNA, each ribose sugar normally has a 2′-OH (hydroxyl) group which is a main hotspot for nucleolytic degradation; hence, removal of 2′OH makes the sgRNAs resistant to cellular nucleases (Hendel et al., 2015). In a 2′-O-methyl modification, the hydrogen of the ribose 2′-OH is replaced by a methyl group ($–CH_3$), forming a 2′-O-Me (methoxy) group instead of the native 2′-OH. However, complete modification of 2′OH everywhere may reduce efficiency or specificity. Thus, balanced and minimal modification tends to work better (Basila et al., 2017).

Another modification is the phosphorothioate linkage; here one of the non-bridging oxygen atoms is replaced by a sulphur atom in the phosphate backbone. Usually the sulphur–phosphate bond is more resistant to enzymatic degradation by nucleases that cleave phosphodiester bonds (Ryan et al., 2018).

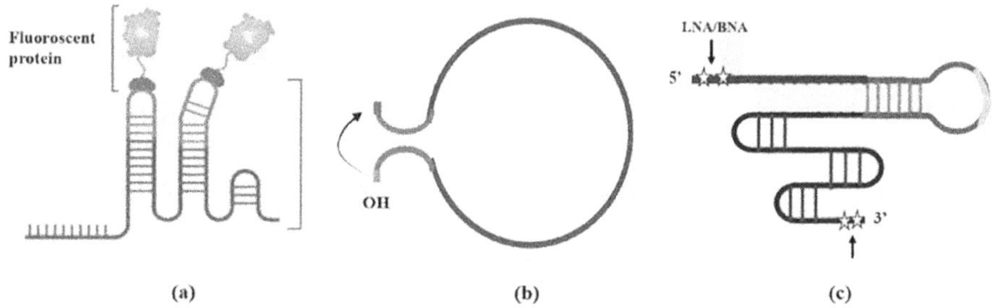

FIGURE 10.3 Illustrations showing advanced gRNA engineering for fluorescent tagged and chemical based sgRNAs. (a) Fluorogenic sgRNA: fluorescence-responsive sgRNA used for visualization and tracking of CRISPR activity (b) Circular sgRNA: covalently circularized sgRNA designed to improve nuclease resistance and controllable activation (c) LNA/BNA-based sgRNA: chemically modified sgRNA incorporating LNA/BNA nucleotides to improve stability and genome editing performance.

Sakinova et al. selectively used incorporated 2′-fluoro (2′-F), 2′-O-methyl (2′-OMe), locked nucleic acid (LNA), or BNA bases into critical regions of crRNA/tracrRNA (Figure 10.3c). In crRNA, they have chosen spacer and RAR, while in tracrRNA, the anti-repeat and stem-loop region. The seed region (1–10 nucleotides of spacer) are usually avoided for modifications. In 2′-F modification, the fluorine atom replaces the 2′-OH on the ribose sugar that strongly stabilize RNA duplex by maintaining the A-form conformation and base pairing. In LNA, the ribose ring between the 2′-oxygen and 4′-carbon is locked by a methylene bridge for better thermal and nuclease stability. The 2′-F and LNA modifications yielded the lowest degradation rates and longest half-life than unmodified gRNA or 2′-O-methyl-only guides. Dual RNA (crRNA/tracrRNA) platform was best suited for chemical modifications as compared to fused sgRNA as they are less tolerant (Sakovina et al., 2022). A novel modification called 2′-O-methyl-3′-phosphonoacetate was shown to increase the sgRNA stability in cells and achieve better editing yields as compared to phosphothiorate. The 3′-phosphonoacetate is a backbone modification introduced at the 3′ end of the sgRNA that replaces the normal phosphodiester linkage with a phosphonoacetate ester. This makes the backbone non-hydrolysable by most exonucleases as they fail to recognize the altered 3′ terminal linkage and hence, the degradation stalls (Ryan et al., 2022).

Studies also utilized circular gRNA, as the linear sgRNA is unstable and more prone to degradation. They used self-splicing ribozymes (RNA enzymes) that cut and join the RNA in a way that forms a circle (Figure 10.3b). In *Escherichia coli*, a 251-nucleotide design produced a stable circular guide which increased the activity of both cytidine base editors and adenine base editors, performing better than linear sgRNAs (Liu et al., 2022). A similar strategy was later tested using the Tornado system in mammalian cells, where the sgRNA is flanked by *Twister* ribozymes. These ribozymes cut themselves out, and the ends of the RNA are then joined into a circle by the cell's own RNA ligase, RtcB. However, in mammalian cells, the circular sgRNAs were successfully produced, but showed lower on-target editing (Wang et al., 2022).

10.4 DATABASES FOR THE GUIDE RNA DESIGNS

Computational tools for gRNA design provide essential support for predicting on-target efficiency, minimizing off-target activity, and optimizing CRISPR performance across diverse Cas systems. The CRISPOR, CHOPCHOP, CRISPR RGEN tools, and E-CRISP are quite helpful to handle hundreds of organisms (Li et al.,2023). Also, there are specific databases such as CRISPR-PLANT, CRISPR-P, and CRISPR-GE for plants; DRSC Find CRISPRs for *Drosophila*; and EuPaGDT for eukaryotic pathogens (Table 10.1). The very first step in editing is to design sgRNA by using suitable tool. For designing sgRNAs, PAM is one of the crucial factors, for which there are tools like CHOPCHOP that provide an option to customize PAM types. CRISPOR and CHOPCHOP incorporate multiple scoring models which are quite useful in predicting sgRNA efficiency and specificity (Li et al., 2023). CHOPCHOP provides sgRNA/crRNA design for DNA and RNA editing. It ranks guide sequence based on on-target score, off-target likelihood, melting temperature, and genomic context around the PAM region to guide efficient sgRNA selection (Labun et al., 2019).

NHEJ is a common repair pathway after the creation of DSBs and responsible for the generation of indels. Studies highlighted that homologous sequences present around DSBs decide the nonrandomness and biasness of NHEJ repair; this is known as microhomology-mediated end joining (Bae et al., 2014b). To handle large indel dataset, FORECasT and inDelphi predictive tools are useful (Allen et al., 2019).

Tools like E-CRISP identify targetable regions and evaluate each sgRNA for genome-wide uniqueness using strict seed-region matching (Heigwer et al., 2014). CRISPRseek is a customizable R-package that allows users to search off-targets, adjust mismatch penalties, and analyse CRISPR outcomes programmatically (Zhu et al., 2014). There are tools such as GUIDES (Graphical User Interface for DNA-Editing Screens) which help in the construction of genome-wide mutation libraries (Meier et al., 2017). CRISPRlnc and CRISPRz collect experimentally validated sgRNAs provided by publication sources and mutagenesis data (Chen et al., 2019; Varshney et al., 2016). Other than sgRNAs, there are databases such as PnB Designer for pegRNA used in prime editing (Siegner et al., 2021). A list of computational tools available for gRNA design is provided in (Table 10.1).

TABLE 10.1
List of Computational Tools Available for Guide RNA Designs

Tool	Features	Organisms	Limitations	Website Link	References
			Diversified Tools		
CRISPOR	On-target/off-target scoring, primers for vector construction, shows number of mismatches	More than 100 species	Often underestimates off-targets	https://crispor.gi.ucsc.edu/	Haeussler et al. (2016)
CRISPRdirect	Simple guide design, off-target checking	More than 200 species	Limited customization; not suitable for large-scale libraries	https://crispr.dbcls.jp/	Naito et al. (2015)
CHOPCHOP	Supports multiple Cas variants, visualizes genomic location of targets	More than 100 species	Predictions may vary across species, reducing accuracy in complex genomes	https://chopchop.cbu.uib.no/	Montague et al. (2014)
E-CRISP	Create genome-scale libraries	More than 50 species	Employs simplified mismatch rules that can miss PAM-distal or noncanonical off-target sites, restricted to few animals	https://www.e-crisp.org/	Heigwer et al. (2014)
CRISPR RGEN	Predicts off-targets number through Cas-OFFinder and outframe scores by Microhomology Predictor	More than 100 species	Off-target prediction is basic and mismatch-based, lacking advanced bulge or context-aware models	https://www.genome.arizona.edu/crispr2/	Bae et al. (2014b)
CRISTA	Used for RNA thermodynamics, ranking and detecting off-targets	More than 100 species	Computationally heavy and slow for large genomes due to complex machine learning models	https://crista.tau.ac.il/	Abadi et al. (2017)
CCTop	Searches single/multi-queries, indicates mismatch limits, and predicts sgRNA efficiency and off-target sites	More than 100 species	Uses simple mismatch counting for off-target analysis, missing bulges and chromatin effects	https://www.flyrnai.org/tools/snp_crispr/web/	Stemmer et al. (2015)
SNP-CRISPR	Designs sgRNA for targeting SNPs or indel variants	9 plant and animal species	Only designs guides around SNPs and has limited off-target prediction strength beyond local regions	https://www.flyrnai.org/tools/snp_crispr/web/	Chen et al. (2020)
PnB Designer	Designs pegRNAs for prime editors and sgRNAs for base editors	6 species	Focused only on prime/base editing and does not evaluate actual repair outcomes or large indels	https://fgcz-shiny.uzh.ch/PnBDesigner/	Seigner et al. (2021)

(Continued)

TABLE 10.1 (*Continued*)
List of Computational Tools Available for Guide RNA Designs

Tool	Features	Organisms	Limitations	Website Link	References
			Plant-Specific Tools		
CRISPR-P 2.0	Plant-specific, on/off-target scoring, gRNA sequence analysis	49 plant species	May not support noncanonical PAMs	http://crispr.hzau.edu.cn/CRISPR2/	Lei et al. (2014)
CRISPR-GE	Provides toolkits, primer for vector construction, target amplification, and sequencing result analysis	More than 40 plant species	Depends on predefined reference builds	https://skl.scau.edu.cn/	Xie et al. (2017)
CRISPR-PLANT V2	Provides selection of chromosome and locations	7 crop plant species	Limited support for newer Cas systems	https://www.genome.arizona.edu/crispr2/	Minkenberg et al. (2018)
			Animal-Specific Tools		
GUIDES	Helps in designing CRISPR knockout libraries	Human and mouse	Dataset-biased predictions that don't generalize well across species or cell types	https://github.com/sanjanalab/GUIDES	Meier et al. (2017)
DRSC Find CRISPRs	Incorporates off-target stringency from 3 to 5 mismatches, separate target region and potential off-targets	*Drosophila*	Provides no on-target efficiency scoring and uses only basic mismatch rules for off-target checks, specific to Drosophila only	https://fgcz-shiny.uzh.ch/PnBDesigner/	Housden et al. (2016)
WU-CRISPR	Provides machine learning algorithm based on experimental data, incorporates custom sequences from 26 to 30,000 bp	Human and mouse	Uses an outdated prediction model that lacks chromatin context and performs weak off-target analysis	https://crispr.wustl.edu/	Wong et al. (2015)
GPP sgRNA Designer	Up to 200 transcript or gene IDs, maximizes on-target activity and minimizes off-target activity, score on-target efforts	Human, mouse, and rat	Focused on pooled screens rather than precise editing	https://portals.broadinstitute.org/gpp/public/analysis-tools/sgrna-design	Fusi et al. (2015)
CRISPRz	Validated sgRNA database, specific for a variety of cell lines and zebrafish	Zebrafish, human, and mouse	Uses simplistic scoring with minimal off-target screening, giving low accuracy for guide quality	https://research.nhgri.nih.gov/CRISPRz/	Varshney et al. (2016)
FORECasT	Anticipates editing outcomes	Human	Can show cell-type bias and inaccurate for complex/large indels	https://elixir.ut.ee/forecast/guide	Allen et al. (2019)

(*Continued*)

TABLE 10.1 (*Continued*)
List of Computational Tools Available for Guide RNA Designs

Tool	Features	Organisms	Limitations	Website Link	References
AsCRISPR	Design sgRNA, identifies allele-specific genetic elements	Human and mouse	Works only when a clear allele-discriminating SNP is present near PAM/seed	https://www.liebertpub.com/	Zhao et al. (2020)
inDelphi	Anticipates editing outcomes	Human	Cell line–dependent predictions that fail to generalize in many tissues/species	https://indelphi.giffordlab.mit.edu/guide	Shen et al. (2018)
			Others		
sgRNA Scorer 2.0	Machine learning–based on target prediction	Multiple	No off-target analysis, mostly for SpCas9	https://frederick.cancer.gov/resources/repositories/sgrnascorer	Chari et al. (2017)
GuideScan2	Efficient genome-wide sgRNA design, custom genomes, supports mismatches/bulges, batch library design	Multiple	High CPU/memory for large genomes, web version limited for arbitrary sgRNAs, still needs experimental validation	https://www.guidescan.com/	Schmidt et al. (2025)
CRIBAR	CRISPR for imaging/tracking rather than editing	Multiple	Relies on sequence-only scoring, giving limited accuracy for real editing outcomes	http://genome.ucf.edu/CRIBAR/input_form	Chen et al. (2025)
EuPaGDT	Broad compatibility for eukaryotic pathogen genomes	Eukaryotic pathogens	Its over-stringent off-target filtering often removes many usable guides and slows genome searches	https://grna.ctegd.uga.edu/	Peng and Tarleton (2015)
CaSilico	Supports Cas12/13/14 crRNA design on any user-provided FASTA, checking GC, self-complementarity, PAM/PFS, accessibility, conserved targets, and off-target hits	Any	Off-target checking relies on user-provided sequences, not built-in genomes	https://github.com/mrb20045/CaSilico	Asadbeigi et al. (2022)
PrimedSherlock	Automates crRNA and primer design for Cas12 diagnostics, covering RNA/viral genomes and all known strains via multiple-sequence input	Any	Designed for Cas12 diagnostics, not for general genome editing sgRNA design, requires good curation of "on-target" and "off-target" sequence sets (manual curation recommended)	https://github.com/JamesGerardMann/PrimedSherlock	Mann and Pitts (2022)

Some *in silico* tools are omitted as their website links are currently unavailable.

10.5 DELIVERY METHODS FOR GUIDE RNAS

CRISPR-Cas genome editing relies on the precise delivery of the Cas nuclease along with its gRNA to the target cells. The efficiency, specificity, and safety of genome editing are heavily dependent on how the sgRNA is going to be delivered. The strategies can be broadly based on the type of cargo (DNA, RNA, or RNP) and the mode of transport (physical, chemical, or biological) (Kocsisova & Coneva, 2023). Recent innovations focus on transient, non-integrative systems to improve editing and reduce off-target effects.

10.5.1 Physical/Direct Methods for Delivery

Physical delivery methods introduce CRISPR components into cells by applying external mechanical or electrical forces that transiently disrupt cellular barriers. Electroporation is a widely used method, where the electric field causes mechanical disruption of cell membrane and creates transient pores for the entry of macromolecules. In comparison with DNA, the RNPs electroporation gives better editing efficiency, reduced off-targets, and is less toxic to different cell lines. In addition to this, microfluidic devices with constricted channels engineered to squeeze cells, resulting in higher delivery efficiency compared to electroporation (Yin et al., 2017). Next approach is particle bombardment, where CRISPR components are delivered directly into plant cells using high-velocity gold/tungsten microprojectiles coated with DNA, RNA, or RNPs. This approach has high success rate in cereals, can work for any species, and generate DNA-free plants if delivered in the form of RNPs. Studies found the detectable traces of RNPs delivered through bombardment in plants up to 14 days (Poddar et al., 2023). Bombardment can be used with immature embryos, callus, or explants. The applicability is limited as gene gun and other consumables are costly; also, bombardment can damage the cell or chromosomes. Delivery of RNPs has not been optimized to the same extent as DNA delivery methods; therefore, efficiency is often lower than that achieved with Agrobacterium-mediated transformation (Kocsisova & Coneva, 2023). One of the most precise physical delivery techniques is microinjection, where plasmid DNA, mRNA, or pre-assembled RNP complexes can be injected into cells like zygotes or embryos. Microinjection of sgRNAs and Cas9 mRNAs into mouse embryo cells results in high levels of biallelic variation in postnatal mice, indicating successful delivery (Yang et al., 2014).

10.5.2 Chemical Methods for Delivery

Chemical methods rely on engineered molecules or materials that can form stable complexes with CRISPR cargo to facilitate their entry into cells. RNP-based delivery involves pre-assembling the purified recombinant Cas protein with its *in vitro* transcribed gRNA and delivering this protein–RNA complex directly into cells. PEG-mediated delivery is a widely used method, which involves enzymatic degradation of plant cell wall to produce protoplasts, followed by treatment with PEG to facilitate the uptake of RNPs. Here, the editing efficiency can be rapidly assessed 1–3 days post-transfection (Ramakrishnan et al., 2025). Although the technique gives effective results, there are challenges in protoplast isolation, culture, and regeneration. Additionally, the chances of introducing somaclonal variations increase across diverse plant systems. A new study introduced PEG-mediated RNP transfection in rice zygotes that increased the editing efficiency from 14% to 64% in mature plants derived from transfected zygotes (Toda et al., 2019). Nowadays, nano-carriers (NCs) are made for the delivery of plasmids (up to 10 kb), mRNAs, and editing reagents (RNPs) (Hejabi et al., 2022). In animal systems, nucleic acids are coated with cationic materials to facilitate nanoparticle formation. During application, these are endocytosed by cells through mechanisms like receptor-mediated endocytosis and phagocytosis (Yin et al., 2017). NC like single-walled carbon nanotubes can cross the cell wall, plasma membrane, and lipid bilayer of chloroplasts (Wong et al., 2016). Likewise, in mitochondria, the NC-assisted process of delivery is utilized (Arya et al., 2021).

Studies reported the use of peptide-carriers (PC) to deliver RNPs into the cell by traversing the cell wall during infiltration (Terada et al., 2020). The cell-penetrating peptides were made with the fusion of homopolymer or copolymer of positively charged amino acid residues such as arginine, lysine, and histidine with the peptide of interest (Kotadiya et al., 2024). Studies also reported PC-based selective delivery of plasmid DNA into the cell organelles like chloroplast and mitochondria (Miyamoto et al., 2023).

10.5.3 Biological Methods for Delivery

Biological delivery systems harness living organisms like bacteria or viral components to transport CRISPR machinery into host cells through natural infection or genetic transformation mechanisms. These methods rely on biological interactions rather than physical force or chemical carriers, enabling efficient delivery and systemic movement in many plant and animal systems. *Agrobacterium* delivers T-DNA carrying the CRISPR cassette into the plant nucleus, enabling stable or transient expression. This method is widely used in plant research and is good for low-copy integrations. But the process is time-consuming as it requires an intervening regeneration step based on tissue culture, which needs species-specific optimization. Also, the regenerated plants must undergo rigorous segregation to obtain transgene-free plants to address regulatory concerns (Kocsisova & Coneva, 2023). Viral vectors are commonly used to transfer genes into the living cells. In plants, virus-induced gene editing (VIGE) has emerged as a promising approach for genome editing. Host range and incorporation of editing cargo into the viral genome are two deciding factors for the successful delivery through VIGE. The method involves insertion of viral genome in a T-DNA containing vector and delivery via *Agrobacterium* through leaf infiltration. When T-DNA is expressed transiently, the viral genome having editing reagents forms viral particles and moves systematically to other parts of the plant. When these viral particles reach the meristematic cells and edit the genome, the changes can be propagated to embryos. Since these embryos do not have any foreign DNA, the resulting plants are known as transgene-free edited plants (Kocsisova & Coneva, 2023). Despite successful implementation, this technique has limitations such as limited size capacity for the insertion cassette, choice of vector, and the systemic movement ability of recombinant constructs. In animals, there are three main classes of vectors such as retroviral, adenoviral, and adeno-associated viruses (AAVs) which are studied extensively and tested. The retroviral vectors can integrate their DNA into cells that do not undergo cell division, adenoviral vectors deliver dsDNA for transient expression and do not incorporate it into the host cell, while AAV vectors can transduce dividing and nondividing cells with genomic integrations (Yin et al., 2017).

10.6 EVALUATION OF THE GUIDE RNA SPECIFICITY AND OFF-TARGETS

A persistent challenge in all CRISPR systems is the specificity of gRNAs and the generation of unintended cleavage events that compromise data interpretation, genome integrity, and therapeutic safety. To screen sgRNA efficiency, the field now relies on an integrated framework of computational prediction tools and experimental genome-wide off-target mapping methods.

Off-target prediction software can be categorized into alignment-based tools such as CasOT, Cas-OFFinder, FlashFry, and CRISFlash, which identify putative off-target regions of sgRNAs, and scoring-based models such as the MIT score (score generated from the mismatch position of sgRNA to evaluate off-targets), CCTop (Consensus Constrained TOPology prediction), CROP-IT (CRISPR-Cas9 Off-target Prediction and Identification Tool), CFD (Cutting frequency determination), DeepCRISPR, etc. which predict off-target activity more precisely (Guo et al., 2023) (Table 10.2).

To further validate these, experimental tools are available and they are divided into three main categories: cell-free methods, cell culture–based methods, and *in vivo* detection. The cell-free methods recruit extracted genomic DNA from cells to identify cleavages. Methods such as Digenome-seq

TABLE 10.2
***In Silico* Prediction Tools and Experimental Methods for Guide RNA Specificity and Off-Target Detection**

Tools	Features	Limitations	References
		In Silico Prediction Tools	
Cas-OFFinder	High tolerance for sgRNA length, PAM types, number of mismatches or bulges	Biased as mostly sgRNA-dependent off-target prediction, needs experimental proof	Bae et al. (2014a)
CasOT	Predicts off-target sites, custom adjustments of PAM sequence and mismatch number	Limited off-target scoring; only counts mismatches without considering bulges or chromatin context	Xiao et al. (2014)
FlashFry	Characterizes thousands of targets in a short time period, high throughput, information for GC contents and on/off-targets	High memory usage for large genomes; off-target predictions are basic and mismatch-based	McKenna and Shendure (2018)
Crisflash	Faster sgRNA design and off-target prediction	Computationally intensive for multiple genomes; limited support for newer Cas variants	Jacquin et al. (2019)
MIT algorithm	Generates score based on position effects of mismatches in sgRNA sequence	Simple mismatch-based scoring; does not account for bulges, chromatin, or repair outcomes	Haeussler et al. (2016)
CCTop	Depends on the mismatches distance to the PAM	Uses basic mismatch counting; ignores bulges and chromatin accessibility effects	Stemmer et al. (2015)
CROP-IT	Predicts sgRNA on-target efficiency, highlights potential off-target sites, and supports batch guide design for CRISPR experiments	Focused on efficiency prediction only; limited off-target evaluation and no repair outcome forecasting	Singh et al. (2015)
CFD algorithm	Based on experimental dataset	Only predicts off-target cleavage likelihood; does not evaluate on-target efficiency or repair outcomes	Doench et al. (2016)
DeepCRISPR	Utilizes deep learning for off-target, considers epigenetic features	Requires high computational resources; performance may drop for non-model organisms or unseen genome contexts	Chuai et al. (2018)
Elevation tools	Have DNA accessibility information to predict potential off-targets	Limited to human exome (GRCh38)	Listgarten et al. (2018)
		Experimental Validation Methods	
CLEAVE-seq	High-sensitivity, *in vitro* assay that maps Cas nuclease on- and off-target cuts and supports parallel testing of multiple guides on plant DNA	May over-predict sites because it's *in vitro*, needs high-quality DNA, and doesn't reflect *in vivo* chromatin conditions	Young et al. (2019)
Digenome-seq	Incubates Cas9/sgRNA (RNP) complex, DSBs detected by WGS, off-target screening	Requires high sequencing coverage, high sequencing cost, requires a reference genome, chromatin states omitted in the off-target assessment	Kim et al. (2016)
DIG-seq	Updated version of Digenome-seq, uses chromatin instead of purified DNA, high accuracy for off-targets	Lower sensitivity in inaccessible regions	Kim and Kim (2018)

(Continued)

TABLE 10.2 (*Continued*)
***In Silico* Prediction Tools and Experimental Methods for Guide RNA Specificity and Off-Target Detection**

Tools	Features	Limitations	References
Extru-seq	Live cells mixed with purified Cas9/sgRNA complex and genomic DNA of cells released through mechanical lysis, low false positive and false negative rates	Difficult to detect large deletions and translocations	Kwon et al. (2023)
SITE-seq	Enriches cleaved genomic sites via selective biotinylation and streptavidin pulldown before sequencing, reducing background and sequencing needs compared to Digenome-seq	Low accuracy of finding off-target sites	Cameron et al. (2017)
CIRCLE-seq	Detects genome-wide off-targets without WGS by shearing and circularizing DNA, Cas9/sgRNA does selective linearizing followed by sequencing, reduces background noise	False positive rate is high, needs targeted sequencing for further validation	Tsai et al. (2017)
WGS	Compares genome sequences before and after editing, uncovers desired and unwanted editing events	Expensive in case of low-frequency off-sites determination	Smith et al. (2014)
Cas9 ChIP-seq	Identifies Cas9 binding sites, uses dCas9 as this can bind to DNA without causing DSBs and easily detach from edited sites	Low validation, depends on antibody specificity and chromatin accessibility	Kuscu et al. (2014)
IDLV	Labels intracellular DSBs, integrated sites amplified for high-throughput sequencing, detects off-targets in cells which are difficult to transfect	Low sensitivity and a high false-positive rate resulting from nonspecific integration and amplification events	Ferrari et al. (2022) and Martin et al. (2016)
GUIDE-seq	Incorporation of double-stranded oligos (dsODNs) with known sequences into DSBs during NHEJ, dsODNs provide template for targeted Polymerase chain reaction (PCR) and sequencing, more sensitive	Lowers detection rate 30%–50% due to low delivery efficiency of dsODNs	Tsai et al. (2015)
LAM-HTGTS	Cells are edited with Cas9 to create known "bait" and unknown "prey" DSBs; prey sites ligate to bait, followed by biotin-based linear amplification, adaptor ligation, and nested PCR for Next-generation sequencing (NGS), enabling detection of chromosomal translocations	Only detects DSBs having translocations, sensitivity to off-target detection is still uncertain, limited chromatin accessibility	Hu et al. (2016)
BLESS	Captures *in situ* DSBs, ligate biotinylated linkers to cleavage sites, false positive rate <1%	Captures DSBs that are present at the fixation time	Crosetto et al. (2013)
BLISS	Breaks labelling *in situ* and sequencing, adapter ligation of DSB ends, linearly amplified and sequenced, higher sensitivity than BLESS	Resolution is limited and cannot distinguish Cas-induced vs. natural breaks	Yan et al. (2017)
Discover-seq	Utilizes MRE11, an endogenous DNA repair protein, captures DSBs	Lower sensitivity for low-frequency off-targets	Wienert et al. (2020)
GUIDE-tag	An improved GUIDE-seq method using Cas9-mSA RNP to recruit biotinylated dsODNs at DSBs, increasing DSB detection sensitivity over DISCOVER-seq	Low incorporation rates of biotin-dsDNA	Liang et al. (2022)

(digested genome sequencing), DIG-seq (Digenome-seq using cell-free chromatin DNA), Extru-seq, SITE-seq (selective enrichment and identification of tagged genomic DNA ends by sequencing), and CIRCLE-seq (circularization for *in vitro* reporting of cleavage effects by sequencing) are included here. The second category is cell culture–based methods, where the off-target would be assessed inside the cells as the intra-nuclear content affects the behaviour of editors. This category includes WGS (whole-genome sequencing), Cas9 ChIP-seq (chromatin immunoprecipitation then high-throughput sequencing), IDLVs (integrase defective lentiviral vectors), GUIDE-seq (genome-wide, unbiased identification of DSBs enabled by sequencing), LAM-HTGTS (linear amplification-mediated high-throughput genome-wide sequencing), BLESS (breaks labelling, enrichment on streptavidin, and next-generation sequencing), and BLISS (breaks labelling *in situ* and sequencing). Third category is based on *in vivo* detection; here the off-targets can be directly assessed in living cells/tissues. Techniques like GUIDE-tag and Discover-seq (discovery of *in situ* Cas off-targets and verification by sequencing) belong to this category (Guo et al., 2023). The *in silico* prediction tools and experimental methods for gRNA specificity and off-target detection are indexed in (Table 10.2).

10.7 CONCLUSION

gRNAs form the programmable core of CRISPR systems, directing nuclease binding and cleavage with sequence-specific precision. Their structural elements including the spacer region, scaffolding hairpins, and secondary/tertiary motifs govern binding affinity. Structure-based approaches including optimized scaffolds, ligand-responsive modules, and multiplexed architectures have expanded the functional versatility of gRNAs, while chemical modifications now offer enhanced stability and nuclease resistance for both plant and animal applications. Alongside these developments, computational platforms and empirical off-target profiling methods together make the designing of highly specific gRNAs easy and effective by minimizing unintended edits.

Despite these advances, several key challenges remain. There is a need of improving the predictability of gRNA behaviour across species, integrating epigenomics and RNA-structure features into next-generation design algorithms. With the emergence of smaller and nontraditional Cas systems, there is a need of new gRNA designs to match their unique structures and requirements. In parallel, delivery remains a major challenge especially in plants and complex tissues. This highlights the need for more efficient, tissue-specific, and DNA-free delivery platforms, including improved nanoparticle carriers, viral replicons, and programmable protein–RNA assemblies. All these aspects will contribute to the generation of efficient gRNAs that can sustain longer and improve genome editing.

REFERENCES

Abadi, S., Yan, W. X., Amar, D., & Mayrose, I. (2017). A machine learning approach for predicting CRISPR–Cas9 cleavage efficiencies and patterns underlying its mechanism of action. *PLoS Computational Biology,* 13(10), e1005807. https://doi.org/10.1371/journal.pcbi.1005807

Allen, F., Crepaldi, L., Alsinet, C., Strong, A. J., Kleshchevnikov, V., De Angeli, P., Páleníková, P., Khodak, A., Kiselev, V., Kosicki, M., Bassett, A. R., Harding, H., Galanty, Y., Muñoz-Martínez, F., Metzakopian, E., Jackson, S. P., & Parts, L. (2019). Predicting the mutations generated by repair of Cas9-induced double-strand breaks. *Nature Biotechnology,* 37(1), 64–72. https://doi.org/10.1038/nbt.4317

Allen, D., Rosenberg, M., & Hendel, A. (2021). Using synthetically engineered guide RNAs to enhance CRISPR genome editing systems in mammalian cells. *Frontiers in Genome Editing,* 2, 617910. https://doi.org/10.3389/fgeed.2020.617910

Anzalone, A. V., Koblan, L. W., & Liu, D. R. (2020). Genome editing with CRISPR–Cas nucleases, base editors, transposases, and prime editors. *Nature Biotechnology,* 38(7), 824–844. https://doi.org/10.1038/s41587-020-0561-9

Arya, S. S., Tanwar, N., & Lenka, S. K. (2021). Prospects of nano- and peptide-carriers to deliver CRISPR cargos in plants to edit across and beyond central dogma. *Nanotechnology for Environmental Engineering,* 6(2), 22. https://doi.org/10.1007/s41204-021-00118-z

Asadbeigi, A., Norouzi, M., Vafaei Sadi, M. S., Saffari, M., & Bakhtiarizadeh, M. R. (2022). CaSilico: A versatile CRISPR package for in silico CRISPR RNA designing for Cas12, Cas13, and Cas14. *Frontiers in Bioengineering and Biotechnology,* 10, 957131. https://doi.org/10.3389/fbioe.2022.957131

Bae, S., Park, J., & Kim, J. S. (2014a). Cas-OFFinder: A fast and versatile algorithm that searches for potential off-target sites of Cas9 RNA-guided endonucleases. *Bioinformatics,* 30(10), 1473–1475. https://doi.org/10.1093/bioinformatics/btu048

Bae, S., Kweon, J., Kim, H. S., & Kim, J. S. (2014b). Microhomology-based choice of Cas9 nuclease target sites. *Nature Methods,* 11(7), 705–706. https://doi.org/10.1038/nmeth.3015

Basila, M., Kelley, M. L., & Smith, A. V. B. (2017). Minimal 2′-O-methyl phosphorothioate linkage modification pattern of synthetic guide RNAs for increased stability and efficient CRISPR–Cas9 gene editing avoiding cellular toxicity. *PLoS ONE,* 12(11), e0188593. https://doi.org/10.1371/journal.pone.0188593

Beloglazova, N., Kuznedelov, K., Flick, R., Datsenko, K. A., Brown, G., Popovic, A., Lemak, S., Semenova, E., Severinov, K., & Yakunin, A. F. (2015). CRISPR RNA binding and DNA target recognition by purified Cascade complexes from *Escherichia coli. Nucleic Acids Research,* 43(1), 530–543. https://doi.org/10.1093/nar/gku1285

Briner, A. E., Donohoue, P. D., Gomaa, A. A., Selle, K., Slorach, E. M., Nye, C. H., Haurwitz, R. E., Beisel, C. L., May, A. P., & Barrangou, R. (2014). Guide RNA functional modules direct Cas9 activity and orthogonality. *Molecular Cell,* 56(2), 333–339. https://doi.org/10.1016/j.molcel.2014.09.019

Cameron, P., Fuller, C. K., Donohoue, P. D., Jones, B. N., Thompson, M. S., Carter, M. M., Gradia, S., Vidal, B., Garner, E., Slorach, E. M., Lau, E., Banh, L. M., Lied, A. M., Edwards, L. S., Settle, A. H., Capurso, D., Llaca, V., Deschamps, S., Cigan, M., Young, J. K., & May, A. P. (2017). Mapping the genomic landscape of CRISPR–Cas9 cleavage. *Nature Methods,* 14(6), 600–606. https://doi.org/10.1038/nmeth.4284

Chari, R., Yeo, N. C., Chavez, A., & Church, G. M. (2017). sgRNA Scorer 2.0: A species-independent model to predict CRISPR/Cas9 activity. *ACS Synthetic Biology,* 6(5), 902–904. https://doi.org/10.1021/acssynbio.6b00343

Chaudhary, N., Im, J. K., Nho, S. H., & Kim, H. (2021). Visualizing live chromatin dynamics through CRISPR-based imaging techniques. *Molecules and Cells,* 44(9), 627–636. https://doi.org/10.14348/molcells.2021.2254

Chen, B., Hu, J., Almeida, R., Liu, H., Balakrishnan, S., Covill-Cooke, C., Lim, W. A., & Huang, B. (2016). Expanding the CRISPR imaging toolset with *Staphylococcus aureus* Cas9 for simultaneous imaging of multiple genomic loci. *Nucleic Acids Research,* 44(8), e75. https://doi.org/10.1093/nar/gkv1533

Chen, C. L., Rodiger, J., Chung, V., Viswanatha, R., Mohr, S. E., Hu, Y., & Perrimon, N. (2020). SNP-CRISPR: A web tool for SNP-specific genome editing. *G3: Genes, Genomes, Genetics,* 10(2), 489–494. https://doi.org/10.1534/g3.119.400904

Chen, W., Zhang, G., Li, J., Zhang, X., Huang, S., Xiang, S., Hu, X., & Liu, C. (2019). CRISPRlnc: A manually curated database of validated sgRNAs for lncRNAs. *Nucleic Acids Research,* 47(D1), D63–D68. https://doi.org/10.1093/nar/gky904

Chen, X., Rahaman, M. M., Naseri, A., & Zhang, S. (2025). CRIBAR: A fast and flexible sgRNA design tool for CRISPR imaging. *Bioinformatics Advances,* 5(1). https://doi.org/10.1093/bioadv/vbaf022

Cho, S. W., Kim, S., Kim, Y., Kweon, J., Kim, H. S., Bae, S., & Kim, J. S. (2014). Analysis of off-target effects of CRISPR/Cas-derived RNA-guided endonucleases and nickases. *Genome Research,* 24(1), 132–141. https://doi.org/10.1101/gr.162339.113

Chuai, G., Ma, H., Yan, J., Chen, M., Hong, N., Xue, D., Zhou, C., Zhu, C., Chen, K., Duan, B., Gu, F., Qu, S., Huang, D., Wei, J., & Liu, Q. (2018). DeepCRISPR: Optimized CRISPR guide RNA design by deep learning. *Genome Biology,* 19(1), 80. https://doi.org/10.1186/s13059-018-1459-4

Cong, L., Ran, F. A., Cox, D., Lin, S., Barretto, R., Habib, N., Hsu, P. D., Wu, X., Jiang, W., Marraffini, L. A., & Zhang, F. (2013). Multiplex genome engineering using CRISPR/Cas systems. *Science,* 339(6121), 819–823. https://doi.org/10.1126/science.1231143

Crosetto, N., Mitra, A., Silva, M. J., Bienko, M., Dojer, N., Wang, Q., Karaca, E., Chiarle, R., Skrzypczak, M., Ginalski, K., Pasero, P., Rowicka, M., & Dikic, I. (2013). Nucleotide-resolution DNA double-strand break mapping by next-generation sequencing. *Nature Methods,* 10(4), 361–365. https://doi.org/10.1038/nmeth.2408

Dahlman, J. E., Abudayyeh, O. O., Joung, J., Gootenberg, J. S., Zhang, F., & Konermann, S. (2015). Orthogonal gene knockout and activation with a catalytically active Cas9 nuclease. *Nature Biotechnology,* 33(11), 1159–1161. https://doi.org/10.1038/nbt.3390

Dang, Y., Jia, G., Choi, J., Ma, H., Anaya, E., Ye, C., Shankar, P., & Wu, H. (2015). Optimizing sgRNA structure to improve CRISPR–Cas9 knockout efficiency. *Genome Biology,* 16(1), 280. https://doi.org/10.1186/s13059-015-0846-3

De Saeger, J. (2025). A guide to guides: An overview of SpCas9 sgRNA scaffold variants and modifications. *SynBio,* 3(4), 19. https://doi.org/10.3390/synbio3040019

Doench, J. G., Fusi, N., Sullender, M., Hegde, M., Vaimberg, E. W., Donovan, K. F., Smith, I., Tothova, Z., Wilen, C., Orchard, R., Virgin, HW., Listgarten, J., & Root, D. E. (2016). Optimized sgRNA design to maximize activity and minimize off-target effects of CRISPR–Cas9. *Nature Biotechnology,* 34(2), 184–191. https://doi.org/10.1038/nbt.3437

Dong, H., Su, B., Chen, Z., & Song, X. (2025). A tRNA–gRNA array-based CRISPR system for efficient and precise marker-free multiple gene editing of *Trichoderma reesei* to enhance cellobiose production. *Industrial Crops and Products,* 231, 121169. https://doi.org/10.1016/j.indcrop.2025.121169

Doudna, J. A., & Charpentier, E. (2014). The new frontier of genome engineering with CRISPR–Cas9. *Science,* 346(6213), 1258096. https://doi.org/10.1126/science.1258096

Ferrari, S., Jacob, A., Cesana, D., Laugel, M., Beretta, S., Varesi, A., Unali, G., Conti, A., Canarutto, D., Albano, L. and Calabria, A. (2022). Choice of template delivery mitigates the genotoxic risk and adverse impact of editing in human hematopoietic stem cells. *Cell Stem Cell, 29*(10), 1428–1444.

Ferry, Q. R., Lyutova, R., & Fulga, T. A. (2017). Rational design of inducible CRISPR guide RNAs for de novo assembly of transcriptional programs. *Nature Communications,* 8(1), 14633. https://doi.org/10.1038/ncomms14633

Fu, Y., Sander, J. D., Reyon, D., Cascio, V. M., & Joung, J. K. (2014). Improving CRISPR–Cas nuclease specificity using truncated guide RNAs. *Nature Biotechnology,* 32(3), 279–284. https://doi.org/10.1038/nbt.2808

Fusi, N., Smith, I., Doench, J., & Listgarten, J. (2015). In silico predictive modeling of CRISPR/Cas9 guide efficiency. *bioRxiv*, 021568. https://doi.org/10.1101/021568

Guo, C., Ma, X., Gao, F., & Guo, Y. (2023). Off-target effects in CRISPR/Cas9 gene editing. *Frontiers in Bioengineering and Biotechnology,* 11, 1143157. https://doi.org/10.3389/fbioe.2023.1143157

Haeussler, M., Schönig, K., Eckert, H., Eschstruth, A., Mianné, J., Renaud, J. B., Schneider-Maunoury, S., Shkumatava, A., Teboul, L., Kent, J., Joly, J. S., & Concordet, J. P. (2016). Evaluation of off-target and on-target scoring algorithms and integration into the guide RNA selection tool CRISPOR. *Genome Biology,* 17(1), 148. https://doi.org/10.1186/s13059-016-1012-2

Hao, Y., Li, J., Li, Q., Zhang, L., Shi, J., Zhang, X., Aldalbahi, A., Wang, L., Fan, C., & Wang, F. (2020). Programmable live-cell CRISPR imaging with toehold-switch-mediated strand displacement. *Angewandte Chemie International Edition,* 59(46), 20612–20618. https://doi.org/10.1002/anie.202009062

Heigwer, F., Kerr, G., & Boutros, M. (2014). E-CRISP: Fast CRISPR target site identification. *Nature Methods,* 11(2), 122–123. https://doi.org/10.1038/nmeth.2812

Hejabi, F., Abbaszadeh, M. S., Taji, S., O'Neill, A., Farjadian, F., & Doroudian, M. (2022). Nanocarriers: A novel strategy for the delivery of CRISPR/Cas systems. *Frontiers in Chemistry,* 10, 957572. https://doi.org/10.3389/fchem.2022.957572

Hendel, A., Bak, R. O., Clark, J. T., Kennedy, A. B., Ryan, D. E., Roy, S., Steinfeld, I., Lunstad, B. D., Kaiser, R. J., Wilkens, A. B., Bacchetta, R., Tsalenko, A., Dellinger, D., Bruhn, L., & Porteus, M. H. (2015). Chemically modified guide RNAs enhance CRISPR–Cas genome editing in human primary cells. *Nature Biotechnology,* 33(9), 985–989. https://doi.org/10.1038/nbt.3290

Housden, B. E., Hu, Y., & Perrimon, N. (2016). Design and generation of *Drosophila* single guide RNA expression constructs. *Cold Spring Harbor Protocols, 2016*(9), pdb.prot090779. https://doi.org/10.1101/pdb.prot090779

Hsu, P. D., Scott, D. A., Weinstein, J. A., Ran, F. A., Konermann, S., Agarwala, V., & Zhang, F. (2013). DNA targeting specificity of RNA-guided Cas9 nucleases. *Nature Biotechnology,* 31(9), 827–832. https://doi.org/10.1038/nbt.2647

Hu, J., Meyers, R. M., Dong, J., Panchakshari, R. A., Alt, F. W., & Frock, R. L. (2016). Detecting DNA double-stranded breaks in mammalian genomes by linear amplification–mediated high-throughput genome-wide translocation sequencing. *Nature Protocols,* 11(5), 853–871. https://doi.org/10.1038/nprot.2016.043

Hu, L. F., Li, Y. X., Wang, J. Z., Zhao, Y. T., & Wang, Y. (2023). Controlling CRISPR–Cas9 by guide RNA engineering. *Wiley Interdisciplinary Reviews: RNA,* 14(1), e1731. https://doi.org/10.1002/wrna.1731

Jacquin, A. L., Odom, D. T., & Lukk, M. (2019). Crisflash: Open-source software to generate CRISPR guide RNAs against genomes annotated with individual variation. *Bioinformatics,* 35(17), 3146–3147. https://doi.org/10.1093/bioinformatics/btz019

Jain, P. K., Ramanan, V., Schepers, A. G., Dalvie, N. S., Panda, A., Fleming, H. E., & Bhatia, S. N. (2016). Development of light-activated CRISPR using guide RNAs with photocleavable protectors. *Angewandte Chemie International Edition,* 55(40), 12440–12444. https://doi.org/10.1002/anie.201606123

Jiang, F., Zhou, K., Ma, L., Gressel, S., & Doudna, J. A. (2015). A Cas9–guide RNA complex preorganized for target DNA recognition. *Science,* 348(6242), 1477–1481. https://doi.org/10.1126/science.aab1452

Jinek, M., Chylinski, K., Fonfara, I., Hauer, M., Doudna, J. A., & Charpentier, E. (2012). A programmable dual-RNA–guided DNA endonuclease in adaptive bacterial immunity. *Science,* 337(6096), 816–821. https://doi.org/10.1126/science.1225829

Kavuri, N. R., Ramasamy, M., Qi, Y., & Mandadi, K. (2022). Applications of CRISPR/Cas13-based RNA editing in plants. *Cells,* 11(17), 2665. https://doi.org/10.3390/cells11172665

Kelley, M. L., Strezoska, Ž., He, K., Vermeulen, A., & van Brabant Smith, A. (2016). Versatility of chemically synthesized guide RNAs for CRISPR–Cas9 genome editing. *Journal of Biotechnology,* 233, 74–83. https://doi.org/10.1016/j.jbiotec.2016.06.011

Kim, D., & Kim, J. S. (2018). DIG-seq: A genome-wide CRISPR off-target profiling method using chromatin DNA. *Genome Research,* 28(12), 1894–1900. https://www.genome.org/cgi/doi/10.1101/gr.236620.118

Kim, D., Kim, S., Kim, S., Park, J., & Kim, J. S. (2016). Genome-wide target specificities of CRISPR–Cas9 nucleases revealed by multiplex Digenome-seq. *Genome Research,* 26(3), 406–415. https://www.genome.org/cgi/doi/10.1101/gr.199588.115

Kocak, D. D., Josephs, E. A., Bhandarkar, V., Adkar, S. S., Kwon, J. B., & Gersbach, C. A. (2019). Increasing the specificity of CRISPR systems with engineered RNA secondary structures. *Nature Biotechnology,* 37(6), 657–666. https://doi.org/10.1038/s41587-019-0095-1

Kocsisova, Z., & Coneva, V. (2023). Strategies for delivery of CRISPR/Cas-mediated genome editing to obtain edited plants directly without transgene integration. *Frontiers in Genome Editing,* 5, 1209586. https://doi.org/10.3389/fgeed.2023.1209586

Kotadiya, D. D., Patel, P., & Patel, H. D. (2024). Cell-penetrating peptides: A powerful tool for targeted drug delivery. *Current Drug Delivery,* 21(3), 368–388. https://doi.org/10.2174/1567201820666230407092924

Kurata, M., Wolf, N. K., Lahr, W. S., Weg, M. T., Kluesner, M. G., Lee, S., Hui, K., Shiraiwa, M., Webber, B. R., & Moriarity, B. S. (2018). Highly multiplexed genome engineering using CRISPR/Cas9 gRNA arrays. *PLoS ONE,* 13(9), e0198714. https://doi.org/10.1371/journal.pone.0198714

Kuscu, C., Arslan, S., Singh, R., Thorpe, J., & Adli, M. (2014). Genome-wide analysis reveals characteristics of off-target sites bound by the Cas9 endonuclease. *Nature Biotechnology,* 32(7), 677–683. https://doi.org/10.1038/nbt.2916

Kwon, J., Kim, M., Hwang, W., Jo, A., Hwang, G. H., Jung, M., Kim, U. G., Cui, G., Kim, H., Eom, J. H., Hur, J. K., Lee, J., Kim, Y., Kim, J. S., Bae, S., & Lee, J. K. (2023). Extru-seq: A method for predicting genome-wide Cas9 off-target sites with advantages of both cell-based and in vitro approaches. *Genome Biology,* 24(1), 4. https://doi.org/10.1186/s13059-022-02842-4

Labun, K., Montague, T. G., Krause, M., Torres Cleuren, Y. N., Tjeldnes, H., & Valen, E. (2019). CHOPCHOP v3: Expanding the CRISPR web toolbox beyond genome editing. *Nucleic Acids Research,* 47(W1), W171–W174. https://doi.org/10.1093/nar/gkz365

Lei, Y., Lu, L., Liu, H. Y., Li, S., Xing, F., & Chen, L. L. (2014). CRISPR-P: A web tool for synthetic single-guide RNA design of CRISPR system in plants. *Molecular Plant,* 7(9), 1494–1496. https://doi.org/10.1093/mp/ssu044

Li, C., Chu, W., Gill, R. A., Sang, S., Shi, Y., Hu, X., Yang, Y., Zaman, Q. U., & Zhang, B. (2023). Computational tools and resources for CRISPR/Cas genome editing. *Genomics, Proteomics & Bioinformatics,* 21(1), 108–126. https://doi.org/10.1016/j.gpb.2022.02.006

Liang, S. Q., Liu, P., Smith, J. L., Mintzer, E., Maitland, S., Dong, X., Yang, Q., Lee, J., Haynes, C. M., Zhu, L. J., Watts, J. K., Sontheimer, E. J., Wolfe, S. A., & Xue, W. (2022). Genome-wide detection of CRISPR editing in vivo using GUIDE-tag. *Nature Communications,* 13(1), 437. https://doi.org/10.1038/s41467-022-28135-9

Lin, J., Liu, Y., Lai, P., Ye, H., & Xu, L. (2020). Conditional guide RNA through two intermediate hairpins for programmable CRISPR/Cas9 function. *Nucleic Acids Research,* 48(20), 11773–11784. https://doi.org/10.1093/nar/gkaa842

Lin, Q., Jin, S., Zong, Y., Yu, H., Zhu, Z., Liu, G., Kou, L., Wang, Y., Qiu, J. L., Li, J., & Gao, C. (2021). High-efficiency prime editing with optimized, paired pegRNAs in plants. *Nature Biotechnology,* 39(8), 923–927. https://doi.org/10.1038/s41587-021-00868-w

Listgarten, J., Weinstein, M., Kleinstiver, B. P., Sousa, A. A., Joung, J. K., Crawford, J., Gao, K., Hoang, L., Elibol, M., Doench, J. G., & Fusi, N. (2018). Prediction of off-target activities for the end-to-end design of CRISPR guide RNAs. *Nature Biomedical Engineering,* 2(1), 38–47. https://doi.org/10.1038/s41551-017-0178-6

Liu, L., Li, W., Li, J., Zhao, D., Li, S., Jiang, G., Wang, J., Chen, X., Bi, C., & Zhang, X. (2022). Circular guide RNA for improved stability and CRISPR–Cas9 editing efficiency in vitro and in bacteria. *ACS Synthetic Biology,* 12(2), 350–359. https://pubs.acs.org/doi/10.1021/acssynbio.2c00381

Ma, H., Tu, L. C., Naseri, A., Huisman, M., Zhang, S., Grunwald, D., & Pederson, T. (2016). Multiplexed labeling of genomic loci with dCas9 and engineered sgRNAs using CRISPRainbow. *Nature Biotechnology,* 34(5), 528–530. https://doi.org/10.1038/nbt.3526

Makarova, K. S., Wolf, Y. I., Iranzo, J., Shmakov, S. A., Alkhnbashi, O. S., Brouns, S. J. J., Charpentier, E., Cheng, D., Haft, D. H., Horvath, P., Moineau, S., Mojica, F. J. M., Scott, D., Shah, S. A., Siksnys, V., Terns, M. P., Venclovas, Č., White, M. F., Yakunin, A. F., . . . Koonin, E. V. (2019). Evolutionary classification of CRISPR–Cas systems: a burst of class 2 and derived variants. *Nature Reviews Microbiology*, *18*(2), 67–83. https://doi.org/10.1038/s41579-019-0299-x

Mann, J. G., & Pitts, R. J. (2022). PrimedSherlock: A tool for rapid design of highly specific CRISPR–Cas12 crRNAs. *BMC Bioinformatics,* 23(1), 428. https://doi.org/10.1186/s12859-022-04968-5

Martin, F., Sánchez-Hernández, S., Gutiérrez-Guerrero, A., Pinedo-Gomez, J., & Benabdellah, K. (2016). Biased and unbiased methods for the detection of off-target cleavage by CRISPR/Cas9. *International Journal of Molecular Sciences,* 17(9), 1507. https://doi.org/10.3390/ijms17091507

McKenna, A., & Shendure, J. (2018). FlashFry: A fast and flexible tool for large-scale CRISPR target design. *BMC Biology,* 16(1), 74. https://doi.org/10.1186/s12915-018-0545-0

Meier, J. A., Zhang, F., & Sanjana, N. E. (2017). GUIDES: sgRNA design for loss-of-function screens. *Nature Methods,* 14(9), 831–832. https://doi.org/10.1038/nmeth.4423

Minkenberg, B., Zhang, J., Xie, K., & Yang, Y. (2018). CRISPR-PLANT v2: An online resource for highly specific guide RNA spacers. *Plant Biotechnology Journal,* 17(1), 5. https://doi.org/10.1111/pbi.13025

Montague, T. G., Cruz, J. M., Gagnon, J. A., Church, G. M., & Valen, E. (2014). CHOPCHOP: A CRISPR/Cas9 and TALEN web tool for genome editing. *Nucleic Acids Research,* 42(W1), W401–W407. https://doi.org/10.1093/nar/gku410

Naito, Y., Hino, K., Bono, H., & Ui-Tei, K. (2015). CRISPRdirect: Software for designing CRISPR/Cas guide RNA with reduced off-target sites. *Bioinformatics,* 31(7), 1120–1123. https://doi.org/10.1093/bioinformatics/btu743

Nelson, J. W., Randolph, P. B., Shen, S. P., Everette, K. A., Chen, P. J., Anzalone, A. V., An, M., Newby, G. A., Chen, J. C., Hsu, A., & Liu, D. R. (2022). Engineered pegRNAs improve prime editing efficiency. *Nature Biotechnology,* 40(3), 402–410. https://doi.org/10.1038/s41587-021-01039-7

Nishimasu, H., Ran, F. A., Hsu, P. D., Konermann, S., Shehata, S. I., Dohmae, N., Ishitani, R., Zhang, F., & Nureki, O. (2014). Crystal structure of Cas9 in complex with guide RNA and target DNA. *Cell,* 156(5), 935–949. https://doi.org/10.1016/j.cell.2014.02.001

Paul, B., & Montoya, G. (2020). CRISPR–Cas12a: Functional overview and applications. *Biomedical Journal,* 43(1), 8–17. https://doi.org/10.1016/j.bj.2019.10.005

Peng, D., & Tarleton, R. (2015). EuPaGDT: A web tool tailored to design CRISPR guide RNAs for eukaryotic pathogens. *Microbial Genomics,* 1(4), e000033. https://doi.org/10.1099/mgen.0.000033

Poddar, S., Tanaka, J., Running, K. L., Kariyawasam, G. K., Faris, J. D., Friesen, T. L., Cho, M. J., Cate, J. H. D., & Staskawicz, B. (2023). Optimization of exogenous-DNA-free Cas9 RNP-mediated gene editing in wheat. *Frontiers in Plant Science,* 13, 1084700. https://doi.org/10.3389/fpls.2022.1084700

Ramakrishnan, M., Kaul, R., Sharma, A., Ahmad, Z., Vijayakanth, V., Keerthana, K., Gao, Z., Zhou, M., & Wei, Q. (2025). CRISPR RNP-mediated transgene-free genome editing in plants. *Plant, Cell & Environment.* https://doi.org/10.1111/pce.70176

Riesenberg, S., Helmbrecht, N., Kanis, P., Maricic, T., & Pääbo, S. (2022). Improved gRNA secondary structures allow editing of resistant target sites. *Nature Communications,* 13(1), 489. https://doi.org/10.1038/s41467-022-28137-7

Ruble, B. K., Yeldell, S. B., & Dmochowski, I. J. (2015). Caged oligonucleotides for studying biological systems. *Journal of Inorganic Biochemistry,* 150, 182–188. https://doi.org/10.1016/j.jinorgbio.2015.03.010

Ryan, D. E., Diamant-Levi, T., Steinfeld, I., Taussig, D., Visal-Shah, S., Thakker, S., Lunstad, B. D., Kaiser, R. J., McCaffrey, R., Ortiz, M., Townsend, J., Welch, W. R. W., Singh, M., Curry, B., Dellinger, D. J., & Bruhn, L. (2022). Phosphonoacetate modifications enhance guide RNA stability. *Biochemistry,* 62(24), 3512–3520. https://doi.org/10.1021/acs.biochem.1c00768

Ryan, D. E., Taussig, D., Steinfeld, I., Phadnis, S. M., Lunstad, B. D., Singh, M., Vuong, X., Okochi, K. D., McCaffrey, R., Olesiak, M., Roy, S., Yung, C. W., Curry, B., Sampson, J. R., Bruhn, L., & Dellinger, D. J. (2018). Improving CRISPR–Cas specificity with chemical modifications in sgRNAs. *Nucleic Acids Research,* 46(2), 792–803. https://doi.org/10.1093/nar/gkx1199

Sakovina, L., Vokhtantsev, I., Vorobyeva, M., Vorobyev, P., & Novopashina, D. (2022). Improving stability and specificity of CRISPR/Cas9 by selective gRNA modification. *International Journal of Molecular Sciences,* 23(21), 13460. https://doi.org/10.3390/ijms232113460

Schmidt, H., Zhang, M., Chakarov, D., Bansal, V., Mourelatos, H., Sánchez-Rivera, F. J., Lowe, S. W., Ventura, A., Leslie, C. S., & Pritykin, Y. (2025). Genome-wide CRISPR guide RNA design with GuideScan2. *Genome Biology,* 26(1), 1–25. https://doi.org/10.1186/s13059-025-03488-8

Shen, M. W., Arbab, M., Hsu, J. Y., Worstell, D., Culbertson, S. J., Krabbe, O., Cassa, C. A., Liu, D. R., Gifford, D. K., & Sherwood, R. I. (2018). Predictable and precise template-free CRISPR editing. *Nature,* 563(7733), 646–651. https://doi.org/10.1038/s41586-018-0686-x

Siegner, S.M., Karasu, M.E., Schröder, M.S., Kontarakis, Z. and Corn, J.E. (2021). PnB Designer: a web application to design prime and base editor guide RNAs for animals and plants. *BMC Bioinformatics*, *22*(1), 101.

Singh, R., Kuscu, C., Quinlan, A., Qi, Y., & Adli, M. (2015). Cas9–chromatin binding enables accurate off-target prediction. *Nucleic Acids Research,* 43(18), e118. https://doi.org/10.1093/nar/gkv575

Smith, C., Gore, A., Yan, W., Abalde-Atristain, L., Li, Z., He, C., Wang, Y., Brodsky, R. A., Zhang, K., Cheng, L., & Ye, Z. (2014). Whole-genome sequencing reveals high specificity of CRISPR/Cas9. *Cell Stem Cell,* 15(1), 12–13. https://doi.org/10.1016/j.stem.2014.06.011

Stemmer, M., Thumberger, T., del Sol Keyer, M., Wittbrodt, J., & Mateo, J. L. (2015). CCTop: A flexible CRISPR/Cas9 target prediction tool. *PLoS ONE,* 10(4), e0124633. https://doi.org/10.1371/journal.pone.0124633

Sternberg, S. H., LaFrance, B., Kaplan, M., & Doudna, J. A. (2015). Conformational control of DNA cleavage by CRISPR–Cas9. *Nature,* 527(7576), 110–113. https://doi.org/10.1038/nature15544

Terada, K., Gimenez-Dejoz, J., Miyagi, Y., Oikawa, K., Tsuchiya, K., & Numata, K. (2020). Artificial cell-penetrating peptides with long-term internalization. *ACS Biomaterials Science & Engineering,* 6(6), 3287–3298. https://pubs.acs.org/doi/10.1021/acsbiomaterials.0c00182

Toda, E., Koiso, N., Takebayashi, A., Ichikawa, M., Kiba, T., Osakabe, K., Osakabe, Y., Sakakibara, H., Kato, N., & Okamoto, T. (2019). An efficient DNA- and selectable-marker-free genome-editing system in rice. *Nature Plants,* 5(4), 363–368. https://doi.org/10.1038/s41477-019-0386-z

Tsai, S. Q., Nguyen, N. T., Malagon-Lopez, J., Topkar, V. V., Aryee, M. J., & Joung, J. K. (2017). CIRCLE-seq: A sensitive in vitro screen for CRISPR off-targets. *Nature Methods,* 14(6), 607–614. https://doi.org/10.1038/nmeth.4278

Tsai, S. Q., Zheng, Z., Nguyen, N. T., Liebers, M., Topkar, V. V., Thapar, V., Wyvekens, N., Khayter, C., Iafrate, A. J., Le, L. P., Aryee, M. J., & Joung, J. K. (2015). GUIDE-seq enables genome-wide profiling of off-target cleavage. *Nature Biotechnology,* 33(2), 187–197. https://doi.org/10.1038/nbt.3117

Varshney, G. K., Zhang, S., Pei, W., Adomako-Ankomah, A., Fohtung, J., Schaffer, K., Carrington, B., Maskeri, A., Slevin, C., Wolfsberg, T., Ledin, J., Sood, R., & Burgess, S. M. (2016). CRISPRz: A database of zebrafish validated sgRNAs. *Nucleic Acids Research,* 44(D1), D822–D826. https://doi.org/10.1093/nar/gkv998

Wang, M., Xu, J., Meng, J., & Huang, X. (2022). Synthetic circular gRNA-mediated biological functions of CRISPR-(d)Cas9. *Frontiers in Cell and Developmental Biology,* 10, 863431. https://doi.org/10.3389/fcell.2022.863431

Wienert, B., Wyman, S. K., Yeh, C. D., Conklin, B. R., & Corn, J. E. (2020). CRISPR off-target detection with DISCOVER-seq. *Nature Protocols,* 15(5), 1775–1799. https://doi.org/10.1038/s41596-020-0309-5

Wong, M.H., Misra, R.P., Giraldo, J.P., Kwak, S.Y., Son, Y., Landry, M.P., Swan, J.W., Blankschtein, D. and Strano, M.S. (2016). Lipid exchange envelope penetration (LEEP) of nanoparticles for plant engineering: a universal localization mechanism. *Nano Letters*, *16*(2), 1161–1172.

Wong, N., Liu, W., & Wang, X. (2015). WU-CRISPR: Characteristics of functional guide RNAs for the CRISPR/Cas9 system. *Genome Biology,* 16(1), 218. https://doi.org/10.1186/s13059-015-0784-0

Xiao, A., Cheng, Z., Kong, L., Zhu, Z., Lin, S., Gao, G., & Zhang, B. (2014). CasOT: A genome-wide Cas9/gRNA off-target searching tool. *Bioinformatics,* 30(8), 1180–1182. https://doi.org/10.1093/bioinformatics/btt764

Xie, X., Ma, X., Zhu, Q., Zeng, D., Li, G., & Liu, Y. G. (2017). CRISPR-GE: A convenient software toolkit for CRISPR-based genome editing. *Molecular Plant,* 10(9), 1246–1249. https://doi.org/10.1016/j.molp.2017.06.004

Yan, W. X., Mirzazadeh, R., Garnerone, S., Scott, D., Schneider, M. W., Kallas, T., Custodio, J., Wernersson, E., Li, Y., Gao, L., Federova, Y., Zetsche, B., Zhang, F., Bienko, M., & Crosetto, N. (2017). BLISS is a versatile and quantitative method for genome-wide profiling of DNA double-strand breaks. *Nature Communications,* 8(1), 15058. https://doi.org/10.1038/ncomms15058

Yang, H., Wang, H., & Jaenisch, R. (2014). Generating genetically modified mice using CRISPR/Cas-mediated genome engineering. *Nature Protocols,* 9(8), 1956–1968. https://www.nature.com/doifinder/10.1038/nprot.2014.134

Yin, H., Kauffman, K. J., & Anderson, D. G. (2017). Delivery technologies for genome editing. *Nature Reviews Drug Discovery,* 16(6), 387–399. https://dx. doi.org/10.1038/nrd.2016.280

Yin, H., Song, C. Q., Suresh, S., Kwan, S. Y., Wu, Q., Walsh, S., Ding, J., Bogorad, R. L., Zhu, L. J., Wolfe, S. A., Koteliansky, V., Xue, W., Langer, R., & Anderson, D. G. (2018). Partial DNA-guided Cas9 enables genome editing with reduced off-target activity. *Nature Chemical Biology,* 14(3), 311–316. https://dx.doi.org/10.1038/nchembio.2559

Young, J., Zastrow-Hayes, G., Deschamps, S., Svitashev, S., Zaremba, M., Acharya, A., Paulraj, S., Peterson-Burch, B., Schwartz, C., Djukanovic, V., Lenderts, B., Feigenbutz, L., Wang, L., Alarcon, C., Siksnys, V., May, G., Chilcoat, N. D., & Kumar, S. (2019). CRISPR–Cas9 editing in maize: Systematic evaluation of off-target activity and its relevance in crop improvement. *Scientific Reports,* 9(1), 6729. https://doi.org/10.1038/s41598-019-43141-6

Zhan, H., Xiao, L., Li, A., Yao, L., Cai, Z., & Liu, Y. (2020). Engineering cellular signal sensors based on CRISPR–sgRNA reconstruction approaches. *International Journal of Biological Sciences,* 16(8), 1441–1453. https://doi.org/10.7150/ijbs.42299

Zhao, G., Li, J., & Tang, Y. (2020). AsCRISPR: A web server for allele-specific single guide RNA design in precision medicine. *The CRISPR Journal,* 3(6), 512–522. https://doi.org/10.1089/crispr.2020.0071

Zhu, L. J., Holmes, B. R., Aronin, N., & Brodsky, M. H. (2014). CRISPRseek: A Bioconductor package to identify target-specific guide RNAs for CRISPR–Cas9 genome-editing systems. *PLoS ONE,* 9(9), e108424. https://doi.org/10.1371/journal.pone.0108424

11 Epigenome Editing for Gene Regulation and Therapeutics

Laxmipreeya Behera, Ashok Mishra, Rakhi Biswal, Rakesh Kumar Patra, Abinash Mishra, and Kailash Chandra Samal

11.1 INTRODUCTION

Epigenome editing refers to the precise and programmable modification of epigenetic information such as DNA methylation, histone post-translational modifications (PTMs), and higher-order chromatin architecture using engineered, locus-specific molecular tools. These tools, which include catalytically inactive CRISPR/Cas variants (dCas), zinc finger proteins (ZFPs), and transcription activator-like effectors (TALEs), are designed to target specific genomic loci without inducing double-stranded DNA breaks (Thakore *et al.*, 2016). By tethering these DNA-binding platforms to epigenetic effector domains such as DNA methyltransferases, demethylases, histone acetyltransferases, histone methyltransferases, or chromatin remodelers, researchers can selectively modulate the local epigenetic environment. Unlike conventional genome editing, which produces permanent, heritable alterations in nucleotide sequence, epigenome editing modulates gene activity without altering the underlying DNA sequence. This makes epigenome editing inherently reversible and tunable, enabling fine control over gene expression.

Consequently, it presents a potentially safer therapeutic strategy, particularly for diseases where aberrant gene expression arises from epigenetic dysregulation rather than coding-sequence mutations (Laufer & Singh, 2015). A broad spectrum of human diseases including cancer, neurodevelopmental and neurodegenerative disorders, autoimmune diseases, metabolic syndromes, and several developmental abnormalities are now recognized to involve dysfunctional epigenetic regulation. These abnormalities often manifest as inappropriate promoter methylation, enhancer silencing, chromatin compaction, or misregulated histone modifications, all of which alter gene expression patterns crucial for cellular identity and homeostasis.

Epigenome editing provides a powerful approach to correct such defects by reprogramming transcriptional states, restoring normal epigenetic landscapes, and selectively activating or silencing pathogenic loci. Because epigenetic modifications can induce long-lasting changes in gene expression without permanent genomic alteration, epigenome editing is emerging as a promising therapeutic modality. Its precision and reversibility position it as an attractive strategy for the treatment of complex diseases driven by epigenetic imbalance, and it is increasingly being integrated into regenerative medicine, functional genomics, and targeted gene-regulatory therapies (Laufer & Singh, 2015; Kantor *et al.*, 2018.

11.2 MOLECULAR PLATFORMS FOR EPIGENOME EDITING

Epigenome editing relies on engineered molecular systems capable of targeting specific genomic loci and depositing, removing, or modulating epigenetic marks. These platforms integrate programmable DNA-binding domains with effector proteins that modify chromatin states. Among the available tools, CRISPR/dCas-based systems, ZFPs, and TALEs constitute the most widely applied technologies for locus-specific epigenome engineering.

DOI: 10.1201/9781003735113-11

11.2.1 CRISPR/dCas-Based Epigenetic Modulators

The CRISPR/dCas system represents the most versatile and scalable approach for epigenome editing. Catalytically inactivated Cas proteins (dead Cas or dCas) are engineered by mutating their nuclease domains, thereby abolishing DNA cleavage while preserving guide RNA–directed DNA-binding capability (Qi *et al.*, 2013; Gilbert *et al.*, 2013). dCas proteins fused to chromatin modifiers enable locus-specific epigenetic perturbations. When fused to specific chromatin-modifying enzymes or transcriptional regulators, dCas proteins can mediate precise epigenetic modifications at single-gene or enhancer levels without inducing double-strand breaks (Thakore *et al.*, 2016).

11.2.2 DNA Methylation Inducers

Targeted epigenetic modification can be achieved by fusing catalytically inactive Cas9 (dCas9) with de novo DNA methyltransferases such as dCas9–DNMT3A or the cofactor-enhanced dCas9–DNMT3A/3L complex, enabling precise CpG methylation at selected genomic loci. These engineered systems direct methylation to promoter or enhancer regions, resulting in stable transcriptional repression (Vojta *et al.*, 2016), and are particularly useful for silencing pathogenic genes, retrotransposons, or aberrantly active regulatory elements. In contrast, targeted DNA demethylation is accomplished using the dCas9–TET1 catalytic domain, which oxidizes 5-methylcytosine (5mC) into 5-hydroxymethylcytosine and related intermediates, thereby promoting active DNA demethylation. This approach enables the reactivation of previously silenced genes, including tumor suppressors and developmentally regulated loci (Liu *et al.*, 2016).

11.2.3 Histone Modification Editors

Histone PTMs play a central role in regulating chromatin organization and transcriptional activity, and their targeted manipulation can be achieved through fusion of catalytically inactive Cas9 (dCas9) with histone-modifying enzymes. For instance, dCas9–p300 acts as a histone acetyltransferase that deposits H3K27ac, thereby activating enhancers and promoters (Hilton *et al.*, 2015), whereas dCas9–LSD1 removes H3K4me1/2 marks, leading to repression of active or poised enhancers (Kearns *et al.*, 2015). Similarly, fusion of dCas9 with the EZH2 catalytic domain of the PRC2 complex promotes H3K27me3 deposition, resulting in stable transcriptional silencing, while dCas9–G9a induces H3K9me2/3 marks that drive constitutive heterochromatin formation (Stepanov *et al.*, 2021). Collectively, these programmable effectors enable precise, locus-specific remodeling of chromatin states.

11.2.4 Transcriptional Activators and Repressors

CRISPR-based synthetic transcription factors provide a powerful means of regulating gene expression without directly modifying underlying epigenetic marks. Transcriptional activation can be achieved using engineered systems such as dCas9–VP64, dCas9–VPR, and the Synergistic Activation Mediator complex, which promote robust gene expression by recruiting components of the transcriptional machinery to target loci (Chavez *et al.*, 2015; Konermann *et al.*, 2015). Conversely, transcriptional repression is mediated through CRISPR interference, where dCas9 fused to the KRAB repressor domain recruit's heterochromatin-forming factors and promotes enrichment of H3K9me3 at target promoters, leading to effective gene silencing. Together, these programmable systems enable dynamic and reversible control of gene expression.

11.2.5 CRISPRoff/CRISPRon System

Recently developed CRISPRoff and CRISPRon systems have further expanded the capabilities of epigenome editing by enabling programmable and heritable control of gene expression without altering the underlying DNA sequence. CRISPRoff mediates long-term transcriptional silencing by directing targeted DNA methylation along with associated repressive chromatin modifications, producing mitotically stable gene repression that can persist even after the editing machinery is no longer expressed. In contrast, CRISPRon reverses CRISPRoff-induced silencing through TET-based demethylation mechanisms, restoring gene activity at previously repressed loci. Together, these complementary systems establish reversible epigenetic memory, offering precise control over gene regulation—an especially valuable feature for therapeutic applications where the timing, duration, and reversibility of gene modulation are critical.

11.2.6 ZFP- and TALE-Based Epigenetic Editors

Before the advent of CRISPR, engineered DNA-binding proteins—specifically ZFPs and TALEs—represented the primary platforms for targeted epigenetic modification. Although technically more complex to design, these systems continue to offer valuable alternatives, particularly in contexts where CRISPR PAM (Protospacer Adjacent Motif) limitations reduce target accessibility.

11.2.7 ZFP-Based Epigenetic Editors

ZFPs are composed of modular Cys2-His2 motifs, each recognizing a 3-bp DNA sequence. By assembling multiple zinc fingers, proteins can be engineered to bind virtually any DNA sequence. When fused to chromatin effectors such as DNMT3A, KRAB, VP64, p300, or LSD1, ZFP-based editors can induce precise epigenetic modifications at target loci. Although capable of high specificity, ZFP engineering is labor-intensive and requires extensive optimization of finger–finger interactions, limiting scalability.

11.2.8 TALE-Based Epigenetic Editors

TALEs from *Xanthomonas* bacteria possess a modular architecture in which each repeat recognizes a single nucleotide. Their straightforward code results in high targeting accuracy and reduced off-target effects compared to ZFPs. TALE based epigenetic editors provide an alternative platform for targeted gene regulation and include constructs such as TALE–TET1 for DNA demethylation, TALE–DNMT3A for site-specific DNA methylation, and TALE–VPR for transcriptional activation. These systems remain particularly valuable in applications that require precise base-level recognition or when suitable CRISPR PAM sequences are not available at the target locus. Their main advantages include reduced dependence on PAM motifs and highly refined DNA sequence recognition. However, they also present certain limitations, such as large protein size, technical complexity in cloning, and comparatively lower scalability than CRISPR-based systems.

Epigenetic regulation operates through multiple layers of biochemical modification that influence chromatin state and gene expression without altering the underlying DNA sequence. Targeted epigenome editing enables precise manipulation of these regulatory layers, allowing researchers to activate, repress, or reprogram gene expression in a locus-specific manner. Three major mechanisms DNA methylation, histone PTMs, and chromatin architecture collectively determine transcriptional outcomes and can be selectively manipulated using modern epigenome-editing platforms such as CRISPR/dCas, ZFPs, and TALEs.

11.3 DNA METHYLATION DYNAMICS

DNA methylation, particularly the addition of a methyl group to cytosine residues within CpG dinucleotides, is one of the most stable epigenetic marks. Targeted editing of DNA methylation status enables durable yet reversible control of gene expression.

11.3.1 Targeted DNA Methylation and Repression

Engineered methyltransferases directed to specific promoters or enhancers can induce stable transcriptional repression, providing a powerful mechanism for long-term gene silencing. This approach is particularly useful for suppressing retroelements, including transposable elements and endogenous retroviruses, thereby preserving genomic integrity; regulating autosomal gene expression through allele-specific control or silencing of dosage-sensitive genes; and repressing aberrantly active oncogenes to limit tumor progression. Because DNA methylation is mitotically heritable, these silencing effects can persist across multiple cell divisions. In contrast, targeted recruitment of demethylating enzymes such as TET1 removes methyl groups and restores transcriptional activity at previously silenced loci. This strategy enables reactivation of dysregulated imprinted genes, restoration of tumor suppressor function in cancer, and re-engagement of key developmental regulators involved in cell fate reprogramming and physiological gene expression. By precisely modulating methylation and demethylation, targeted epigenetic editing can reshape regulatory landscapes to achieve defined functional states.

11.3.2 Histone Post-Translational Modifications

Histone tails undergo a wide range of PTMs, including acetylation, methylation, phosphorylation, and ubiquitination, which collectively regulate chromatin structure and accessibility. Epigenome-editing platforms fused with histone-modifying enzymes enable precise manipulation of these marks, allowing fine-tuned control of gene regulation. Among key regulatory PTMs, H3K27 acetylation (H3K27ac) is enriched at active enhancers and promoters, where it promotes open chromatin and elevated transcription; its targeted deposition, such as through dCas9–p300, can strongly activate gene expression. H3K4 trimethylation (H3K4me3), a hallmark of active promoters, supports transcription initiation by enhancing promoter competence and RNA polymerase recruitment. In contrast, repressive marks such as H3K9me3 and H3K27me3 define heterochromatin and enforce stable gene silencing, with targeted induction through effectors like G9a or EZH2 effectively suppressing aberrant or pathogenic loci. Because histone modifications operate within interconnected feedback loops and chromatin-modifying networks, targeted editing can reshape transcriptional memory and reconfigure regulatory circuits, potentially producing durable phenotypic changes.

11.3.3 Chromatin Architecture Editing

Beyond DNA methylation and histone modifications, the three-dimensional (3D) organization of the genome exerts substantial influence on gene regulation. Chromatin loops, topologically associated domains (TADs), and insulated neighborhoods govern enhancer–promoter communication and higher-order regulatory interactions. Epigenome-editing technologies have advanced to the point where chromatin structure itself can be directly engineered. For example, dCas9 fused with architectural proteins such as CTCF (CCCTC-binding factor) can re-establish insulator boundaries at genomic regions where their loss leads to gene misregulation. Similarly, dCas9 linked to cohesin loaders or chromatin looping factors can promote or stabilize long-range chromatin interactions, enabling disrupted enhancers to reconnect with their specific target promoters. By reshaping chromatin architecture in this way, targeted epigenome editing holds significant therapeutic and functional potential. It can restore enhancer–promoter communication disrupted in developmental disorders or cancer, rebuild insulator boundaries to prevent inappropriate enhancer activity, and correct TAD alterations associated with structural genomic variants and abnormal gene expression. These structural interventions broaden the scope of epigenome editing beyond chemical modifications, allowing precise reconstruction of the genome's 3D organization required for proper transcriptional regulation.

11.4 DELIVERY PLATFORMS FOR EPIGENOME EDITORS

Delivery remains a major challenge for in vivo therapeutic applications, particularly for epigenome-editing systems that require efficient targeting while minimizing off-target effects and avoiding prolonged editor expression. Viral delivery methods such as AAV (Adeno-Associated Virus) vectors are widely used but face size limitations, as CRISPR/dCas systems are often too large, whereas lentiviral vectors enable stable genomic integration and long-term expression but carry risks of insertional mutagenesis. Nonviral approaches offer alternative strategies, including lipid and polymer nanoparticles, ribonucleoprotein (RNP) complexes for transient activity, and mRNA-mediated delivery, all designed to enhance safety and control expression duration. Virus-like particles, which are non-replicating structures derived from viral proteins, can package epigenome-editing cargo such as Cas9 or dCas9 RNPs without containing viral genomes, thereby reducing mutagenesis risk while maintaining high transduction efficiency and enabling transient delivery suitable for epigenome modulation; they have been applied in locus-specific demethylation and reversible gene reactivation in neurodevelopmental disorders. Emerging nanocarrier platforms further expand delivery options: DNA nanostructures, such as programmable DNA origami frameworks, allow multiplexed targeting of regulatory genomic regions, while self-assembling RNA nanostructures can transport mRNA or guide RNAs with improved cytoplasmic stability and reduced immune activation. Additionally, exosome-mediated delivery uses natural extracellular vesicles to transport mRNA, sgRNA, or RNPs in a tissue-specific manner, with engineering strategies such as membrane peptide display and RNA-binding motifs enabling targeted cellular uptake and efficient loading of dCas9 fusion proteins. Overall, these systems emphasize safety, specificity, and biocompatibility as central goals in advancing therapeutic genome and epigenome editing.

Advanced delivery and control strategies continue to expand the precision and safety of epigenome-editing technologies. Peptide-based delivery systems include cell-penetrating peptides, such as arginine-rich motifs like TAT (Trans-Activator of Transcription) or poly-arginine, which facilitate membrane translocation of cargo, enabling delivery of functional epigenetic editors such as dCas9–DNMT3A for targeted DNA methylation and dCas9–p300 for acetylation-mediated gene activation. Endosomolytic peptides further enhance intracellular delivery by promoting endosomal escape through pH-responsive amphipathic helices. Ligand-targeted delivery systems improve tissue specificity by coating nanocarriers with aptamers, antibodies, receptor ligands, or engineered peptides that selectively bind cell-specific markers, including cancer-associated surface proteins, neuron-specific receptors, or defined immune cell populations, thereby enabling precise epigenome editing in diseased or specialized tissues.

In parallel, CRISPR–artificial transcription factor hybrids with inducible control provide temporal and spatial regulation of gene modulation; small-molecule–responsive systems use ligand-binding domains such as tamoxifen- or abscisic acid–inducible modules, while light-inducible optoepigenetic platforms employ blue light–responsive protein interactions to recruit chromatin modifiers with high spatial precision and reversibility. Protein engineering approaches also enhance delivery efficiency and safety, including split-dCas9 systems that reassemble only within target cells to reduce systemic off-target effects, and miniaturized epigenetic effector domains designed for compact packaging, particularly for vector-based delivery. Cell-based delivery platforms offer another emerging strategy, where engineered therapeutic cells release epigenome editors through membrane nanotubes, exosomes, or synaptic contacts, exemplified by immune cells programmed to modulate tumor microenvironments.

11.5 APPLICATIONS IN GENE REGULATION

Controlled manipulation of endogenous loci enables precise tuning of gene expression without introducing exogenous transgenes. This approach is particularly valuable for investigating the function of regulatory elements and understanding transcriptional networks within their native chromatin context.

By modulating gene activity directly at endogenous sites, researchers can dissect gene regulation with greater physiological relevance. Epigenome-editing technologies also facilitate the reprogramming of cell fate. Activation of lineage-specific transcription factors using tools such as dCas9–VPR or p300 has been widely applied in induced pluripotent stem cell reprogramming. At the same time, targeted silencing of differentiation inhibitors further supports cellular reprogramming. These strategies provide powerful means to manipulate developmental pathways and explore mechanisms of cellular identity. Another major application lies in studying enhancers and other regulatory elements. CRISPR activation and interference screening approaches allow systematic mapping of enhancer function, while targeted epigenetic perturbations enable functional genomics studies at unprecedented resolution. This provides deeper insight into chromatin architecture and gene-regulatory hierarchies.

In cancer therapy, epigenome editing offers promising strategies for both gene reactivation and suppression. Tumor suppressor genes can be reactivated through targeted demethylation, while oncogenes can be silenced using transcriptional repressors or DNA methyltransferases. In addition, editing aberrant enhancer landscapes that drive malignant gene expression represents a novel therapeutic direction. Neurological and neurodevelopmental disorders also represent key targets for epigenetic intervention. Disease-associated gene expression patterns can be modulated through targeted epigenetic reprogramming, such as reactivating silenced genes, adjusting regulatory protein levels, or modifying expression of mutant alleles. These approaches aim to restore normal gene regulation without altering the underlying DNA sequence. Similarly, epigenome editing shows potential in autoimmune and inflammatory disorders by modifying regulatory T cell–associated epigenetic states and reprogramming cytokine expression. Such targeted modulation of immune gene regulation may help restore immune balance and reduce pathological inflammation.

Monogenic diseases can also benefit from epigenetic therapies. For example, activating alternative or compensatory gene pathways such as fetal gene expression programs may alleviate disease symptoms where direct mutation correction is difficult. Likewise, selectively silencing pathogenic alleles offers a therapeutic alternative when precise genome editing is technically challenging. In viral diseases, epigenome editing enables targeted silencing of integrated viral genomes, thereby suppressing viral replication. Conversely, controlled reactivation of latent viral sequences can support clearance strategies, particularly when combined with immune-mediated elimination. Rapid technological advances continue to expand the capabilities of epigenome-editing platforms. Miniaturized and split-dCas systems improve delivery efficiency, especially for viral vectors with limited cargo capacity. Multiplexed epigenome editing allows simultaneous targeting of multiple enhancers or chromatin marks, enabling combinatorial reprogramming of entire regulatory networks. RNA-targeted epigenetic editing, particularly using dCas13-based systems, permits modification of RNA methylation states and expands regulation to the epitranscriptomic level. Hybrid systems that combine base editing with epigenetic modulation further enable dual control over genetic and epigenetic information.

Despite these advances, significant technical and biological challenges remain. One major concern is off-target DNA binding in CRISPR/dCas-based systems. Although catalytically inactive Cas proteins do not generate double-strand breaks, they may still bind partially homologous sequences. Such unintended recruitment of DNA methyltransferases, demethylases, or histone-modifying enzymes can lead to unintended epigenetic changes, resulting in inappropriate gene activation or silencing. Because chromatin networks are highly sensitive, even minor off-target epigenetic perturbations may produce substantial biological consequences, particularly in therapeutically relevant tissues. These limitations highlight the need for improved targeting specificity, safety evaluation, and long-term functional assessment in epigenome-editing applications.

11.6 CHALLENGES AND LIMITATIONS OF EPIGENOME EDITING

The durability of targeted epigenetic modifications remains a primary challenge for clinical translation due to the highly variable persistence of marks like CpG methylation or H3K9me3, which are often countered by endogenous chromatin-maintenance systems (Bird, 2002; Hathaway *et al.*, 2012). This stability is dictated by a complex interplay of locus context, cell type–specific resetting machinery, and the baseline chromatin state of the target region (Cedar & Bergman, 2009; Smith & Meissner, 2013; Allis & Jenuwein, 2016), ultimately complicating the design of predictable therapies (O'Geen *et al.*, 2017). Furthermore, delivery remains a significant bottleneck; the large size of dCas9–effector complexes often exceeds the capacity of AAV vectors (Liao *et al.*, 2020), while alternatives like lentiviral vectors carry risks of insertional mutagenesis and nonviral methods like LNPs struggle with precise in vivo biodistribution (Zuris *et al.*, 2015). These hurdles are exacerbated by the interconnected nature of the epigenome, where a single modification can trigger compensatory crosstalk—such as targeted demethylation recruiting histone modifiers or repressive marks inducing de novo DNA methylation (Du *et al.*, 2015). Consequently, isolated analysis of single marks is insufficient, necessitating multi-omic integration (ATAC-seq, ChIP-seq, and transcriptomics) to accurately interpret functional biological outcomes (Buenrostro *et al.*, 2013; Roadmap Epigenomics Consortium, 2015).

11.7 ETHICAL AND REGULATORY CONSIDERATIONS IN EPIGENOME EDITING

Epigenome editing occupies a unique ethical and regulatory space because, unlike traditional genome editing, it modulates gene expression without altering the underlying DNA sequence. This "genetically non-permanent" intervention is frequently viewed as a safer, more ethically acceptable alternative, particularly for therapeutic applications requiring reversibility or controlled temporal effects. By targeting epigenetic marks like DNA methylation or histone modifications, these somatic therapies can transiently manage disease pathways or neurological disorders where dose and timing are crucial, all while minimizing the risks associated with irreversible genetic changes. However, the assumption of non-permanence must be evaluated with caution, particularly regarding the potential for inadvertent germline transmission. While most epigenetic marks are reset during gametogenesis, some modifications can escape this reprogramming, raising the possibility that engineered changes could influence future generations without their consent. This risk complicates the regulatory distinction between somatic and germline interventions, necessitating stringent delivery controls and reproductive risk assessments to prevent unintended phenotypic effects in offspring. Beyond inheritance, the long-term impact on the cellular chromatin landscape remains incompletely understood. Even if the initial molecular tool is degraded, artificially recruited effectors may trigger feedback loops that lead to persistent transcriptional reprogramming or "epigenetic drift." Evidence from animal models suggests that environmental stresses and artificial methylation states can sometimes propagate through the germline as epimutations. Although this remains controversial in humans, the potential for such transgenerational effects requires a precautionary approach and longitudinal studies using high-resolution profiling. Consequently, epigenome editing challenges existing regulatory frameworks that were originally designed for classical gene therapy. Global trends are shifting toward multilayered oversight that evaluates not just DNA sequence integrity, but also off-target epigenetic changes and the mechanisms for controlled deactivation. As these technologies advance, international consensus-building is essential to address the molecular, clinical, and societal impacts, ensuring equitable access to these complex and potentially expensive therapeutic innovations.

Advances in epigenome editing are poised to transform both basic biological research and therapeutic development, yet several technological frontiers must be addressed to fully realize its clinical potential. One major direction is the engineering of high-fidelity dCas systems and orthologous Cas variants with improved DNA-binding specificity, reduced off-target recruitment of chromatin modifiers, and minimized immunogenicity (Chen *et al.*, 2020). The expanding catalog of compact Cas proteins—including Cas12a, SaCas9, and other naturally occurring CRISPR orthologs—offers new possibilities for optimizing editing efficiency, vector compatibility, and cell-type adaptability (Zetsche *et al.*, 2015). Such high-precision dCas platforms will be essential for therapeutic intervention where epigenetic perturbation must be tightly controlled. A second area of innovation involves the development of cell type– and tissue-specific epigenetic editors, achieved through integration of targeted constructs with tissue-restricted promoters, inducible regulatory elements, or microRNA-responsive switches.

These strategies minimize ectopic activity and reduce the risk of aberrant chromatin remodeling, particularly in sensitive systems such as neural, cardiac, and hematopoietic tissues (Dominguez *et al.*, 2016). Context-specific delivery is a prerequisite for ensuring safety in translational applications. Emerging applications also highlight the potential of epigenome-wide multiplexed engineering, wherein multiple genomic loci are targeted simultaneously to modulate complex gene-regulatory networks implicated in polygenic disorders. Multiplexed CRISPR/dCas platforms—enabled by guide RNA architectures, RNA scaffolds, and multivalent dCas fusion systems—permit coordinated activation or repression of entire signaling pathways rather than single genes (Chavez *et al.*, 2016; Morita *et al.*, 2016). This systems-level manipulation mirrors the natural integration of epigenetic mechanisms and is particularly advantageous for diseases such as cancer, metabolic disorders, and neurodegeneration.

Integration of epigenome editing with single-cell multi-omics technologies represents another major frontier. Single-cell ATAC-seq, RNA-seq, and spatial multi-omics now provide high-resolution insights into chromatin accessibility, transcriptional heterogeneity, and epigenetic lineage dynamics (Stuart & Satija, 2019). Coupling these tools with targeted epigenetic perturbations may enable personalized therapeutic design by identifying patient-specific regulatory vulnerabilities and tracking the stability or reversibility of edited states across individual cells (Zhang *et al.*, 2023). Finally, the clinical translation of epigenome-editing therapeutics is expected to expand significantly across neurological, oncological, and hematopoietic disorders, where epigenetic dysregulation plays a central role. Preclinical studies using targeted DNA methylation, histone modification, and CRISPRoff-based silencing have demonstrated stable, long-term gene regulation without permanent genomic alteration (Nuñez *et al.*, 2021). Nevertheless, stringent evaluation of safety, durability, delivery efficiency, off-target chromatin modifications, and immunogenicity remains paramount (Boehm & Hahn, 2011). As delivery platforms improve and regulatory frameworks evolve, epigenome editing is likely to become a foundational modality in next-generation gene and cell therapies.

11.8 CONCLUSION

Epigenome editing represents a transformative frontier in precision medicine, offering the ability to modulate gene expression without the risks associated with permanent DNA sequence alterations. By leveraging the inherent flexibility of the chromatin landscape, these technologies provide a unique therapeutic middle ground where dose, timing, and reversibility can be fine-tuned to treat complex diseases, from neurodevelopmental disorders to chronic inflammation. However, the path to clinical translation is paved with significant challenges, including the variable durability of epigenetic marks, the technical hurdles of delivering large molecular complexes, and the potential for unintended transgenerational inheritance. Addressing these complexities requires a holistic approach that integrates high-resolution multi-omic profiling with stringent regulatory oversight to ensure both safety and efficacy. As our understanding of the interconnected nature of chromatin crosstalk matures, epigenome editing is poised to shift from a specialized research tool to a controllable and durable therapeutic platform, redefining how we approach the treatment of genetically driven pathologies.

REFERENCES

Allis, C. D., & Jenuwein, T. (2016). The molecular hallmarks of epigenetic control. *Nature Reviews Genetics,* 17(8), 487–500. https://doi.org/10.1038/nrg.2016.59

Bird, A. (2002). DNA methylation patterns and epigenetic memory. *Genes & Development,* 16(1), 6–21. https://doi.org/10.1101/gad.947102

Boehm, J. S., & Hahn, W. C. (2011). Towards systematic functional characterization of cancer genomes. *Nature Reviews Genetics,* 21(12), 487–498. https://doi.org/10.1038/nrg3013

Buenrostro, J. D., Giresi, P. G., Zaba, L. C., *et al.* (2013). Transposition of native chromatin for fast and sensitive epigenomic profiling of open chromatin (ATAC-seq). *Nature Methods,* 10(12), 1213–1218. https://doi.org/10.1038/nmeth.2688

Cedar, H., & Bergman, Y. (2009). Linking DNA methylation and histone modification: Patterns and paradigms. *Nature Reviews Genetics,* 10(5), 295–304. https://doi.org/10.1038/nrg2540

Chavez, A., Scheiman, J., Vora, S., et al. (2015). Highly efficient Cas9-mediated transcriptional programming. *Nature Methods,* 12(4), 326–328.

Chavez, A., Tuttle, M., Pruitt, B. W., et al. (2016). Comparison of Cas9 activators in multiple species. *Nature Methods,* 13(7), 563–567. https://doi.org/10.1038/nmeth.3871

Chen, X., Rinsma, M., Janssen, J. M., et al. (2020). Probing the impact of chromatin conformation on genome editing tools. *Nucleic Acids Research,* 48(9), 5033–5048. https://doi.org/10.1093/nar/gkaa231

Dominguez, A. A., Lim, W. A., & Qi, L. S. (2016). Beyond editing: Repurposing CRISPR–Cas9 for precision genome regulation and interrogation. *Nature Reviews Molecular Cell Biology,* 17(1), 5–15. https://doi.org/10.1038/nrm.2015.2

Du, J., Johnson, L. M., Jacobsen, S. E., *et al.* (2015). DNA methylation pathways and their crosstalk with histone methylation. *Nature Reviews Molecular Cell Biology,* 16(9), 519–532. https://doi.org/10.1038/nrm4043

Gilbert, L. A., Larson, M. H., Morsut, L., et al. (2013). CRISPR-mediated modular RNA-guided regulation of transcription in eukaryotes. *Cell,* 154(2), 442–451.

Hathaway, N. A., Bell, O., Hodges, C., et al. (2012). Dynamics and memory of heterochromatin in living cells. *Cell,* 149(7), 1447–1460. https://doi.org/10.1016/j.cell.2012.03.052

Hilton, I. B., D'Ippolito, A. M., Vockley, C. M., et al. (2015). Epigenome editing by a CRISPR–Cas9-based acetyltransferase activates genes from promoters and enhancers. *Nature Biotechnology,* 33(5), 510–517.

Kantor, B., Tagliafierro, L., Gu, J., *et al.* (2018). Epigenetic engineering for neuroscience: What does the future hold? *Frontiers in Genetics,* 9, 610. https://doi.org/10.3389/fgene.2018.00610

Kearns, N. A., Pham, H., Tabak, B., et al. (2015). Functional annotation of enhancers and promoters using the dCas9-p300 epigenome editing tool. *Nature Biotechnology,* 33(5), 555–560.

Konermann, S., Brigham, M. D., Trevino, A. E., et al. (2015). Genome-scale transcriptional activation by an engineered CRISPR–Cas9 complex. *Nature,* 517(7536), 583–588.

Laufer, B. I., & Singh, S. M. (2015). Strategies for precision modulation of gene expression by epigenome editing: An overview. *Epigenetics & Chromatin,* 8(1), 34. https://doi.org/10.1186/s13072-015-0023-7

Liao, H. K., Hatanaka, F., Araoka, T., *et al.* (2020). In vivo target gene activation via CRISPR/Cas9-mediated trans-epigenetic modulation. *Cell,* 181(6), 1509–1523.e15. https://doi.org/10.1016/j.cell.2020.04.019

Liu, X. S., Wu, H., Krzisch, M., et al. (2016). Editing DNA methylation in the mammalian genome. *Cell,* 167(1), 233–247.e17.

Morita, S., Noguchi, H., Horii, T., *et al.* (2016). Targeted DNA demethylation in vivo using dCas9–TET3CD. *Nature Communications,* 7, 12017. https://doi.org/10.1038/ncomms12017

Nuñez, J. K., Chen, J., Pommier, G. C., *et al.* (2021). Genome-wide programmable transcriptional memory by CRISPR-based epigenome editing. *Cell,* 184(9), 2503–2519. https://doi.org/10.1016/j.cell.2021.03.047

O'Geen, H., Henry, I. M., Bhakta, M. S., *et al.* (2017). A genome-wide analysis of Cas9 binding specificity using ChIP-seq and targeted sequence capture. *Nucleic Acids Research,* 45(12), e118.

Qi, L. S., Larson, M. H., Gilbert, L. A., et al. (2013). Repurposing CRISPR as an RNA-guided platform for gene expression control. *Nature Biotechnology,* 31(12), 1149–1156.

Roadmap Epigenomics Consortium. (2015). Integrative analysis of 111 reference human epigenomes. *Nature,* 518(7539), 317–330. https://doi.org/10.1038/nature14248

Smith, Z. D., & Meissner, A. (2013). DNA methylation: Roles in mammalian development. *Nature Reviews Genetics,* 14(3), 204–220. https://doi.org/10.1038/nrg3354

Stepanov, G. A., Semina, S. E., & Kulbachinskiy, A. V. (2021). Targeted epigenome editing tools for gene regulation and studies of chromatin modifications. *Biochemistry (Moscow),* 86(4), 416–432.

Stuart, T., & Satija, R. (2019). Integrative single-cell analysis. *Nature Reviews Genetics,* 20(5), 257–272. https://doi.org/10.1038/s41576-019-0093-7

Thakore, P. I., Black, J. B., Hilton, I. B., *et al.* (2016). Editing the epigenome: Technologies for programmable transcription and epigenetic modulation. *Nature Methods,* 13(2), 127–137.

Vojta, A., Dobrinic, P., Tadic, V., et al. (2016). Repurposing the CRISPR–Cas9 system for targeted DNA methylation. *Nucleic Acids Research,* 44(12), 5615–5628.

Zetsche, B., Gootenberg, J. S., Abudayyeh, O. O., *et al.* (2015). Cpf1 is a single-RNA-guided endonuclease of a class 2 CRISPR–Cas system. *Cell,* 163(3), 759–771. https://doi.org/10.1016/j.cell.2015.09.038

Zhang, H., Xiang, S., Li, Y., *et al.* (2023). Single-cell epigenomic profiling for precision medicine. *Nature Biotechnology,* 41(3), 318–330. https://doi.org/10.1038/s41587-022-01593-8

Zuris, J. A., Thompson, D. B., Shu, Y., *et al.* (2015). Cationic lipid-mediated delivery of proteins enables efficient protein-based genome editing in vitro and in vivo. *Nature Biotechnology,* 33(1), 73–80. https://doi.org/10.1038/nbt.3081

12 Single-Cell Genome Editing Technologies

Marylin Taye, Dhritismita Deka, and Tuward J.

12.1 INTRODUCTION

The deficiencies of bulk editing methods, which mask important cellular heterogeneity that influences biological outcomes, are addressed by single-cell genome editing (ScGE). Although bulk CRISPR techniques aid in the identification of gene functions, they are unable to identify uncommon but important cellular states in complex diseases like cancer, where a variety of tumor cell populations change as a result of treatment. The need for editing platforms targeted at individual cells is highlighted by high-resolution multi-omics studies that demonstrate the impact of different cell subpopulations on disease progression and treatment resistance (Wang et al. 2022a). This change is especially noticeable in oral squamous cell carcinoma (OSCC), where PD-1/PD-L1 signaling was not sufficiently represented by conventional bulk knockout screenings (Elumalai and Ezhilarasan 2022; Soni and Pandey 2024). Bulk measurements fail to identify microclusters with immune-evasive signatures and stress-related epigenetic adaptations, but single-cell analyses do (Hu et al. 2023). These findings indicate that non-responding edited clones in treatment-resistant epithelial tumors exhibit unique transcription factors and traits detectable only through single-cell tracking (Chai et al. 2020; Menon et al. 2025).

ScGE is a major breakthrough relative to traditional CRISPR approaches, as it allows in-depth biological diversity inside each cell. ScGE and CRISPR enable accurate monitoring of cellular alterations, immune evasion, and differentiation processes (Liuet al. 2025). It deals with population evolutionary processes and verifies gene changes by profiling malignancy and tracking resistant lineage using the latest techniques and single-cell sequencing to improve cancer treatments, making it much easier to correct mutations in grafts produced from stem cells for regenerative therapy, thus inducing precise cell identity–based medicine model that informs clonal evolution and multiple risks to tissues. Table 12.1 represents the evaluation of single-cell and gene editing in biological measures.

12.2 BUILDING BLOCKS OF SINGLE-CELL EDITING

Components of CRISPR like modular nucleases (SpCas9) with gRNA necessitate ScGE via standard procedures (Cheng et al. 2013). Its (CRISPR's) versatile nature is exemplified by Cas9 protein, connected to transcriptional machinery for gene expression regulation (Dang et al. 2015). For better accuracy reduction of off-targets, gRNA and editing of nucleases are pivotal in sgRNA structure to increase the effectiveness of cas9 knockdown strategies (Graham and Root 2015), for which recent works focused on improving guidelines to boost precision. Moreover, new breakthroughs in enhanced stimuli-responsive carriers with better specificity in tissues have resounded the urge to improve single-cell editing techniques.

To lower cellular toxicity and improve tracking, base editors are an excellent approach that connects faulty Cas proteins with degrading enzymes to allow nucleotide (C–T or A–G)conversions without double-strand breaks (DSBs) (Cui et al. 2018). Prime editing utilizes a Cas9 nickase and enzyme (reverse transcriptase) for accurate editing of sequences of genes to repair mutations such as those in the *DYSF* gene through a "search-and-replace" approach (Jiang and Doudna 2017).

DOI: 10.1201/9781003735113-12

TABLE 12.1
Evaluation of Single-Cell and Gene Editing in Biological Measures

Biological Dimension	Bulk Gene Editing Detection	What Bulk Gene Editing Hides and Leaves Out	Single-Cell Gene Editing: What It Reveals	Implications for Disease and Therapy	References
Genetic variation	Average population-wide mutation response	Rare tolerant clones with unique variation loads	Edit-specific effects in individual cells; after clonal selection	Early detection of resistant genes prevents therapy escape	Goldman et al. (2019)
Epigenetic state	The average of all edited cells' epigenetic signatures	Buffering and trends in methylation in minor groups caused by stress	Histone code changes and edited chromatin remodeling at single-cell precision	Identifies the epigenetic pathways leading to drug resistance and renewal	Hu et al. (2023) and Gaj et al. (2016)
Transcriptional response	Average efficiency of transcriptome knockdown	Multiple transcription rewiring subclusters right after editing	Variations in pathway activation between cells (e.g., immune suppression, metabolic rewiring)	Permits cellular identity-related functional CRISPR readouts	Soni and Pandey (2024) and McKenzie et al. (2022)
Cell-state identity	"Single outcome" translation	Tissue-like, Epithelial–Mesenchymal Transition (EMT)-like, or hidden T-reg-like states that endure due to CRISPR modification	Results of state-based modification tracked over time	Explains why some modified cells divide while others relapse	Sowmya et al. (2024) and Geyer et al. (2024)
Clonal evolution under pressure	Overall treatment response to editing	Survival of micro-lineages that later dominate relapse	Fitness trajectories of individual clones during editing-mediated selection	Enables prediction of therapy failure and adaptive resistance	Elumalai and Ezhilarasan (2022) and Xue et al. (2025)
Tumor-immune interaction	CRISPR proves to be effective for immune-active tumors	Immune-evasive circuit retention (e.g., TIGIT-high/PD-L1-high microclusters)	Effects of personalized editing in tumor-immune synapses	Real CRISPR defects in the design of immune therapy	Bode et al. (2021) and Raikwar et al. (2019)
Therapeutic translation	Assumes a consistent therapeutic benefit	Indirect growth, unreliable clones, and mosaic risk	Testing of safe clones before therapy development	Speeds up the progress of precision cell therapy while ensuring safety	Menon et al. (2025) and Liu et al. (2025)

Moreover, effective delivery and analysis procedures correct interruptions in complicated phenotypes. For delivery, current techniques utilize ribonucleoprotein (RNP) complexes, viral vectors, and nonviral nanoparticles, designed for applications in plants and cancer therapy. Large-scale analysis of modified cells for transcriptional and chromatin changes is made possible by powerful readout techniques like CROP-seq and Perturb-ATAC allowing for a system that combines flexible editors and customized guides with single-cell analysis.

12.3 ISOLATION AND MANIPULATION OF SINGLE CELLS

Single-cell analysis mainly depends on the isolation and manipulation of single cells that needs isolation from adjacent cell populations for functional analysis and validation. Physical, microfluidic, and field-mediated methods are the three fundamental handling methods currently used in research. These approaches explore how machinery accuracy and localized impact single-cell growth and survival (Tanet al. 2010).

Electrochemical and fluorescent reporters are two of the most widely used biosensor systems for classifying and monitoring the effects of gene editing and altered cells in real time without harming or destroying other healthy cells/tissues (Onan et al. 2025). Technologies like Bio-Rad's digital droplet assays make it easier to check how well CRISPR works before it is used on a larger scale (Bio-Rad 2024). With tremendous advances in microfluidics and single-cell readout strategies, single-cell isolation has developed into a multistage pipeline that includes the isolation, delivery, validation, and expansion of edited cells, creating a complete ecosystem for genome via biomimetic micro carriers and three-dimensional scaffolds used in protecting transformed cells growth during regenerative therapy (Onan et al. 2025).

12.4 DELIVERY STRATEGIES: CASE EXAMPLES

The development of gene editing and therapeutic molecule delivery strategies focuses on a variety of techniques intended to safely and effectively introduce these agents directly into targeted living cells to improve therapeutic outcomes.

12.4.1 Physical Methods

12.4.1.1 Electroporation in Single-Cell Chips

Single-cell microfluidic chip electroporation allows the delivery of CRISPR RNPs and DNA into numerous cell types, incorporating mouse zygotes and human T cells, by temporary membrane openings created by electronic pulses (Liu et al. 2017). Proper regulation of pulse parameters assures cell membrane resealing, which is a key for maintaining cell viability, while strong pulses can lead to the death of cells. Recent studies have shown successful delivery of Cas9 mRNA, sgRNA, and donor DNA using this method (Park et al. 2024; Fajrial et al. 2020). Real-time monitoring differentiates living cells from the dead ones, by factors such as crucial electric tension and duration of the pulse vital for optimizing cell viability. Microinjections are another specific gene editing technique used in animal embryos, ensuring direct delivery of genetic materials into nucleus and/or cytosol. This method enables early editing of genome for creating genetically modified embryos. Additionally, engaging cell-penetrating peptides (CPPs) can mitigate death of cells in association with electroporation through facilitating safer transportation of Cas9 protein and escort RNAs into appropriate cell targets. This method involves transporting CRISPR/Cas9 into one-cell rat embryos to create genetically modified embryos by targeted gene insertions or deletions. Researchers must deliver the Cas9 protein and lead RNAs into the targeted cells, often using electroporation, which can be harmful. To reduce this, CPPs are used as helper molecules to allow safe entry via cell membrane.

12.4.2 Chemical Methods: Application of Lipid Nanoparticles Designed for Delicate Stem Cells

Lipid nanoparticle (LNP) delivery systems are designed to deliver genetic materials like siRNA and mRNA into cells, including stem cells. LNPs shield these materials from degradation by enzymes and make use of endocytosis for entry into the cell. Inside the cells, LNPs are confined in endosomes, which gradually become acidic as they mature. This acidic nature causes ionization of lipids in LNPs to become positively charged, followed by interaction with the negatively charged endosomal membrane, resulting in its rupture and releasing of the components into the cytosol. The released components can then be translated into proteins or interact with the genome. Optimized LNPs are efficient in delivering CRISPR Cas9 and other RNA therapies for gene editing in multipotent adult stem cells and induced pluripotent stem cells (iPSCs), increasing editing efficiency and survival of the cell (Zong et al. 2023; Hou et al. 2021).

12.4.3 Low-MOI Lentiviral Vectors for Clonal Editing in Cancer Cell Lines

Lentiviral vectors, which merge into host DNA and transduce both dividing and nondividing cells, including cell lines like HepG2 and OSCC, are utilized for stable genome editing (Lane-Reticker et al. 2023). Stable genotypic and phenotypic correlations are achieved in clonal CRISPR editing by using a low multiplicity of infection (MOI; approximately 0.3), which enables single-cell clones to acquire a single viral integrant (Li et al., 2018; Bystrykh et al., 2014). Although low-MOI transduction is commonly preferred to minimize multiple integrations, higher MOIs may be used in certain barcoded CRISPR applications to improve clone recovery and identification efficiency. This approach conserves the stability of Cas9, sgRNAs, and shRNAs without cellular impairment. Elevated knockout efficiency in demanding cell populations can be achieved by merging lentiviral delivery with electroporation, ameliorating clonal stability and genetic traceability in performance evaluation pertaining to oral and hepatic cancers.

12.4.4 Some Developments in Nanoparticle-CRISPR in Breast Cancer

Polymeric lipid nanoparticles (PLNPs) effectively deliver CRISPR/Cas9 components to solid tumors (Zhang et al. 2022), particularly targeting treatment-resistant cancer cell clones (Wang et al. 2022b). In A375 melanoma cells, PLNP/DNA complexes aimed at *PLK-1* achieved 47.4% transfection efficiency, compared to only 3.09% with Lipofectamine 2000. *PLK-1,* which is overexpressed in therapy-resistant breast cancers, regulates chromosomal instability and contributes to paclitaxel resistance. Silencing PLK-1 increased susceptibility to cytotoxicity from paclitaxel, while its up regulation correlated with lower patient survival rates (Zhang et al. 2022). Additionally, PLNP-mediated CRISPR delivery showed a 16.1% mutation efficiency, demonstrating functional gene editing in vivo. PLNP/DNA targeting of *PLK-1* resulted in significant tumor growth inhibition, reducing tumor volume by about 67%, and outperformed other delivery methods in silencing *PLK-1* protein expression, which was linked to diminished tumor size and weight.

PLK-1 targeting induces G2-M phase cell-cycle arrest followed by apoptotic cell death, with CRISPR-mediated PLK-1 editing demonstrating a fivefold reduction in cell viability compared to untreated controls, evident from DAPI (4′,6-diamidino-2-phenylindole) / Annexin V stainingXTT [2,3-bis(2-methoxy-4-nitro-5-sulfophenyl)-2H-tetrazolium-5-carboxanilide] assay, and achieving significant G2-M arrest within 48 hours(Wang et al. 2022b). This loss also activates DNA damage response pathways, such as DNA damage and checkpoint proteins. The performance of *PLK-1* nanoparticle-based method depends on their abundance in resistant cancer cells (Worthington and Forsberg 2022; Krishnamurthy et al. 2015). CRISPR/Cas9-mediated disruption of PLK-1 yields significantly improved tumor inhibition compared to traditional siRNA methods that offer only transient gene silencing. Studies on PLK-1 inhibitors like volasertib reveal substantial reductions in tumor size

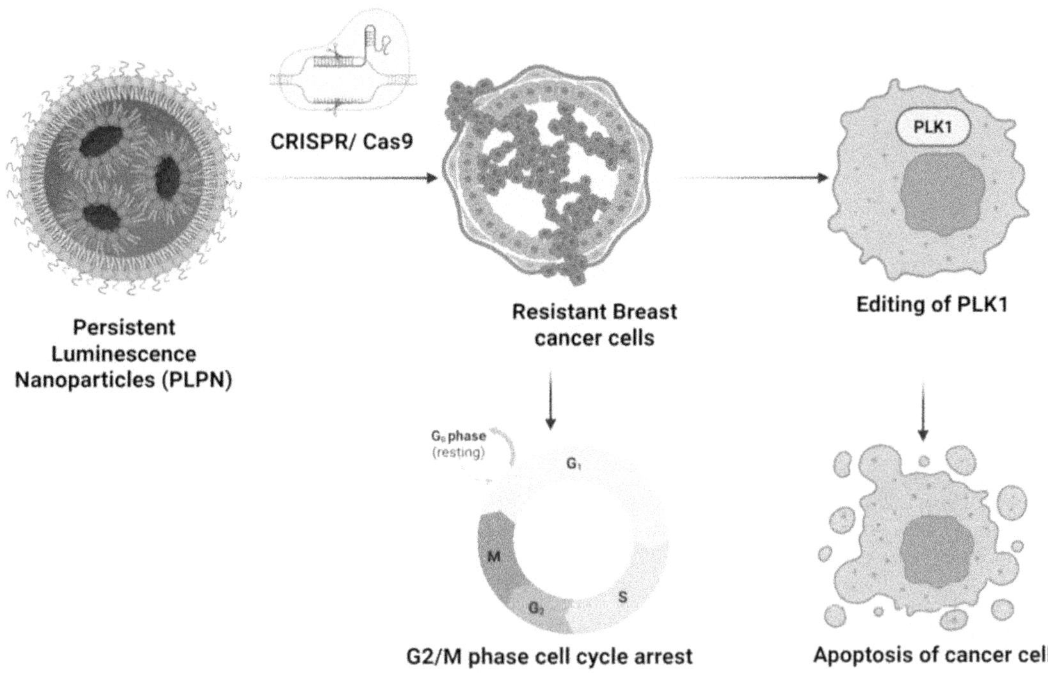

FIGURE 12.1 Schematic representation of PLNP (Polymer–Lipid Nanoparticles) -mediated CRISPR/Cas9 delivery targeting the *PLK1* gene to induce cell-cycle arrest and apoptosis in breast cancer cells.

and weight in patient-derived xenograft models, supporting clinical application. The integration of CRISPR precision with PLNP delivery presents a novel therapeutic pathway for PLK-1-dependent drug-resistant breast cancer, circumventing conventional chemotherapy approaches (Rosenblum et al. 2020). The illustration of PLNPs that facilitate effective delivery of Cas9/sgPLK-1 to therapy-resistant breast cancer cells, achieving high transfection rates, PLK-1 silencing, G2-M cell-cycle arrest, apoptosis, and notable tumor growth inhibition in vivo is shown in Figure 12.1.

12.5 APPLICATIONS IN ONCOLOGY

12.5.1 Tracking Resistant Subclones in CAL-27 and SCC-25

A total of 918 context-specific fitness genes were found through genome-wide CRISPR/Cas9 screening of 21 OSCC cell lines, mostly from Asian patients with a history of chewing betel quid. The screening identified unique genetic vulnerabilities, including the functions of *YAP1* and *WWTR1* in the Hippo pathway, and identified essential genes across lineages. These genes showed compensatory roles in some OSCC lines and mutually exclusive dependencies in others (Quan et al. 2022). Furthermore, *WWTR1*-dependent OSCC lines showed a notable enrichment of immune-related hallmarks associated with biomarkers that indicate favorable reactions to immune checkpoint inhibitors. According to the results, betel quid extract stimulates NF-κB signaling in oral epithelial cells, which is consistent with the crucial function of the NF-κB pathway in betel quid–associated OSCC models (Chai et al. 2020).

12.5.2 Single-Cell CRISPR for Checkpoint Regulators

Through single-cell CRISPR immune screening, transcriptional regulators of immune checkpoint molecules have been found, and *MLLT6* has been shown to be crucial for PD-L1 expression and presentation in cancer cells. CD8 T cell–mediated lysis is diminished when MLLT6 is lost

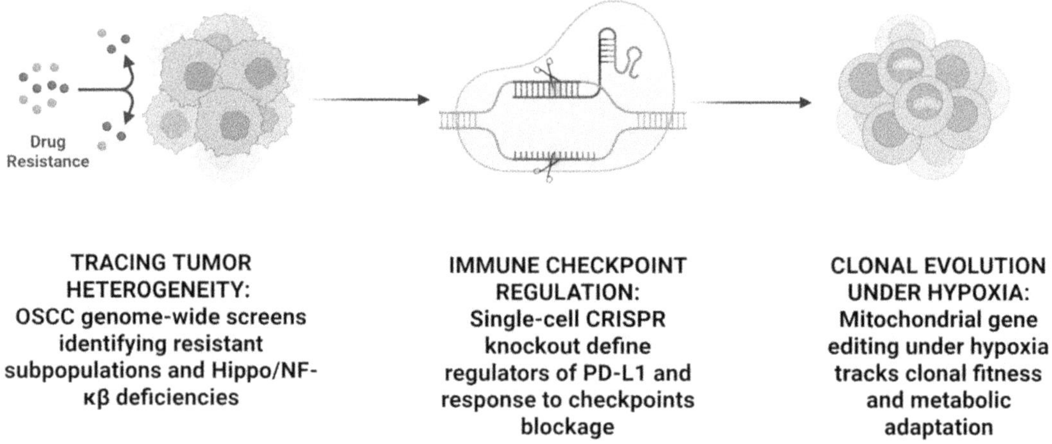

FIGURE 12.2 Illustration of CRISPR applications in tracing tumor heterogeneity, regulating immune checkpoints, and studying clonal evolution under hypoxic conditions.

(Hou et al. 2021). Additional research revealed that editing genes like *Asf1a, PRMT1, RIPK1, MAN2A1, IFNGR2*, and *JAK1* increases sensitivity to PD-L1 or anti-PD-1 treatments. Multimodal single-cell CRISPR platforms such as Perturb-seq and CROP-seq, which measure gene expression and perturbation effects at the individual cell level, are used to investigate the relationship between immune evasion mechanisms and variability in PD-L1 expression (Xu et al. 2022).

12.5.3 Mitochondrial Edits and Clonal Fitness

CRISPR editing of mitochondrial regulators under stress conditions aids in assessing mitochondrial function vital for tumor clonal fitness in hypoxic environments. Hypoxia promotes the selection of invasive clones and induces genomic instability by inhibiting homologous recombination and favoring error-prone non-homologous end joining (Kunachowicz et al. 2025). HIF-mediated responses in hypoxic areas support cancer stem cell maintenance and stabilize stemness factors. By employing CRISPR on regulators like *PINK1, Parkin, DRP1, and OPA1,* researchers can observe clonal dynamics as modified populations adapt to low oxygen. Tracking these edited clones helps identify therapeutic targets for resistance in hypoxia-adapted populations through single-cell barcode integration (Kunachowicz et al. 2025). However, the tumor heterogeneity, immune checkpoint regulation, and clonal evolution under hypoxic conditions are investigated using CRISPR-based techniques (Figure 12.2).

12.6 APPLICATIONS IN STEM CELLS AND REGENERATION

Pluripotent stem cells can be edited at the single-cell stage using patient-derived induced pluripotent stem cells (iPSCs) for the correction of rare genetic disorders. Single-cell editing of patient-derived iPSCs shows therapeutic potential for rare genetic disorders through repairing disease-causing mutations. Using CRISPR technology and optimized homology-directed repair templates, editing ability in iPSCs reaches 17.6% (He et al. 2024). Distinct barcode sequences assist in differentiating edited cells. A remarkable application was the*AKT1 E17K* mutation which is linked to Proteus syndrome, which shows 50% mosaic mutation rates in organoids, thereby depicting clinical mutation patterns. Modified iPSCs can help in mechanistic research and potential therapeutic applications, while holding on to pluripotency and expressing disease-related (Worthington and Forsberg 2022).

12.6.1 CRISPR Barcoding to Map Stem Cell Lineage

CRISPR barcoding has revolutionized developmental cell tracking by enhancing lineage tracing by DNA barcodes, allowing for the reconstruction of differentiated hierarchies from single-cell data. A transgenic mouse model uses doxycycline-inducible Cas9 to produce up to 44,000 unique barcode sequences via stochastic CRISPR mutations. The LinTIMaT statistical method boosts accuracy in reconstructing lineage trees by integrating single-cell transcriptomic data with mutational barcode data. This approach has progenitor–progeny relationships in both fish and mammalian hematopoiesis, thereby enhancing the comprehension of hematopoietic differentiation courses (Zafaret al. 2020).

12.6.2 Single-Cell Editing to Model Carcinogen-Driven Dysbiosis in Oral Organoids

Chewing betel nut notably alters the oral microbiome, causing organoid-level toxicity and promoting the development of a carcinogenic microenvironment. This habit causes genotoxicity and damage to mitochondrial DNA in oral epithelial cells, which contributes to oral dysbiosis, characterized by reduced bacterial diversity and elevated pathogenic taxa in long-term consumers. Organoid models, produced using CRISPR to knock out specific genes from patient-derived tissues, enhance the evaluation of genetic pre-suspension to carcinogenic metabolites. They also assist the simulation of epithelial-to-mesenchymal transitions while observing vital genomic alterations during the transformation process related with carcinogens. To prevent undesired mosaicism in the development of edited stem cells into organoids, strict clonal validation procedures are necessary. Efficient strategies include sorting of FACS-based single-cell, morphological monitoring during colony growth, pre-differentiation genotyping through Sanger sequencing, and post-expansion karyotyping for chromosomal stability. Maintaining a 50% mutation rate in the *AKT1 E17K* organoid model that replicates Proteus syndrome requires careful assessment of mutation status and minimization of in vitro culture duration to improve editing efficiency prior to differentiation (Bhargava et al. 2022).

12.7 INTEGRATING WITH MULTI-OMICS PLATFORMS

In order to analyze how genome edits influence cellular phenotypes and functional results at the single-cell level and to regulate in-depth analyses of phenotype and genotype relationships, multi-omics integration is important.

12.7.1 scRNA-seq Combined with Genome Editing

Single-cell RNA sequencing (scRNA-seq), when merged with genome editing techniques such as CRISPR/Cas9, divulges the systematic transcriptional landscape of altered cells, and uncovers modifications in signaling networks and unpredicted transcriptional feedbacks (Huang et al. 2023). scRNA-seq can distinguish successfully edited cells from cells with off-target effects or failed modifications, while tracking changes in immune, metabolic, and stress-response pathways alone with stress responses. It is effective in identifying rare operational status post-editing, including compensatory gene expressions and cell identity shifting in oncology (Mahat et al. 2024). Disruption of genes such as *MYC* or *TP53* reveals secondary cellular processes, including apoptosis and senescence, which vary depending on the cancer type and editing outcome (Tornesello 2024; Hernandez et al. 2017).

12.7.2 CITE-seq and Proteomics Integration

Cellular Indexing of Transcriptomes and Epitopes by sequencing (CITE-seq) measures transcriptomes and abundance of the surface proteins in cell editing and at the same time enhances the multi-omics analysis. Because various regulatory mechanisms cause dissimilarities between abundance of mRNA and protein expression, this integration is vital (Lenkiewicz and Kaminska 2025).

CITE-seq demonstrates how immune cell surface marker genomic modifications influence protein presentation, sequentially affecting immunological functions. Weighted Nearest Neighbor analysis is a technique that enables to identify various cell populations with diverse protein-to-RNA ratios (Song et al. 2025; Wu et al. 2024). Spatial CITE-seq technique can be applied to intact tissues to show how editing alters cell interactions and functioning of tissues.

12.7.3 Lineage Barcoding with CRISPR Recorders

By heritable mutations at particular location of genome, CRISPR-based molecular barcoding systems make permanent genetic "time stamps" that let clonal evolution to be traced. As the cells divide, these mutational barcodes make up and conceal lineage histories in the genomes of descended cells. Researchers can particularly rebuild cellular descent and profile transcriptomic and proteomic states by combining these barcodes with high-throughput single-cell sequencing (Huang et al. 2023). Thousands of cells can be traced during development or the course of a disease from sophisticated platforms like CARLIN and scGESTALT. In order to deduce developmental trajectories and for identifying regulatory transitions, methods like decadence inference algorithms, such as LinTIMaT, along with barcode mutations, use maximum-likelihood techniques (Rajet al. 2018). Tracing embryonic developmental decisions and exploring the evolution of cancer are crucial applications. The ScGE workflow that incorporates multi-omics platforms is shown in Figure 12.3.

12.7.4 Computational Integration of Multi-Omics Single-Cell Datasets

Significant computational problems arise at single-cell resolution in aligning transcriptomes, proteomes, and genome edits due to the complex and high-dimensional multi-omics data. This data often comprises about 30,000 genes and 200 protein markers, together with various epigenetic measurements (Jiang et al. 2025; Lee et al. 2024). Techniques such as Mutual Nearest Neighbors (MNN), Harmony, and Canonical Correlation Analysis (CCA) are used to correct batch effects while maintaining biological signal integrity. Advanced deep learning of architectures such as variational autoencoders and graph neural networks (GNNs) eases the learning of joint representations from numerous data modalities. Tools such as Smmit aid in integrating samples and detecting biological signals while obliging patient-specific variations (Zou et al. 2021). Customized bioinformatics workflows must be developed for identifying edited cells, differentiating modification types, clustering cells by phenotype, and conducting multi-omics differential analysis. As datasets expand toward millions of cells, cloud computing and GPU-based solutions from tools such as Seurat, Scanpy, and Pegasus become vital for scalable analyses. Furthermore, a comparative evaluation of five multi-omics technologies for ScGE, including analytical performance and application suitability, is presented in Table 12.2.

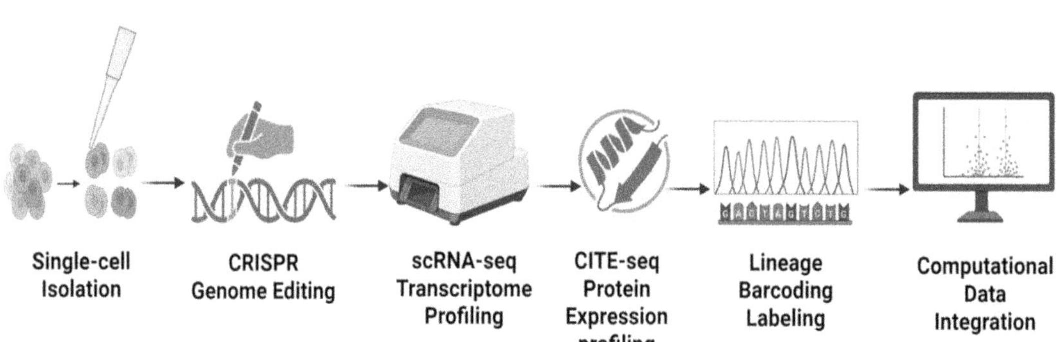

FIGURE 12.3 Workflow for integrating single-cell CRISPR genome editing with multi-omics profiling and computational data analysis.

TABLE 12.2
Comparative Evaluation of Five Multi-Omics Technologies for Single-Cell Genome Editing: Analytical Performance and Application Suitability

Technology	Sensitivity	Specificity	Dynamic Range	Processing Time	Dropout Rate (%)	Best Use Case
scRNA-seq andediting	1 transcript/cell	Approximately 98%	7 orders	24–72 hours	15–30	Cell-state discovery after editing
CITE-seq (protein andRNA)	10–100 proteins/cell	95%–99%	3–4 orders	48–96 hours	5–15	Immune protein profiling after editing
CRISPR lineage barcoding	Detects clones, more than or approximately 10 cells	More than 99%	Very high	1–2 weeks	2–5	Clonal evolution and fate tracking
Spatial transcriptomics	10–100 transcripts/spot	95%–99%	4 orders	48–120 hours	20–35	Spatial mapping of edited cells in tissue
scATAC-seq andmulti-omics	Approximately 1,000 reads/cell	More than 98%	2–3 orders	48–96 hours	40–60	Chromatin accessibility after editing

12.8 ETHICAL, SAFETY, AND TRANSLATION ISSUES

ScGE methods require a systematic review of safety hazards, ethical concerns, and regulatory pathways, with a special emphasis on proliferative stem cells and reproductive contexts.

12.8.1 Off-Target Risks Amplified in Stem Cell Therapies

Stem cell editing poses significant risks due to the potential for inadvertent mutations in a single cell to propagate. CRISPR/Cas9 can induce off-target double-strand breaks (DSBs), and even high-fidelity variants may result in unintended alterations, large chromosomal rearrangements, and chromothripsis. In hematopoietic stem and progenitor cell therapies, even one modified cell can spread harmful mutations, while mosaicism creates a composite population that convolutes clinical results. Regulatory bodies mandate comprehensive genomic safety assessments, incorporate whole-genome sequencing, and targeted sequencing at estimated off-target sites, detection of large chromosomal deformities, and clonal expansion assays.

12.8.2 Ethical and Regulatory Concerns with Germline Editing

In comparison to somatic cell editing, germline gene editing raises different ethical questions because the changes are transmissible and have impact on future generations who are not able to give assent. Global ethical analysis was initiated by the 2018 Chinese dispute surrounding CRISPR/Cas9 editing of human embryos brought to light issues such as potential genetic asymmetry, long-term multigenic effects, and indefinite discrepancy between therapeutic outcomes and amplification. Regulatory reactions differ: clinical germline editing for cell proliferation is generally considered reprehensible, whereas basic research may be sanctioned under proper authorized supervision. Before any clinical application can proceed, the WHO encourages promote strict regulations and public consent.

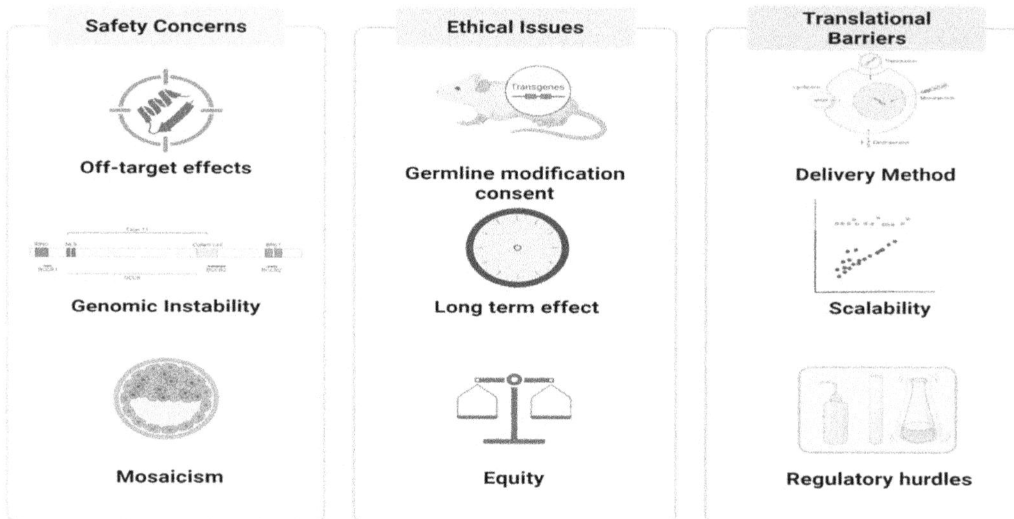

FIGURE 12.4 Overview of the primary safety, ethical, and translational challenges associated with the clinical implementation of CRISPR technology.

12.8.3 Clinical Translations

Regulatory agencies' rigorous preclinical data prerequisites are one of the many challenges in the ScGE's refinement from lab research to authorized cellular therapies. These include validating the efficacy and precision of editing, guaranteeing that no undesirable mutations occur, maintaining genomic stability, etc. The process is further hindered by the complexities of producing autologous cell therapies, which requires patient-specific workflows from cell analysis to treatment(Zhang et al. 2025).Substantial challenges include exorbitant, precise manufacturing rules and the prerequisite for effective long-term results and adverse event tracking (Raigani et al. 2025). As the regulatory frameworks are evolving to analyze these problems, standardization of methods and guaranteeing consistent findings across various interventions continue to be challenging endeavors. The schematic overview of safety, ethical, and translational challenges in ScGE for regenerative and germline applications is shown in Figure 12.4.

12.9 FUTURE PROSPECTS

12.9.1 Spatial Transcriptomics Integration

Researchers can map genetic alterations within intact tissue structure by merging ScGE with spatial transcriptomics, which shows cell–cell interactions that frequently conceal dissociation-based scRNA-seq (Zheng and Yang 2023). Gene expression tracing at specific tissue coordinates supported by platforms like ten times Visium and seqFISH, which enables modified cells in their environmental settings. The analysis of tissue morphology and molecular states is made simple by methods such as GNNs and vision transformers (Schmoldt et al. 1975). These techniques help explain how modified stem cell gives out in tissue repair and how modified immune cells are involved with the tumor microenvironment. Further studies should prioritize long-term impacts of designed modifications on tissue structure and function along with real-time spatial analysis.

12.9.2 AI-Designed sgRNAs for Precision and Efficiency

Through the prediction of on-target editing efficiency and off-target probabilities with greater accuracy than conventional sequence-based scoring methods, AI is changing the guide RNA design. Sequence features and chromatin availability patterns are used by machine learning models that are trained on large datasets of sgRNAs tested in multiple contexts (Sherkatghanad et al. 2023; Tahsin et al. 2025). CNNs, RNNs, and transformers are few examples of deep learning architectures that perform better than physics-based predictors. Various designs and datasets are used by tools such as CRISPR on and DeepSpCas9. AI models examine variables such as sequence similarity, chromatin accessibility, methylation of DNA, and celltype–specific gene expression that impact wrong predictions (Lee 2023). These models comprise cellstate–dependent elements for ScGE, allowing the development of tailored sgRNA libraries intended to increase target editing while decreasing off-target effects.

12.9.3 Personalized Regeneration via Ex Vivo Single-Cell Editing

ScGE is a standard autologous therapy which ensures specific correction of mutations that may lead to disease. Patient-derived iPSCs can be differentiated into appropriate cell types for a variety of diseases after CRISPR/Cas9 corrections. The goal of cutting-edge editing methods such as base and prime editing is to boost accuracy and reduce the risk of mutagenesis while monitoring modified cells use a variety of long-term techniques (McTague et al. 2021). Unusual drug-resistant subclones can be found and targeted by using specialized genome editing techniques through profiling of distinct transcriptomic, proteomic, and genomic characteristics of a patient's tumor (Li et al. 2025). This method guarantees better stem cell graft combination and permits repeated treatment improvements based on multi-omics profiling. Adaptive interventions based on cellular behavior are made possible through real-time monitoring after editing. Strong computational infrastructure, individual response prediction models, adaptable regulatory frameworks, and ethical standards to guarantee fair use are all important elements of this approach (Singh et al. 2025). Personalized medicine is becoming increasingly possible due to developments in single-cell editing and AI-driven analytics, which enable precise outcome assessment, diagnosis, and intervention.

12.10 CONCLUSION

ScGEemergence has seen a pivotal shift toward cell-resolved highly automated interventions, linking genomic edits with transcriptional, proteomic, epigenetic, and clonal outcomes across heterogeneous populations with accuracy. Via CRISPR system integrations like delivery techniques, prime editors, this framework reveals immune-evasive states, resistant subclones, and stemcell dynamics that bulk approaches consistently mask. However, off-target issues, mosaicism, difficulties in manufacturing, and ethicalities restrict its therapeutic potential in regenerative medicine cancer biology, thus making reliable computational analysis and stringent safety procedures paramount to address these challenges. New insights in ex vivo, scalable multi-omics analytics, autologous single-cell editing techniques, and AI-based sgRNA design have the potential to advance the field, and in an ethically controlled, gene editing treatments.

REFERENCES

Bhargava, Nupur, Priya Thakur, Thulasi P. Muruganandam, Shashank Jaitly, Pragya Gupta, Neelam Lohani, Sangam G. Goswami, et al. 2022. "Development of an Efficient Single-Cell Cloning and Expansion Strategy for Genome Edited Induced Pluripotent Stem Cells." *Molecular Biology Reports* 49 (8): 7887–98. doi:10.1007/s11033-022-07621-9.

Bio-Rad. (2024). *Genome edit detection assays*. Available at: https://www.bio-rad.com/digital-assays/assays-create/genome. Assessed on 20 November 2025.

Bode, Daniel, Alyssa H. Cull, Juan A. Rubio-Lara, and David G. Kent. 2021. "Exploiting Single-Cell Tools in Gene and Cell Therapy." *Frontiers in Immunology* 12 (July): 702636. doi:10.3389/fimmu.2021.702636.

Bystrykh, Leonid V., Gerald De Haan, and Evgenia Verovskaya. 2014. "Barcoded Vector Libraries and Retroviral or Lentiviral Barcoding of Hematopoietic Stem Cells." In *Hematopoietic Stem Cell Protocols*, edited by Kevin D. Bunting and Cheng-Kui Qu, 1185:345–60. Methods in Molecular Biology. New York, NY: Springer New York. doi:10.1007/978-1-4939-1133-2_23.

Chai, Annie W. Y., Pei S. Yee, Stacey Price, Shi M. Yee, Hui M. Lee, Vivian K. Tiong, Emanuel Gonçalves, et al. 2020a. "Genome-Wide CRISPR Screens of Oral Squamous Cell Carcinoma Reveal Fitness Genes in the Hippo Pathway." *eLife* 9 (September): e57761. doi:10.7554/eLife.57761.

Cheng, Albert W., Haoyi Wang, Hui Yang, Linyu Shi, Yarden Katz, Thorold W. Theunissen, Sudharshan Rangarajan, et al. 2013. "Multiplexed Activation of Endogenous Genes by CRISPR-on, an RNA-Guided Transcriptional Activator System." *Cell Research* 23 (10): 1163–71. doi:10.1038/cr.2013.122.

Cui, Yingbo, Jiaming Xu, Minxia Cheng, Xiangke Liao, and Shaoliang Peng. 2018. "Review of CRISPR/Cas9 sgRNA Design Tools." *Interdisciplinary Sciences: Computational Life Sciences* 10 (2): 455–65. doi:10.1007/s12539-018-0298-z.

Dang, Ying, Gengxiang Jia, Jennie Choi, Hongming Ma, Edgar Anaya, Chunting Ye, Premlata Shankar, et al. 2015. "Optimizing sgRNA Structure to Improve CRISPR-Cas9 Knockout Efficiency." *Genome Biology* 16 (1): 280. doi:10.1186/s13059-015-0846-3.

Elumalai, Perumal, and Devaraj Ezhilarasan. 2022. "Emerging Applications of CRISPR/Cas9 Gene Editing Technology in Reversing Drug Resistance in Oral Squamous Cell Carcinoma." *Oral Oncology* 134 (November): 106100. doi:10.1016/j.oraloncology.2022.106100.

Fajrial, Apresio K., Qing Q. He, Nurul I. Wirusanti, Jill E. Slansky, and Xiaoyun Ding. 2020. "A Review of Emerging Physical Transfection Methods for CRISPR/Cas9-Mediated Gene Editing." *Theranostics* 10 (12): 5532–49. doi:10.7150/thno.43465.

Gaj, Thomas, Shannon J. Sirk, Sai-lan Shui, and Jia Liu. 2016. "Genome-Editing Technologies: Principles and Applications." *Cold Spring Harbor Perspectives in Biology* 8 (12): a023754. doi:10.1101/cshperspect.a023754.

Geyer, Maximilian, Fabian Geyer, Ute Reuning, Sarah Klapproth, Klaus-Dietrich Wolff, and Markus Nieberler. 2024. "CRISPR/Cas9-Mediated Knock out of ITGB6 in Human OSCC Cells Reduced Migration and Proliferation Ability." *Head & Face Medicine* 20 (1): 37. doi:10.1186/s13005-024-00437-x.

Goldman, Samantha L., Matthew MacKay, Ebrahim Afshinnekoo, Ari M. Melnick, Shuxiu Wu, and Christopher E. Mason. 2019. "The Impact of Heterogeneity on Single-Cell Sequencing." *Frontiers in Genetics* 10 (March): 8. doi:10.3389/fgene.2019.00008.

Graham, Daniel B., and David E. Root. 2015. "Resources for the Design of CRISPR Gene Editing Experiments." *Genome Biology* 16 (1): 260. doi:10.1186/s13059-015-0823-x.

He, Siyu, Yuefei Zhu, Shradha Chauhan, Daniel N. Tavakol, Jong H. Lee, Rayna B.-L. Berris, Cong Xu, et al. 2024. "Human Vascular Organoids with a Mosaic *AKT1* Mutation Recapitulate Proteus Syndrome." *Bioengineering*. doi:10.1101/2024.01.26.577324.

Hernandez, Brenda Y., Xuemei Zhu, Marc T. Goodman, Robert Gatewood, Paul Mendiola, Katrina Quinata, and Yvette C. Paulino. 2017. "Betel Nut Chewing, Oral Premalignant Lesions, and the Oral Microbiome." edited by Brenda A Wilson. *PLoS ONE* 12 (2): e0172196. doi:10.1371/journal.pone.0172196.

Hou, Xucheng, Tal Zaks, Robert Langer, and Yizhou Dong. 2021. "Lipid Nanoparticles for mRNA Delivery." *Nature Reviews Materials* 6 (12): 1078–94. doi:10.1038/s41578-021-00358-0.

Hu, Yuhua, Feng Shen, Xi Yang, Tingting Han, Zhuowen Long, Jiale Wen, Junxing Huang, et al. 2023a. "Single-Cell Sequencing Technology Applied to Epigenetics for the Study of Tumor Heterogeneity." *Clinical Epigenetics* 15 (1): 161. doi:10.1186/s13148-023-01574-x.

Huang, Dezhi, Naya Ma, Xinlei Li, Yang Gou, Yishuo Duan, Bangdong Liu, Jing Xia, et al. 2023. "Advances in Single-Cell RNA Sequencing and Its Applications in Cancer Research." *Journal of Hematology & Oncology* 16 (1): 98. doi:10.1186/s13045-023-01494-6.

Jiang, Fuguo, and Jennifer A. Doudna. 2017. "CRISPR–Cas9 Structures and Mechanisms." *Annual Review of Biophysics* 46 (1): 505–29. doi:10.1146/annurev-biophys-062215-010822.

Jiang, Ziming, Haoxuan Zhang, Yibo Gao, and Yingli Sun. 2025. "Multi-Omics Strategies for Biomarker Discovery and Application in Personalized Oncology." *Molecular Biomedicine* 6 (1): 115. doi:10.1186/s43556-025-00340-0.

Krishnamurthy, Sangeetha, Rajendran Vaiyapuri, Liangfang Zhang, and Juliana M. Chan. 2015. "Lipid-Coated Polymeric Nanoparticles for Cancer Drug Delivery." *Biomaterials Science* 3 (7): 923–36. doi:10.1039/C4BM00427B.

Kunachowicz, Dominika, Paulina Tomecka, Mikołaj Sędzik, Jarosław Kalinin, Jacek Kuźnicki, and Nina Rembiałkowska. 2025. "Influence of Hypoxia on Tumor Heterogeneity, DNA Repair, and Cancer Therapy: From Molecular Insights to Therapeutic Strategies." *Cells* 14 (14): 1057. doi:10.3390/cells14141057.

Lane-Reticker, Sarah Kate, Emily A. Kessler, Audrey J. Muscato, Sarah Y. Kim, John G. Doench, Kathleen B. Yates, et al. 2023. "Protocol for in Vivo CRISPR Screening Using Selective CRISPR Antigen Removal Lentiviral Vectors." *STAR Protocols* 4 (1): 102082. doi:10.1016/j.xpro.2023.102082.

Lee, Minhyeok. 2023. "Deep Learning in CRISPR-Cas Systems: A Review of Recent Studies." *Frontiers in Bioengineering and Biotechnology* 11 (July): 1226182. doi:10.3389/fbioe.2023.1226182.

Lee, Su-Hyeon, Junha Park, and Byungjin Hwang. 2024. "Multiplexed Multimodal Single-Cell Technologies: From Observation to Perturbation Analysis." *Molecules and Cells* 47 (12): 100147. doi:10.1016/j.mocell.2024.100147.

Lenkiewicz, Anna M., and Bozena Kaminska. 2025. "Application of Cellular Indexing of Transcriptomes and Epitopes by Sequencing (CITE-Seq) to Precisely Dissect the Brain Tumor Immune Microenvironment." In *Cancer Immunosurveillance*, edited by Alejandro López-Soto and Alicia R. Folgueras, 2930:31–43. Methods in Molecular Biology. New York, NY: Springer US. doi:10.1007/978-1-0716–4558-1_3.

Li, Jinxin, Jiatao Hu, Yiren Yang, Hanzhong Zhang, Ying Liu, Yu Fang, Le Qu, et al. 2025. "Drug Resistance in Cancer: Molecular Mechanisms and Emerging Treatment Strategies." *Molecular Biomedicine* 6 (1): 111. doi:10.1186/s43556-025-00352-w.

Li, Ya, Weifeng Zhang, Junli Zhao, Sai Li, Linlin Shan, Jiuling Zhu, Yan Li, et al. 2018. "Establishing a Dual Knock-out Cell Line by Lentivirus Based Combined CRISPR/Cas9 and Loxp/Cre System." *Cytotechnology* 70 (6): 1595–1605. doi:10.1007/s10616-018-0252-2.

Liu, Chang, Li Zhang, Hao Liu, and Kun Cheng. 2017. "Delivery Strategies of the CRISPR-Cas9 Gene-Editing System for Therapeutic Applications." *Journal of Controlled Release* 266 (November): 17–26. doi:10.1016/j.jconrel.2017.09.012.

Liu, Dongqi, Di Cao, and Renzhi Han. 2025. "Recent Advances in Therapeutic Gene-Editing Technologies." *Molecular Therapy* 33 (6): 2619–44. doi:10.1016/j.ymthe.2025.03.026.

Mahat, Dig B., Nathaniel D. Tippens, Jorge D. Martin-Rufino, Sean K. Waterton, Jiayu Fu, Sarah E. Blatt, and Phillip A. Sharp. 2024. "Single-Cell Nascent RNA Sequencing Unveils Coordinated Global Transcription." *Nature* 631 (8019): 216–23. doi:10.1038/s41586-024-07517-7.

McKenzie, Rebecca E, Emma M. Keizer, Jochem N. A. Vink, Jasper Van Lopik, Ferhat Büke, Vera Kalkman, Christian Fleck, et al. 2022. "Single Cell Variability of CRISPR-Cas Interference and Adaptation." *Molecular Systems Biology* 18 (4): e10680. doi:10.15252/msb.202110680.

McTague, Amy, Giada Rossignoli, Arianna Ferrini, Serena Barral, and Manju A. Kurian. 2021. "Genome Editing in iPSC-Based Neural Systems: From Disease Models to Future Therapeutic Strategies." *Frontiers in Genome Editing* 3 (March): 630600. doi:10.3389/fgeed.2021.630600.

Menon, A. Vipin, Bicna Song, Lumen Chao, Diksha Sriram, Pamela Chansky, Ishnoor Bakshi, Jane Ulianova, and Wei Li. 2025. "Unraveling the Future of Genomics: CRISPR, Single-Cell Omics, and the Applications in Cancer and Immunology." *Frontiers in Genome Editing* 7 (April): 1565387. doi:10.3389/fgeed.2025.1565387.

Onan, Deniz, Melike Özder, Meryem İ. Sipahi, Nazlıcan Poyraz, Ceylin Apaydın, Gülşah Erel-Akbaba, and Hasan Akbaba. 2025a. "Microfluidics Based Particle and Droplet Generation for Gene and Drug Delivery Approaches." *Journal of Biomedical Materials Research Part B: Applied Biomaterials* 113 (2): e35530. doi:10.1002/jbm.b.35530.

Park, Insu, Seungyeop Choi, Youngwoo Gwak, Jingwon Kim, Gyeongjun Min, Danyou Lim, and Sang Woo Lee. 2024. "Microfluidic Electroporation Arrays for Investigating Electroporation-Induced Cellular Rupture Dynamics." *Biosensors* 14 (5): 242. doi:10.3390/bios14050242.

Quan, Mingji, Yumi Oh, Sung-Yup Cho, Ju H. Kim, and Hyeong-Gon Moon. 2022. "Polo-Like Kinase 1 Regulates Chromosomal Instability and Paclitaxel Resistance in Breast Cancer Cells." *Journal of Breast Cancer* 25 (3): 178. doi:10.4048/jbc.2022.25.e28.

Raigani, Mozhghan, Zohre Eftekhari, Ahmad Adeli, and Fatemeh Kazemi-Lomedasht. 2025. "Advancing Gene Editing Therapeutics: Clinical Trials and Innovative Delivery Systems across Diverse Diseases." *Molecular Therapy Nucleic Acids* 36 (3): 102666. doi:10.1016/j.omtn.2025.102666.

Raikwar, Sudhanshu P., Nidhi S. Kikkeri, Ragha Sakuru, Daniyal Saeed, Haris Zahoor, Keerthivaas Premkumar, Shireen Mentor, et al. 2019. "Next Generation Precision Medicine: CRISPR-Mediated Genome Editing for the Treatment of Neurodegenerative Disorders." *Journal of Neuroimmune Pharmacology* 14 (4): 608–41. doi:10.1007/s11481-019-09849-y.

Raj, Bushra, James A. Gagnon, and Alexander F. Schier. 2018. "Large-Scale Reconstruction of Cell Lineages Using Single-Cell Readout of Transcriptomes and CRISPR–Cas9 Barcodes by scGESTALT." *Nature Protocols* 13 (11): 2685–2713. doi:10.1038/s41596-018-0058-x.

Rosenblum, Daniel, Anna Gutkin, Ranit Kedmi, Srinivas Ramishetti, Nuphar Veiga, Ashley M. Jacobi, Mollie S. Schubert, et al. 2020. "CRISPR-Cas9 Genome Editing Using Targeted Lipid Nanoparticles for Cancer Therapy." *Science Advances* 6 (47): eabc9450. doi:10.1126/sciadv.abc9450.

Schmoldt, A., Benthe, H. F., & Haberland, G. 1975. Digitoxin metabolism by rat liver microsomes. *Biochemical Pharmacology*, 24(17): 1639–1641. https://doi.org/10.1016/0006-2952(75)90094-5.

Sherkatghanad, Zeinab, Moloud Abdar, Jeremy Charlier, and Vladimir Makarenkov. 2023. "Using Traditional Machine Learning and Deep Learning Methods for On- and off-Target Prediction in CRISPR/Cas9: A Review." *Briefings in Bioinformatics* 24 (3): bbad131. doi:10.1093/bib/bbad131.

Singh, Rohan, Hamid E. Orimi, Praveen K. R. Pedabaliyarasimhuni, Corinne A. Hoesli, and Moncef Chioua. 2025. "AI-Driven Quality Monitoring and Control in Stem Cell Cultures: A Comprehensive Review." *Biotechnology Journal* 20 (8): e70100. doi:10.1002/biot.70100.

Song, Hye-Wong, Jody Martin, Xiaoshan Shi, and Aaron J. Tyznik. 2025. "Key Considerations on CITE-Seq for Single-Cell Multiomics." *Proteomics* 25 (21–22): 206–13. doi:10.1002/pmic.202400011.

Soni, Jyoti, and Rajesh Pandey. 2024. "Single Cell Genomics Based Insights into the Impact of Cell-Type Specific Microbial Internalization on Disease Severity." *Frontiers in Immunology* 15 (May): 1401320. doi:10.3389/fimmu.2024.1401320.

Sowmya, S. V., Dominic Augustine, Shazia Mushtaq, Hosam A. Baeshen, Heba Ashi, Reem N. Hassan, Mohammed Alshahrani, and Shankargouda Patil. 2024. "Revitalizing Oral Cancer Research: Crispr-Cas9 Technology the Promise of Genetic Editing." *Frontiers in Oncology* 14 (June): 1383062. doi:10.3389/fonc.2024.1383062.

Tahsin, Anika, Zarin Tasnim, Muneera Chowdhury, Kangkhita Hassin, Galib H. Meraz, Kazi F. Aziz, and Swakkhar Shatabda. 2025. "CRISPR-Embedding: CRISPR/Cas9 off-Target Activity Prediction Using DNA k -Mer Embedding." *Computational and Structural Biotechnology Reports* 2: 100043. doi:10.1016/j.csbr.2025.100043.

Tan, Swee J., Qingsen Li, and Chwee T. Lim. 2010. "Manipulation and Isolation of Single Cells and Nuclei." *Methods in Cell Biology* 98:79–96. doi:10.1016/S0091–679X(10)98004-7.

Tornesello, Maria. 2024. "TP53 Mutations in Cancer: Molecular Features and Therapeutic Opportunities (Review)." *International Journal of Molecular Medicine* 55 (1): 7. doi:10.3892/ijmm.2024.5448.

Wang, Tingjie, Ningxin Dang, Guangbo Tang, Zihang Li, Xiujuan Li, Bingyin Shi, Zhong Xu, et al. 2022a. "Integrating Bulk and Single-cell RNA Sequencing Reveals Cellular Heterogeneity and Immune Infiltration in Hepatocellular Carcinoma." *Molecular Oncology* 16 (11): 2195–2213. doi:10.1002/1878-0261.13190.

Wang, Zeyu, Shasha Zhao, Xiaolin Lin, Guanglong Chen, Jiawei Kang, Zhongping Ma, Yiming Wang, et al. 2022b. "Application of Organoids in Carcinogenesis Modeling and Tumor Vaccination." *Frontiers in Oncology* 12 (March): 855996. doi:10.3389/fonc.2022.855996.

Worthington, Atesh K., and E. Camilla Forsberg. 2022a. "A CRISPR View of Hematopoietic Stem Cells: Moving Innovative Bioengineering into the Clinic." *American Journal of Hematology* 97 (9): 1226–35. doi:10.1002/ajh.26588.

Wu, Xiangyu, Xin Yang, Yunhan Dai, Zihan Zhao, Junmeng Zhu, Hongqian Guo, and Rong Yang. 2024. "Single-Cell Sequencing to Multi-Omics: Technologies and Applications." *Biomarker Research* 12 (1): 110. doi:10.1186/s40364-024-00643-4.

Xu, Yanxin, Chen Chen, Yaxin Guo, Shengyun Hu, and Zhenqiang Sun. 2022. "Effect of CRISPR/Cas9-Edited PD-1/PD-L1 on Tumor Immunity and Immunotherapy." *Frontiers in Immunology* 13 (March): 848327. doi:10.3389/fimmu.2022.848327.

Xue, Feng, Hai Yang, Pengyan Xu, Shuman Zhang, Nathalie Britzen-Laurent, Li-Li Bao, Robert Grützmann, et al. 2025. "CRISPR/Cas9 Screening Highlights PFKFB3 Gene as a Major Contributor to 5-Fluorouracil Resistance in Esophageal Cancer." *Cancers* 17 (10): 1637. doi:10.3390/cancers17101637.

Zafar, Hamim, Chieh Lin, and Ziv Bar-Joseph. 2020. "Single-Cell Lineage Tracing by Integrating CRISPR-Cas9 Mutations with Transcriptomic Data." *Nature Communications* 11 (1): 3055. doi:10.1038/s41467-020-16821-5.

Zhang, Jie, Lei Wang, Qianqian Song, Mingbing Xiao, Jie Gao, Xiaolei Cao, and Wenjie Zheng. 2022. "Organoids in Recapitulating Tumorigenesis Driven by Risk Factors: Current Trends and Future Perspectives." *International Journal of Biological Sciences* 18 (7): 2729–43. doi:10.7150/ijbs.70406.

Zhang, Tianjiao, Zhongqian Zhao, Hongfei Zhang, Zhenao Wu, Fang Wang, and Guohua Wang. 2025. "Multi-Omics Single-Cell Data Alignment and Integration with Enhanced Contrastive Learning and Differential Attention Mechanism." edited by Laura Cantini. *Bioinformatics* 41 (8): btaf443. doi:10.1093/bioinformatics/btaf443.

Zheng, Yue, and Xiaofeng Yang. 2023. "Spatial RNA Sequencing Methods Show High Resolution of Single Cell in Cancer Metastasis and the Formation of Tumor Microenvironment." *Bioscience Reports* 43 (2): BSR20221680. doi:10.1042/BSR20221680.

Zong, Yan, Yi Lin, Tuo Wei, and Qiang Cheng. 2023. "Lipid Nanoparticle (LNP) Enables mRNA Delivery for Cancer Therapy." *Advanced Materials* 35 (51): 2303261. doi:10.1002/adma.202303261.

Zou, Bin, Tongda Zhang, Ruilong Zhou, Xiaosen Jiang, Huanming Yang, Xin Jin, and Yong Bai. 2021. "deep-MNN: Deep Learning-Based Single-Cell RNA Sequencing Data Batch Correction Using Mutual Nearest Neighbors." *Frontiers in Genetics* 12 (August): 708981. doi:10.3389/fgene.2021.708981.

13 Computational Tools and AI in Genome Editing Design

Sujaan Aehsas and Prabhudatta Behera

13.1 INTRODUCTION

The field of genome editing (GED) has undergone a profound transformation, moving from laborious, low-throughput methods to precise, programmable systems now dominated by CRISPR (clustered regularly interspaced short palindromic repeat)-based technologies (Doudna & Charpentier, 2014). This revolutionary shift has unlocked unprecedented avenues for understanding fundamental gene function and developing targeted therapeutic interventions for a myriad of human diseases (Jinek et al., 2012). However, the sheer volume and complexity of genomic data, combined with the need to accurately predict the on- and off-target effects of guide RNAs (gRNAs), demand computational power that traditional bioinformatics methods often cannot supply. This challenge has necessitated the integration of artificial intelligence (AI), particularly machine learning (ML) and deep learning (DL), to drive the next wave of innovation in GED. The convergence of AI and CRISPR is pivotal, as it moves the process from being largely empirical to being data-driven and predictive. AI models are employed to maximize various critical aspects of the GED process, leading to the development of sophisticated algorithms capable of optimizing gRNA design, evaluating the efficacy of different editing modalities, and ultimately ensuring the safety and precision required for clinical applications.

The promise of this convergence is the realization of precision medicine. By leveraging AI systems to scan a patient's unique genomic information, researchers can pinpoint precise mutations, variations, and biomarkers, enabling the customization of treatment strategies (Wang et al., 2021). This chapter will explore how AI is becoming an indispensable component in empowering the accuracy, speed, and cost-effectiveness of gene editing pipelines. From optimizing the core CRISPR-Cas9 system to enhancing newer technologies like base (BE) and prime editing (PE), AI is fundamentally changing the landscape of genetic engineering, making precise and programmable DNA changes a reality. Nevertheless, as we will discuss below, this rapid advancement is accompanied by significant technical, ethical, and logistical hurdles that must be addressed to fully realize the therapeutic potential of this technology.

13.2 OVERVIEW OF GENOME EDITING

Genome editing involves the targeted modification of an organism's genetic material. Historically, this has involved techniques like zinc finger nucleases and transcription activator-like effector nucleases, which rely on protein-DNA recognition domains (Gaj et al., 2013). Although effective, the laborious and costly process of engineering new protein domains for every target sequence severely limited their widespread adoption. The advent of the CRISPR-Cas9 system changed this paradigm. Derived from a bacterial adaptive immune system, this technology relies on a simple, easily programmable gRNA to direct the Cas9 nuclease to a specific genomic locus, where it induces a double-strand break (DSB) (Jinek et al., 2012). The cell's endogenous DNA repair mechanisms then attempt to fix the break, typically resulting in non-homologous end joining, which causes gene disruption, or homology-directed repair (HDR), which allows for precise sequence insertion or correction using a donor template (Hsu et al., 2013).

 DOI: 10.1201/9781003735113-13

The simplicity of gRNA design has made CRISPR-Cas9 the dominant GED tool, fostering a rapid expansion in both basic research and therapeutic development. Building upon the canonical Cas9 system, researchers have developed alternative editing modalities to overcome limitations such as dependence on DSBs and the low efficiency of HDR in certain cell types. These are as follows:

- **Base Editing (BE)**: This system uses a catalytically impaired Cas9 (nickase) fused to a deaminase enzyme, enabling the direct, irreversible conversion of one base pair to another (e.g., C>T or A>G) without generating a DSB (Gaudelli et al., 2017).
- **Prime Editing (PE)**: Considered an "all-in-one" system, PE employs a Cas9 nickase fused to a reverse transcriptase and uses an extended gRNA, or pegRNA, to directly template the synthesis of new DNA, allowing for all 12 possible base-to-base conversions, as well as targeted insertions and deletions, all without a DSB (Anzalone et al., 2019).
- **Epigenome Editing**: This technique uses catalytically dead Cas9 (dCas9) fused to epigenetic effector domains (e.g., methyltransferases or acetyltransferases) to regulate gene expression without altering the underlying DNA sequence (Hilton et al., 2015).

While these advancements dramatically increase the versatility of GED, they simultaneously multiply the number of design variables and potential off-target liabilities. The success of any GED experiment is fundamentally dependent on designing an editing agent that is both highly efficient (on-target activity) and highly specific (minimal off-target activity). This is where computational and AI-driven approaches become indispensable, providing the necessary predictive power to navigate this complex design space (Doench et al., 2014).

13.3 COMPUTATIONAL APPROACHES IN GENOME EDITING DESIGN

The design of effective genome editing components, particularly the single-guide RNA (sgRNA) for CRISPR-Cas9, is a complex search problem within the vast expanse of a host genome. Computational methods are essential for identifying optimal target sites, predicting editing outcomes, and minimizing unintended edits. Initially, computational approaches relied on rule-based systems and basic bioinformatics algorithms. These methods primarily filtered potential target sites based on sequence context, such as the presence of the required protospacer adjacent motif, and then calculated the potential for off-target binding by searching for sequences with few mismatches to the sgRNA (Hsu et al., 2013). While foundational, these heuristic models often failed to capture the intricate biological and epigenetic factors that govern *in vivo* editing efficiency and specificity. The complexity of target recognition which involves chromatin accessibility, DNA methylation, and the specific cell type necessitated a move toward data-driven ML models. ML algorithms are trained on large, experimentally validated datasets of sgRNA activity and off-target cleavage rates. These models move beyond simple sequence matching to weigh a variety of features that influence editing outcomes (Beak et al., 2021). Key features engineered for these models are as follows:

- **Sequence-Based Features:** Nucleotide composition, position-specific mismatch penalties, and the thermodynamic stability of the RNA:DNA duplex (e.g., minimum free energy calculations).
- **Epigenetic Features:** Chromatin accessibility (e.g., ATAC-seq, DNase-seq data), histone modification marks (e.g., ChIP-seq data), and DNA methylation status.
- **Structural Features:** Local DNA folding energy and secondary structure of the sgRNA itself.

The application of DL, a subfield of ML that uses neural networks with multiple layers, has further refined prediction accuracy. DL models, such as convolutional neural networks (CNNs) and recurrent neural networks (RNNs), have the advantage of learning complex, nonlinear relationships and extracting highly descriptive features directly from the raw DNA and RNA sequences without manual feature engineering (Rafiq et al., 2025). This capability has led to state-of-the-art tools for predicting gRNA efficiency and specificity (Chuai et al., 2018).

This integration of sophisticated computational and learning-based models is not limited to Cas9-mediated cleavage. For newer editors, bespoke computational frameworks are equally critical:

- **Base Editing:** Tools like BE-Hive employ ML models to predict BE efficiency and the pattern of "bystander" mutations (unintended edits in the editing window), accounting for the deaminase's sequence preferences.
- **Prime Editing:** Algorithms like Easy-Prime (Kwon et al., 2021) and Prime Design (Haeussler et al., 2020) are necessary to optimize the complex design parameters of the pegRNA, which include the primer binding site (PBS) and the reverse transcriptase template length, a task infeasible by manual means.

13.4 DATABASES AND BIOINFORMATICS PLATFORMS

The success of any data-driven computational approach in genome editing is intrinsically linked to the quality, quantity, and accessibility of the underlying data. Databases and bioinformatics platforms serve as the essential infrastructure for collecting, standardizing, and presenting the vast experimental data required to train and validate robust AI models (Rafiq et al., 2025). The development of these resources addresses a fundamental challenge: experimental data across different laboratories, cell types, and editing platforms often show high variability (Hsu et al., 2013). A centralized, curated repository is necessary to overcome this fragmentation and enable the training of models that are generalizable across different biological contexts.

13.4.1 Essential Databases for Genome Editing

- **CRISPR-Cas Activity Datasets:** These specialized databases compile experimentally determined on-target cleavage efficiency and off-target indel frequencies for thousands of gRNAs. Examples include datasets from high-throughput screening experiments like the *GeCKO* or *Doench* libraries. The compilation of these diverse datasets is crucial for training predictive models like DeepCRISPR (Chuai et al., 2018).
- **BE Data:** The complexity of BE outcomes, which involve efficiency and bystander rates, necessitates specialized data. BE-dataHIVE, for instance, is a comprehensive database aggregating BE data across multiple studies, enriched with thermodynamic and energy terms for advanced ML model training (Schneider & Minary, 2024).
- **Epigenomic Data:** Since chromatin accessibility is a key determinant of Cas9 activity, databases like the Encyclopedia of DNA Elements (ENCODE) and the Roadmap Epigenomics Project are foundational resources (Wang et al., 2021).

13.4.2 Bioinformatics Platforms and Tools

Beyond raw data, bioinformatics platforms provide the user-friendly interface for researchers to execute computational analyses. These platforms integrate the predictive models developed using the databases, making advanced AI tools accessible to a wider biological community:

- **gRNA Design Tools:** Platforms like CRISPOR (Concordet & Haeussler, 2018) and GuideScan (Perez et al., 2017) are widely used for target site selection. They leverage integrated databases and ML models to score candidate gRNAs based on predicted efficiency and potential off-targets.

- **Variant-to-Edit Mapping:** For precision medicine applications, platforms are being developed to rapidly analyze patient genomic data, identify pathogenic variants, and suggest optimal editing strategies (Wang et al., 2021). This involves mapping disease-causing mutations to the nearest editable target site, factoring in the desired therapeutic outcome.
- **Specialized Editing Calculators:** Tools like CRISTA (CRISPR Target Assessment), DeepHF (Deep High-Fidelity), and the aforementioned BE-Hive provide specialized calculations for specific editor variants (e.g., high-fidelity Cas9 variants or different base editors), allowing users to quickly assess and compare their performance *in silico*.

The long-term success of AI in genome editing requires an ongoing, collaborative effort to create an exhaustive, centralized database of CRISPR outcomes, ensuring that data is FAIR (Findable, Accessible, Interoperable, and Reusable) to continuously enhance predictive model performance and drive collaborative innovation (Rafiq et al., 2025).

13.5 MACHINE LEARNING AND DEEP LEARNING FOR GRNA OPTIMIZATION

The greatest bottleneck in CRISPR-based genome editing lies in the reliable optimization of the gRNA for maximum on-target efficiency and minimal off-target effects. This is a classic prediction problem ideally suited for ML and DL methodologies.

13.5.1 Optimizing On-Target Efficiency

Predicting on-target efficacy requires understanding how the gRNA sequence, its secondary structure, and the local chromatin environment influence Cas9's ability to bind and cleave DNA. Early ML models, such as Support Vector Regression and Random Forest, established the initial predictive framework by analyzing empirically derived features (Doench et al., 2016). However, the complex sequence-context dependencies were best captured by DL.

- **DeepCRISPR:** This comprehensive computational platform unified the prediction of sgRNA on-target knockout efficacy and off-target profiles using a Deep Convolutional De-noising Neural Network (Chuai et al., 2018). By automatically extracting optimal sequence and epigenetic features, it significantly surpassed earlier state-of-the-art tools, demonstrating the power of DL to uncover subtle predictive patterns.
- **CRISTA (CRISPR Target Assessment):** Another DL-based tool, CRISTA, focuses on predicting the activities of different Cas9 variants, helping researchers select the most suitable editor for their specific target site.
- **DeepHF:** This model was specifically developed to optimize the design of high-fidelity Cas9 variants, aiming to maintain high on-target activity while eliminating off-target cleavage (Wang et al., 2025).

13.5.2 Enhancing Specificity and Fidelity

Off-target effects—where the gRNA directs the Cas protein to unintended genomic sites—are a primary safety concern for therapeutic applications. DL models excel at predicting these liabilities:

- **Off-target Scoring:** The prediction of off-target effects involves modeling the tolerance of the Cas9 enzyme to mismatches between the gRNA and the target DNA (Fu et al., 2013). DL models are trained on massive datasets of verified off-target sites, allowing them to model the complex relationship between mismatch number, position, and the resulting cleavage frequency with high accuracy (Rafiq et al., 2025).

TABLE 13.1
Gut Microbiota Composition and AI-Based Computational Insights

Bacteria Type	Role in Gut Health	Computational Insights (AI/ML)
Beneficial bacteria	Support digestion, produce SCFAs, protect gut lining, and compete with pathogens (Hernández Medina et al., 2022).	• ML classifiers identify beneficial-dominant "healthy microbiome signatures." • Autoencoders reveal hidden microbial patterns linked to metabolic health.
Conditional pathogenic bacteria	Normally harmless but can cause disease when overgrown or triggered by stress or antibiotics (Hernández Medina et al., 2022).	• AI models detect abnormal increases of opportunists using anomaly detection. • ML distinguishes harmful vs. harmless strains using genomic features.
Pathogenic bacteria	Cause infections, inflammation, and toxin-mediated damage (Hernández Medina et al., 2022).	• Metagenomic ML tools rapidly identify pathogen genes in gut data. • Predictive models estimate infection risk based on microbiome shifts.

13.5.3 AI in Advanced Editing Systems

ML and DL are equally transformative in optimizing the newer, more complex editing platforms:

- **BE Optimization:** For base editors, models like BE-Hive and DeepCBE/DeepABE (Deep Cytosine/Adenine Base Editing) use ML to predict the efficiency of base conversion and the critical rate of bystander mutations, which is dependent on the deaminase enzyme's activity window and sequence preferences.
- **PE Optimization:** PE design is challenging due to the numerous variables in the pegRNA, including the length of the PBS and the reverse transcription template (RTT). Tools like PrimeDesign and Easy-Prime (Kwon et al., 2021) leverage predictive models trained on high-throughput screens to rapidly identify pegRNA designs that maximize editing and minimize by-products (Anzalone et al., 2019).

The transition from heuristic-based scoring to sophisticated ML-based and DL-based algorithms has been instrumental in moving genome editing design from a trial-and-error process to a precise, predictive discipline. AI is turning out to be the crucial element in ensuring that targeted gene modifications are highly specific and effective, which is a prerequisite for moving these powerful tools from the lab bench into the clinic (Wang et al., 2021).

Computational tools and AI are transforming genome editing by enabling more precise, efficient, and scalable design of gene editing systems. AI-driven approaches accelerate protein engineering and facilitate the discovery and optimization of novel Cas variants, thus broadening the applications of genome editing in medicine, agriculture, and synthetic biology (Ali et al., 2024). Yet, such advances also introduce significant challenges, ethical boundaries, and technical limitations that call for responsible innovation. The classification of gut bacteria and their roles in gut health, with AI/ML-based computational insights, is presented in Table 13.1.

13.6 AI-DRIVEN PROTEIN ENGINEERING AND CAS VARIANT DESIGN

AI is transforming protein engineering by computationally modeling the complex relationship between protein sequences and their functions. These ML and DL models can be used for predictions about how changes in amino acid sequences will affect protein structure, stability, and activity (Gao et al., 2020). This capability is particularly useful in engineering Cas proteins, the

TABLE 13.2
Protein Structural Levels and AI-Based Computational Insights

Level	Structural Features (Short)	Computational Insights (AI/ML)
Primary structure	Linear amino acid sequence determining protein identity and downstream folding (Meng et al., 2025).	• AI sequence models (e.g., protein language models) extract structural patterns directly from raw sequences. • Deep learning improves motif detection and functional site prediction.
Secondary structure	Local backbone folding into α-helices and β-sheets through hydrogen bonding (Meng et al., 2025).	• CNNs, RNNs, and transformers accurately classify helix/sheet regions. • AI captures long-range sequence patterns influencing local folding.
Tertiary structure	Full 3D conformation of a single polypeptide driven by long-range interactions (Meng et al., 2025).	• AlphaFold2 and RoseTTAFold use deep neural networks to predict near-experimental quality 3D structures (Jumper et al., 2021). • Hybrid physics–AI tools refine structural models and side-chain positioning.
Quaternary structure	Assembly of multiple folded subunits into functional complexes (Jumper et al., 2021)	• AI multimer models predict protein–protein interfaces and complex assembly. • ML-based docking tools assess subunit fit and interaction stability.

core enzymes in the CRISPR systems, through identification of mutations that may enhance specificity, efficiency, or expand the range of targetable DNA sequences (Liu et al., 2014). For instance, AI models can rapidly screen millions of potential Cas variants to identify variants displaying improved on-target activities or diminished off-target effects, hence reducing the time required for experimental validation by many orders of magnitude. Generative AI, including the most recent class of protein large language models, pLLMs, can even design completely synthetic Cas proteins or transposases with novel functions (Franciss et al., 2025).

These models learn from huge databases of natural protein sequences to suggest new variants, which may confer better activity, specificity, or delivery system compatibility compared to naturally occurring enzymes. Using such approaches, both synthetic PiggyBac transposases and Cas variants displaying an improved capacity for genome integration or editing were identified, thereby expanding therapeutic and biotechnological options. The protein structural levels and AI-based computational insights are presented in Table 13.2.

13.7 APPLICATIONS IN MEDICINE, AGRICULTURE, AND SYNTHETIC BIOLOGY

13.7.1 Medicine

AI-driven genome editing accelerates gene therapy development for monogenic and polygenic diseases (Srivastav et al., 2025). With the help of AI, highly specific correction of disease-causing mutations has been made possible by optimizing gRNA design and Cas protein variants, having a minimum number of off-target effects. This is important for disorders such as sickle cell anemia, cystic fibrosis, and some cancers. AI also aids in designing personalized therapies by analyzing genomic data from patients to tailor editing strategies, improving therapeutic outcomes and safety.

13.7.2 Agriculture

AI-powered genome editing in agriculture is used to engineer crops for better yield, disease resistance, and nutritional value (Bradbury et al., 2025). AI helps breeders create crops resilient to changes in climate and pests by predicting how genetic modifications will impact plant phenotypes.

For instance, AI models can optimize CRISPR-Cas9 systems with the aim of precision editing of genes responsible for drought tolerance or nutrient uptake, accelerating the development of sustainable agricultural solutions (Liu et al., 2024).

13.7.3 Synthetic Biology

AI is a cornerstone of synthetic biology that enables novel genetic circuit designs, metabolic pathways, and the engineering of organisms (Devi et al., 2025). Large-scale integration of genomic information with functional data allows AI models to predict the consequences of genetic modifications and inform the construction of synthetic systems with predictable properties. This capability is crucial for applications in biofuel production, bioremediation, and the development of microbes with industrial applications.

13.8 CHALLENGES AND LIMITATIONS

Despite the transformative potential of AI-driven genome editing, there are several challenges to be considered. Among these, high quality and diversity are key limitations to training data. AI models are only as good as the data they were trained on, and biases or gaps in genomic datasets lead to incorrect predictions or the continuation of health disparities. Besides, some AI algorithms are not easily interpretable, since the internal structure is a "black box," which raises concerns regarding transparency and accountability (Hassija et al., 2024). Technical challenges include the high computational cost of training AI models and the requirement for large-scale experimental validation of AI-predicted variants. Off-target effects, delivery efficiency, and the complexity of *in vivo* editing remain significant hurdles. Finally, the rapidity with which AI-driven innovation is occurring can outpace regulatory frameworks on the risks of misuse or unintended consequences (Farooq et al., 2025).

13.8.1 Ethical Boundaries

AI and genome editing together raise profound ethical queries. Concretely, concerns include the potential for untargeted genetic alterations, potentially increasing health disparities, and the risk of creating a "genetic underclass" if gene therapies are accessible only to the rich. Ensuring data privacy and security is critical, as genomic data is highly sensitive and can reveal information about an individual and their family. Stronger regulatory frameworks and international cooperation will be necessary to set a code of ethical conduct and avoid fragmentation in regulatory regimes. Transparency, accountability, and inclusiveness are needed at the heart of AI governance related to genome editing. Public debate and interdisciplinary collaboration will be required to weigh scientific progress against ethical and social justice considerations (Kashefi et al., 2024).

13.9 FUTURE DIRECTIONS AND EMERGING TRENDS

Several emerging trends characterize the future of AI-driven genome editing. The development of the virtual cell model powered by AI will, therefore, be able to simulate the functional outcome of genetic modification and guide target selection or predict functional outcomes. These models allow highly accurate and efficient design of genome editing strategies that reduce costly and time-consuming experiments. Other trends involve the integration of AI with high-throughput screening and functional genomics to identify and validate new Cas variants and editing tools rapidly. Improvements in protein language models and generative AI will further extend the synthetic biology toolbox, including the design of completely new enzymes and genetic circuits. Finally, the fusion of AI, CRISPR, and precision medicine will open a way to personalized therapies according to genetic profiles. The integration of the three technologies will revolutionize healthcare, agriculture, and biotechnology to explore more solutions for complex challenges in human health and sustainability.

13.10 CONCLUSION

Computational tools and AI are bringing about a paradigm shift in genome editing, allowing unprecedented precision, efficiency, and scalability in the design of gene editing systems. AI-powered protein engineering and Cas variant design will continue to accelerate the development of new therapies, crops, and synthetic organisms, which have widespread applications in medicine, agriculture, and synthetic biology. Yet these advances need to be accompanied by robust ethical frameworks, transparent governance, and interdisciplinary collaboration to ensure responsible innovation and equitable access. As AI continues to evolve, its integration with genome editing will unlock game-changing advances that will reshape both the future of biotechnology and human health.

REFERENCES

Ali, M., Shabbir, K., Ali, S., Mohsin, M., Kumar, A., Aziz, M., … Sultan, H. M. (2024). A new era of discovery: How artificial intelligence has revolutionized biotechnology. *Nepal Journal of Biotechnology,* 12(1), 1–11.

Anzalone, A. V., Randolph, P. G., Davis, J. R., Sousa, A. A., Koblan, L. W., Levy, J. M., Chen, P. J., Wilson, C., Newby, E. T., Maianti, A. L., Liu, D. R. (2019). Search-and-replace genome editing without double-strand breaks or donor DNA. *Nature*, 576(7785), 149–157. https://doi.org/10.1038/s41586-019-1711-4

Beak, J. Y., Han, J. S., Lee, M., Ghang, B., Park, S. S., Lim, H. S., & Kim, B. S. (2021). CRISTA: A machine learning-based guide RNA scoring tool for CRISPR/Cas9 systems. *Bioinformatics*, 37(10), 1362–1369. https://doi.org/10.1093/bioinformatics/btaa919

Bradbury, A., Clapp, O., Biacsi, A. S., Kuo, P., Gaju, O., Hayta, S., et al. (2025). Integrating genome editing with omics, artificial intelligence, and advanced farming technologies to increase crop productivity. *Plant Communications,* 6(7), 101386. https://doi.org/10.1016/j.xplc.2025.101386

Chuai, G., Ma, H., Yan, J., Chen, M., Hong, M., Xue, D., Li, Y., Zhu, C., Huang, Y., & Liu, Q. (2018). DeepCRISPR: Optimized CRISPR guide RNA design by deep learning. *Genome Biology*, 19(1), 80. https://doi.org/10.1186/s13059-018-1459-4

Concordet, J. P., & Haeussler, M. (2018). CRISPOR: An automated design tool for CRISPR/Cas9, CRISPR/Cpf1 and CRISPR-CBE guide RNAs. *Genome Biology*, 19(1), 215. https://doi.org/10.1186/s13059-018-1549-3

Devi, N. B., Sen, S., & Pakshirajan, K. (2025). Artificial intelligence in synthetic biology. In P.V. Mohanan (Ed.), *Artificial intelligence and biological sciences* (pp. 278–300). CRC Press.

Doench, J. G., Hartenian, E., Graham, D. B., Tothova, Z., Hegde, M., Smith, I., Sullender, M., Ebert, B. L., Houghton, R., & Lander, E. S. (2016). Optimized sgRNA design to maximize activity and minimize off-target effects of CRISPR-Cas9. *Nature Biotechnology*, 34(2), 184–191. https://doi.org/10.1038/nbt.3437

Doudna, J. A., & Charpentier, E. (2014). The new frontier of genome engineering with CRISPR-Cas9. *Science*, 346(6213), 1258096. https://doi.org/10.1126/science.1258096

Farooq, M. U., Lawrie, C. H., & Deng, N. N. (2025). Solving mazes of organelle-targeted therapies with DNA nanomachines. *Advanced Materials,* 37(39), 2508047. https://doi.org/10.1002/adma.202508047

Franciss, N. L. (2025). Generative artificial intelligence in genetics: A comprehensive review. *Deep Learning in Genetics and Genomics*, 2, 237–247.

Fu, Y., Foden, J. A., Ma, W., Lu, Y., Stamnes, J. J., Cotta-Ramusino, L., Guell, X., Rots, D. A., Zhang, F. (2013). High-frequency off-target mutagenesis induced by CRISPR-Cas nucleases in human cells. *Nature Biotechnology*, 31(9), 822–826. https://doi.org/10.1038/nbt.2623

Gaj, T., Scharenberg, C. A., & Liu, P. Q. (2013). ZFN, TALEN, and CRISPR/Cas-based methods for genome engineering. *Trends in Biotechnology*, 31(7), 397–405. https://doi.org/10.1016/j.tibtech.2013.04.004

Gao, W., Mahajan, S. P., Sulam, J., & Gray, J. J. (2020). Deep learning in protein structural modeling and design. *Patterns,* 1(9), 100142. https://doi.org/10.1016/j.patter.2020.100142

Gaudelli, N. M., Komor, A. C., Rees, H. A., Packer, M. S., Badran, A. H., Bryson, D. I., & Liu, D. R. (2017). Programmable base editing of AT to G•C in genomic DNA without DNA cleavage. *Nature*, 551(7681), 464–471. https://doi.org/10.1038/nature24644

Haeussler, M., Yang, Z., Feras, M., Wang, J., Ziyang, J., Li, Y., & Wei, W. (2020). PrimeDesign: A pipeline for streamlined design of prime editing guide RNAs. *Bioinformatics*, 36(15), 4353–4355. https://doi.org/10.1093/bioinformatics/btaa544

Hassija, V., Chamola, V., Mahapatra, A., Singal, A., Goel, D., Huang, K., Hussain, A. (2024). Interpreting black-box models: A review on explainable artificial intelligence. *Cognitive Computation,* 16(1), 45–74.

Hernández Medina, R., Kutuzova, S., Nielsen, K. N., Johansen, J., Hansen, L. H., Nielsen, M., & Rasmussen, S. (2022). Machine learning and deep learning applications in microbiome research. *ISME Communications,* 2, 98.

Hilton, I. B., D'Ippolito, A. M., Kadeer, Z., Kadoch, C., & Qi, L. S. (2015). Epigenome editing using a CRISPR/Cas9-based acetyltransferase. *Nature Biotechnology*, 33(5), 515–523. https://doi.org/10.1038/nbt.3199

Hsu, P. D., Lander, E. S., & Zhang, F. (2013). Development and applications of CRISPR-Cas9 for genome engineering. *Cell*, 157(6), 1262–1278. https://doi.org/10.1016/j.cell.2014.05.025

Jinek, M., Chylinski, K., Fonfara, I., Hauer, A., Doudna, J. A., & Charpentier, E. (2012). A programmable dual-RNA-guided DNA endonuclease in adaptive bacterial immunity. *Science*, 337(6096), 816–821. https://doi.org/10.1126/science.1225829

Jumper, J., Evans, R., Pritzel, A., Green, T., Figurnov, M., Ronneberger, O., et al. (2021). Highly accurate protein structure prediction with AlphaFold. *Nature,* 596, 583–589.

Kashefi, P., Kashefi, Y., & Ghafouri Mirsaraei, A. (2024). Shaping the future of AI: Balancing innovation and ethics in global regulation. *Uniform Law Review,* 29(3), 524–548.

Kwon, Y., Ryu, H., Kim, M., Lee, J., Chung, E., Min, S., & Kim, J. S. (2021). Easy-Prime: A web-based prime-editing guide RNA design tool. *Molecular Therapy - Nucleic Acids*, 24, 822–828. https://doi.org/10.1016/j.omtn.2021.05.006

Liu, B., Jiang, H., & Li, L. (2024). The innovative applications and prospects of artificial intelligence in agricultural research: Gene editing, crop improvement, and intelligent data analysis. *Advances in Resources Research,* 4(4), 624–653.

Liu, L., & Fan, X. D. (2014). CRISPR–Cas system: A powerful tool for genome engineering. *Plant Molecular Biology,* 85(3), 209–218.

Meng, Y., Zhang, Z., Zhou, C., Tang, X., Hu, X., Tian, G., Yang, J., & Yao, Y. (2025). Protein structure prediction via deep learning: An in-depth review. *Frontiers in Pharmacology,* 16, 1498662.

Perez, A. R., Weir, J. S., Volf, V., Rots, D. A., & Liu, Q. (2017). GuideScan: Accurate prediction of on- and off-target activity of Cas9 nucleases. *Bioinformatics*, 33(11), 1673–1679. https://doi.org/10.1093/bioinformatics/btx003

Rafiq, S., Macha, M. A., & Assad, A. (2025). Machine learning and deep learning for genomic data: A data-centric approach to CRISPR/Cas9 gene editing. *International Journal of Data Science and Analytics*, 20(6), 5037–5066. https://doi.org/10.1007/s41060-025-00773-4

Schneider, L., & Minary, P. (2024). Be-dataHIVE: a base editing database. *BMC Bioinformatics*, 25(1), 330. https://doi.org/10.1186/s12859-024-05898-0

Srivastav, A. K., Mishra, M. K., Lillard, J. W., Jr., & Singh, R. (2025). Transforming pharmacogenomics and CRISPR gene editing with the power of artificial intelligence for precision medicine. *Pharmaceutics,* 17(5), 555.

Wu, J. L., Zheng, S. S., Wang, L., Xiang, X., Li, F., Lv, M. L., ... & Miao, H. B. (2025). CRISPR-Cas System-Mediated Genetic Modification in Bacillus spp.: Current Status and Future. *Journal of Agricultural and Food Chemistry*, 73(24), 14760–14775. https://doi.org/10.1021/acs.jafc.5c03140

Wang, H., Li, T., Wang, Z., & Liu, M. (2021). EpiCas-DL: Deep learning-based prediction of CRISPR-Cas9 editing efficiency with epigenomic features. *Bioinformatics*, 37(19), 3123–3129. https://doi.org/10.1093/bioinformatics/btab216

14 AI-Driven Design of CRISPR Immunotherapies

Radhakanta Nag, Prasanti Devi, and Subhashree Pattnaik

14.1 INTRODUCTION

The field of cancer immunotherapy has been revolutionized by modalities such as immune checkpoint blockade (ICB) and chimeric antigen receptor T-cell therapy, yet a significant proportion of patients with solid tumors remain unresponsive (Sammamed & Chen, 2018). The complexity of the tumor microenvironment (TME) and the vast genetic heterogeneity of solid tumors present formidable barriers to these largely empirical approaches (Binnewies et al., 2018). The advent of the CRISPR-Cas9 system has provided an unprecedented tool for precise genomic manipulation, opening new avenues for directly engineering immune cells or modulating tumor cells to enhance immunogenicity (Rupp et al., 2017). However, the transition from using CRISPR as a blunt instrument for gene knockout to a precision scalpel for rational immunotherapy design requires a sophisticated, data-driven framework (Shifrut et al., 2018). This chapter outlines a theoretical and practical bioinformatic workflow for leveraging artificial intelligence (AI) and multi-omic data to design intelligent CRISPR-based immunotherapies, critically addressing the challenges of data integration, model interpretability, and clinical translation (Eraslan et al., 2022).

14.2 THE CONCEPTUAL FRAMEWORK: INTEGRATING MULTI-OMICS AND AI

The rational design of CRISPR immunotherapies hinges on the integration of disparate, high-dimensional biological data types to form a holistic model of tumor-immune interactions (Hasin et al., 2017). Multi-omic data encompassing genomics, transcriptomics, epigenomics, and proteomics provides a multifaceted view of the molecular drivers of cancer and the mechanisms of immune evasion (Wagner et al., 2022). AI, particularly machine learning and deep learning (DL), is uniquely suited to identify complex, nonlinear patterns within these integrated datasets that are intractable to traditional statistical methods (Greener et al., 2022). For instance, AI models can be trained to predict which gene perturbations will synergize to reverse a "cold" tumor to an immunologically "hot" state, or to identify novel, tumor-specific neoantigens that can be targeted by engineered T-cells (Lu et al., 2021). This AI-driven, multi-omic framework moves the field from a one-gene-at-a-time approach to a systems-level strategy for therapeutic design (Huang et al., 2023).

14.3 A PRACTICAL BIOINFORMATIC WORKFLOW FOR AI-DRIVEN CRISPR DESIGN

A robust, reproducible computational pipeline is essential for translating theoretical models into testable therapeutic hypotheses. The following sections detail a concrete workflow from data acquisition to in silico validation.

14.3.1 Data Acquisition and Curation from Public Repositories

The first step involves the systematic acquisition of high-quality, multi-omic data from large-scale public repositories. The Cancer Genome Atlas (TCGA) provides comprehensive genomic,

DOI: 10.1201/9781003735113-14

transcriptomic, and clinical data for over 20,000 primary cancers across 33 cancer types, serving as an invaluable resource for understanding tumor-intrinsic features and their correlation with immune infiltration (Weinstein et al., 2013). Complementarily, the Cancer Dependency Map (DepMap) project employs CRISPR-Cas9 screens to identify genes essential for the fitness of hundreds of cancer cell lines, providing direct functional evidence for potential therapeutic targets (Ghandi et al., 2019). Integrating TCGA data with DepMap dependency scores allows for the prioritization of targets that are not only genetically relevant but also critically important for tumor survival (Tsherniak et al., 2017). Data curation, including batch effect correction, normalization, and clinical annotation, is a critical and often underappreciated step that significantly impacts downstream model performance (Leek et al., 2012).

14.3.2 Target Hypothesis Generation via AI Models

With curated multi-omic data, the next step is to apply AI models to generate high-confidence target hypotheses. Unsupervised learning methods, such as clustering and dimensionality reduction, can identify distinct molecular subtypes of tumors with characteristic immune profiles (Thorsson et al., 2018). Supervised learning models, including random forests and gradient boosting, can be trained to predict key immunotherapy-relevant phenotypes, such as response to ICB or cytotoxic T-cell infiltration, based on the molecular features of the tumor (Jiang et al., 2018). More advanced DL architectures, like graph neural networks, can model the complex protein–protein interaction networks within the TME to identify key network nodes whose perturbation could maximally disrupt the immunosuppressive circuitry (Zheng et al., 2023). The output of this stage is a ranked list of candidate genes whose knockout or activation is predicted to enhance antitumor immunity.

14.3.3 gRNA Design and Off-Target Prediction

Once a target gene is selected, the design of highly specific and efficient single-guide RNAs (sgRNAs) is paramount. Computational tools like CRISPOR leverage sequence composition and thermodynamic rules to score and rank potential sgRNAs based on their on-target efficiency and to predict potential off-target sites across the genome (Haeussler et al., 2016). Recent AI-based tools, such as DeepCRISPR and Elevation, have further improved prediction accuracy by learning the sequence determinants of CRISPR activity from large-scale screening data (Wang et al., 2019; Listgarten et al., 2018). For immunotherapeutic applications, especially ex vivo engineering of T-cells, it is also critical to consider the impact of sgRNA design on delivery efficiency and the potential for chromosomal translocations, which can be mitigated by careful selection of target sites (Roth et al., 2018).

14.3.4 In Silico Validation Through Network Perturbation Analysis

Before committing resources to costly in vitro and in vivo experiments, in silico validation provides a crucial sanity check. This involves simulating the effect of the proposed genetic perturbation within a biological network model. Tools that leverage causal network models, such as CausalPath, can predict the downstream molecular consequences of a gene knockout, revealing potential compensatory mechanisms or unexpected, deleterious effects (Babur et al., 2021). By simulating network perturbations for a top-ranked candidate gene, one can assess whether the predicted outcome aligns with the desired immunomodulatory effect, such as the upregulation of antigen presentation pathways or the downregulation of immunosuppressive cytokines (Keenan et al., 2023). This step adds a layer of systems-level validation that is absent from simple, single-gene dependency analyses.

14.4 CRITICAL CHALLENGES AND FUTURE DIRECTIONS

Despite the promise of this approach, significant challenges remain. The effective integration of diverse, often noisy, multi-omic datasets into a unified model is a major computational hurdle that requires sophisticated data harmonization techniques (Rappoport & Shamir, 2018). Furthermore, the "black box" nature of many complex AI models poses a problem for interpretability; understanding why a model makes a specific prediction is essential for building biological insight and clinical trust (Greener et al., 2022). Emerging fields of explainable AI are developing methods to attribute predictions to specific input features, making AI-driven hypotheses more transparent and testable (Jiménez-Luna et al., 2020). Finally, the ultimate challenge lies in the translation of computational designs into clinically viable therapies, which will require close collaboration between computational biologists, immunologists, and clinicians, as well as robust in vivo model systems for functional validation (LaFleur et al., 2022).

14.5 CONCLUSION

The convergence of CRISPR technology, multi-omic profiling, and AI marks a paradigm shift in cancer immunotherapy development. The workflow presented here—from data-driven target discovery to AI-optimized gRNA design and network-based validation—provides a concrete roadmap for bridging the gap between computational biology and clinical application. While challenges in data integration and model interpretability persist, the continued refinement of these methods promises to unlock a new generation of intelligent, highly personalized CRISPR immunotherapies for the most intractable solid tumors.

REFERENCES

Babur, Ö., Luna, A., Korkut, A., Durupinar, F., Siper, M. C., Dogrusoz, U., … Demir, E. (2021). CausalPath: A causal analysis framework for pathway models. *Bioinformatics*, 37(21), 3799–3806. https://doi.org/10.1093/bioinformatics/btab421

Binnewies, M., Roberts, E. W., Kersten, K., Chan, V., Fearon, D. F., Merad, M., … Krummel, M. F. (2018). Understanding the tumor immune microenvironment (TIME) for effective therapy. *Nature Medicine*, 24(5), 541–550. https://doi.org/10.1038/s41591-018-0014-x

Eraslan, G., Avsec, Ž., Gagneur, J., & Theis, F. J. (2022). Deep learning: New computational modelling techniques for genomics. *Nature Reviews Genetics*, 23(6), 389–403. https://doi.org/10.1038/s41576-021-00432-x

Ghandi, M., Huang, F. W., Jané-Valbuena, J., Kryukov, G. V., Lo, C. C., McDonald, E. R., … Getz, G. (2019). Next-generation characterization of the Cancer Cell Line Encyclopedia. *Nature*, 569(7757), 503–508. https://doi.org/10.1038/s41586-019-1186-3

Greener, J. G., Kandathil, S. M., Moffat, L., & Jones, D. T. (2022). A guide to machine learning for biologists. *Nature Reviews Molecular Cell Biology*, 23(1), 40–55. https://doi.org/10.1038/s41580-021-00407-0

Haeussler, M., Schönig, K., Eckert, H., Eschstruth, A., Mianné, J., Renaud, J. B., … Concordet, J. P. (2016). Evaluation of off-target and on-target scoring algorithms and integration into the guide RNA selection tool CRISPOR. *Genome Biology*, 17(1), 1–12. https://doi.org/10.1186/s13059-016-1012-2

Hasin, Y., Seldin, M., & Lusis, A. (2017). Multi-omics approaches to disease. *Genome Biology*, 18(1), 1–15. https://doi.org/10.1186/s13059-017-1215-1

Huang, L., Wang, Z., Liu, Y., & Wu, Q. (2023). Building predictive models for immunotherapy response with multi-omics integration and deep learning. *Briefings in Bioinformatics*, 24(1), bbac594. https://doi.org/10.1093/bib/bbac594

Jiang, P., Gu, S., Pan, D., Fu, J., Sahu, A., Hu, X., … Han, Z. (2018). Signatures of T cell dysfunction and exclusion predict cancer immunotherapy response. *Nature Medicine*, 24(10), 1550–1558. https://doi.org/10.1038/s41591-018-0136-1

Jiménez-Luna, J., Grisoni, F., & Schneider, G. (2020). Drug discovery with explainable artificial intelligence. *Nature Machine Intelligence*, 2(10), 573–584. https://doi.org/10.1038/s42256-020-00236-4

Keenan, A. B., Jenkins, S. L., Jagodnik, K. M., Koplev, S., He, E., Torre, D., … Ma'ayan, A. (2023). The Library of Integrated Network-Based Cellular Signatures (LINCS) program: A user-guided resource for integrative analysis of cellular responses to perturbations. *Nature Methods*, 20(2), 243–255. https://doi.org/10.1038/s41592-022-01732-8

LaFleur, M. W., Murohashi, M., & Sharpe, A. H. (2022). In vivo CRISPR screens for cancer immunotherapy. *Trends in Cancer*, 8(2), 95–108. https://doi.org/10.1016/j.trecan.2021.10.002

Leek, J. T., Johnson, W. E., Parker, H. S., Jaffe, A. E., & Storey, J. D. (2012). The sva package for removing batch effects and other unwanted variation in high-throughput experiments. *Bioinformatics*, 28(6), 882–883. https://doi.org/10.1093/bioinformatics/bts034

Listgarten, J., Weinstein, M., Kleinstiver, B. P., Sousa, A. A., Joung, J. K., Crawford, J., … Fusi, N. (2018). Prediction of off-target activities for the end-to-end design of CRISPR guide RNAs. *Nature Biomedical Engineering*, 2(1), 38–47. https://doi.org/10.1038/s41551-017-0178-6

Lu, T., Wang, Z., Wang, Z., Xu, J., Li, M., Zhu, J., … Zhang, H. (2021). A machine learning model for predicting response to cancer immunotherapy. *Nature Communications*, 12(1), 1–11. https://doi.org/10.1038/s41467-021-26779-7

Rappoport, N., & Shamir, R. (2018). Multi-omic and multi-view clustering algorithms: Review and cancer benchmark. *Nucleic Acids Research*, 46(20), 10546–10562. https://doi.org/10.1093/nar/gky889

Roth, T. L., Puig-Saus, C., Yu, R., Shifrut, E., Carnevale, J., Li, P. J., … Marson, A. (2018). Reprogramming human T cell function and specificity with non-viral genome targeting. *Nature*, 559(7714), 405–409. https://doi.org/10.1038/s41586-018-0326-5

Rupp, L. J., Schumann, K., Roybal, K. T., Gate, R. E., Ye, C. J., Lim, W. A., & Marson, A. (2017). CRISPR/Cas9-mediated PD-1 disruption enhances anti-tumor efficacy of human chimeric antigen receptor T cells. *Scientific Reports*, 7(1), 1–10. https://doi.org/10.1038/s41598-017-00462-8

Sammamed, M. F., & Chen, L. (2018). A paradigm shift in cancer immunotherapy: From enhancement to normalization. *Cell*, 175(2), 313–326. https://doi.org/10.1016/j.cell.2018.09.035

Shifrut, E., Carnevale, J., Tobin, V., Roth, T. L., Woo, J. M., Bui, C. T., … Marson, A. (2018). Genome-wide CRISPR screens in primary human T cells reveal key regulators of immune function. *Cell*, 175(7), 1958–1971. https://doi.org/10.1016/j.cell.2018.10.024

Thorsson, V., Gibbs, D. L., Brown, S. D., Wolf, D., Bortone, D. S., Ou Yang, T. H., … Shmulevich, I. (2018). The immune landscape of cancer. *Immunity*, 48(4), 812–830. https://doi.org/10.1016/j.immuni.2018.03.023

Tsherniak, A., Vazquez, F., Montgomery, P. G., Weir, B. A., Kryukov, G., Cowley, G. S., … Hahn, W. C. (2017). Defining a cancer dependency map. *Cell*, 170(3), 564–576. https://doi.org/10.1016/j.cell.2017.06.010

Wagner, J., Rapsomaniki, M. A., Chevrier, S., Anzeneder, T., Langwieder, C., Dykgers, A., … Steiger, K. (2022). A single-cell atlas of the tumor and immune ecosystem of human breast cancer. *Cell*, 177(5), 1330–1345. https://doi.org/10.1016/j.cell.2019.03.005

Wang, D., Zhang, C., Wang, B., Li, B., Wang, Q., Liu, D., … Wang, H. (2019). Optimized CRISPR guide RNA design for two high-fidelity Cas9 variants by deep learning. *Nature Communications*, 10(1), 1–14. https://doi.org/10.1038/s41467-019-12281-8

Weinstein, J. N., Collisson, E. A., Mills, G. B., Shaw, K. R., Ozenberger, B. A., Ellrott, K., … Stuart, J. M. (2013). The Cancer Genome Atlas Pan-Cancer analysis project. *Nature Genetics*, 45(10), 1113–1120. https://doi.org/10.1038/ng.2764

Zheng, F., Zhang, Y., & Zhang, Z. (2023). Graph neural networks for predicting cancer drug responses from multi-omics data. *Briefings in Bioinformatics*, 24(1), bbac566. https://doi.org/10.1093/bib/bbac566

15 Deep Learning for Predicting Genome-Editing Outcomes

Efficiency, Specificity, and Repair Pathways

Selorm Adablanu

15.1 INTRODUCTION

CRISPR-Cas systems have dramatically transformed genome engineering in producing targeted DNA modification at high frequency. In CRISPR-Cas9, a single-guide RNA (sgRNA) directs Cas9 to another site along a PAM motif from another DNA site and produces a double-strand break (Jinek et al., 2012; Lee, 2023). In some cases, non-homologous end joining (NHEJ) or homology-directed repair (HDR) is used to insertion, deletion, or template-based editing, but efficiencies vary widely in genomic loci and biological contexts (Kim et al., 2019). But, the avoidance of off-target cleavage remains difficult, since Cas9 does tolerate mismatches at some points (Chuai et al., 2018; Doench et al., 2016; Lin & Wong, 2018).

Earlier computational approaches relied on empirical rules or classical machine-learning models such as MIT and CFD scores and ensemble predictors such as Rule Set 2 (Cao et al., 2025; Du et al., 2025; Lee, 2023). While useful, their performance was inconsistent across datasets and cell types (Mello et al., 2025). Plus, large-scale CRISPR screens and off-target detection assays such as GUIDE-seq and CIRCLE-seq provided the most accurate model for the specificity of cleavage specificity (Cloney, 2019; Shen et al., 2018).

Deep learning has since emerged as a more powerful framework, learning sequence context, structural features, and epigenetic signals directly from high-throughput datasets (Kim et al., 2019; Zhang et al., 2024). These models now outperform earlier computational tools in predicting both on-target efficiency and off-target risk (Liao et al., 2025; Zhang et al., 2024). They also accurately model DNA repair outcomes—such as indel distributions and base/prime editing product profiles—which were previously unpredictable. Similar advances have been reported in other biomedical domains; for example, deep learning has improved CT-based stroke detection while facing analogous challenges in clinical robustness and interpretability (Adablanu et al., 2025).

This chapter reviews advances in deep learning for predicting on-target activity, off-target specificity, and DNA repair outcomes, as well as efforts integrating multi-omics data, transfer learning, and explainable AI. It also highlights emerging directions such as generative design of editors and guides, federated learning for privacy-preserving model development, and automated experimental workflows. Figure 15.1 summarizes the key deep learning inputs, model types, and prediction tasks discussed in this chapter.

Schematic showing typical inputs (sgRNA-target sequence, PAM, chromatin accessibility, epigenomic marks, 3D genome contacts), core deep learning architectures (convolutional neural network (CNN)/recurrent neural network (RNN), transformer/attention, and hybrid multi-omics models), and representative outputs (on-target efficiency scores, off-target risk profiles, and repair outcome distributions for indels, base edits, and prime edits). As shown in Figure 15.1, deep learning

DOI: 10.1201/9781003735113-15

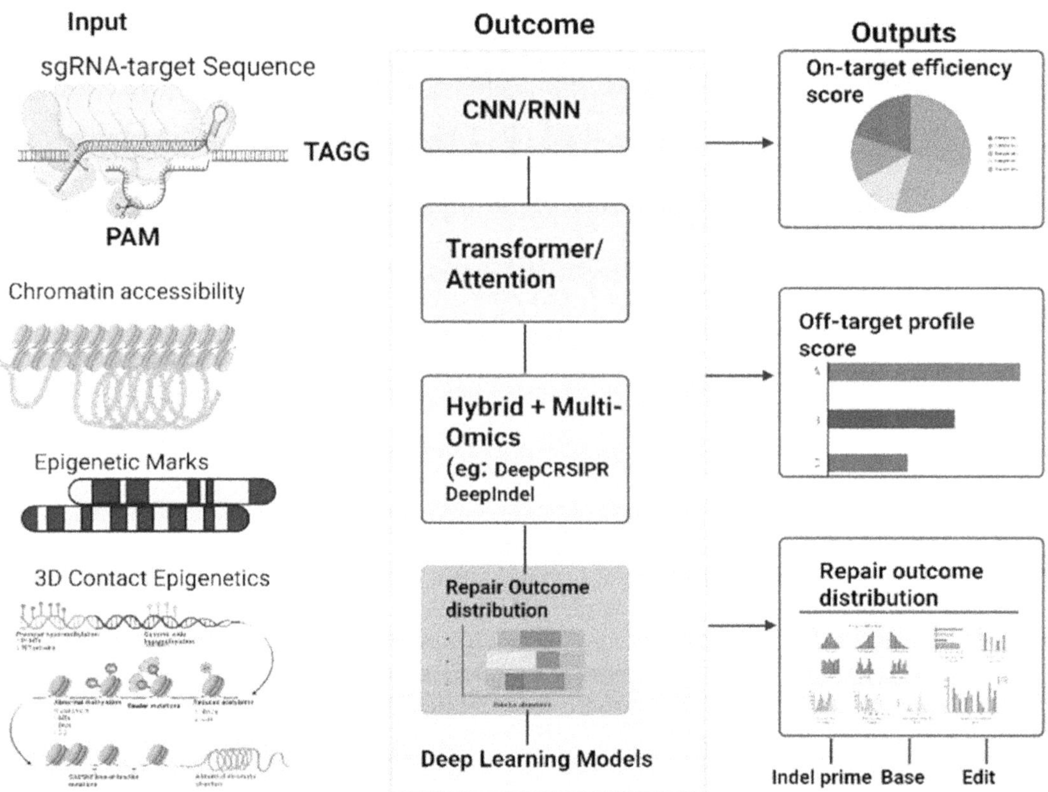

FIGURE 15.1 Deep learning framework for predicting CRISPR outcomes based on sequence features, chromatin accessibility, and epigenetic markers.

models ingest sequence and multi-omics features, map them through CNN, RNN, transformer, or hybrid architectures, and output scores for on-target efficiency, off-target risk, and repair outcome distributions.

15.2 DEEP LEARNING MODELS FOR ON-TARGET EFFICIENCY PREDICTION

Predicting on-target cleavage efficiency is essential for effective genome editing, and deep learning has substantially improved sgRNA design by learning patterns directly from large high-throughput indel and knockout datasets. Early progress came from CNNs, which use sgRNA sequence and local genomic context as inputs. DeepCRISPR was one of the first unified CNN models for on-target and off-target prediction and showed strong performance by learning sequence- and chromatin-associated signals without manual feature engineering (Chuai et al., 2018; Kim et al., 2024; Mathis et al., 2023; Park & Kim, 2023).

Another major contribution was DeepSpCas9, trained on >12,000 sgRNAs from multiple studies (Abbaszadeh & Shahlai, 2025; Kissling et al., 2025). DeepSpCas9 achieved higher correlations with measured indel frequencies than earlier tools ("Human Embryo Research Policy Update," 2018; Kim et al., 2019; Wang et al., 2019). The extended model incorporating chromatin accessibility (DeepSpCas9-CA) further improved predictive accuracy (Lu et al., 2021).

Beyond CNNs, hybrid architectures that integrate RNNs and attention mechanisms have been developed. AttCRISPR combines CNN, LSTM, and attention layers to capture sequential dependencies and highlight nucleotides with the greatest influence on cleavage outcomes (Hsieh et al., 2022), reflecting a growing emphasis on interpretable deep learning.

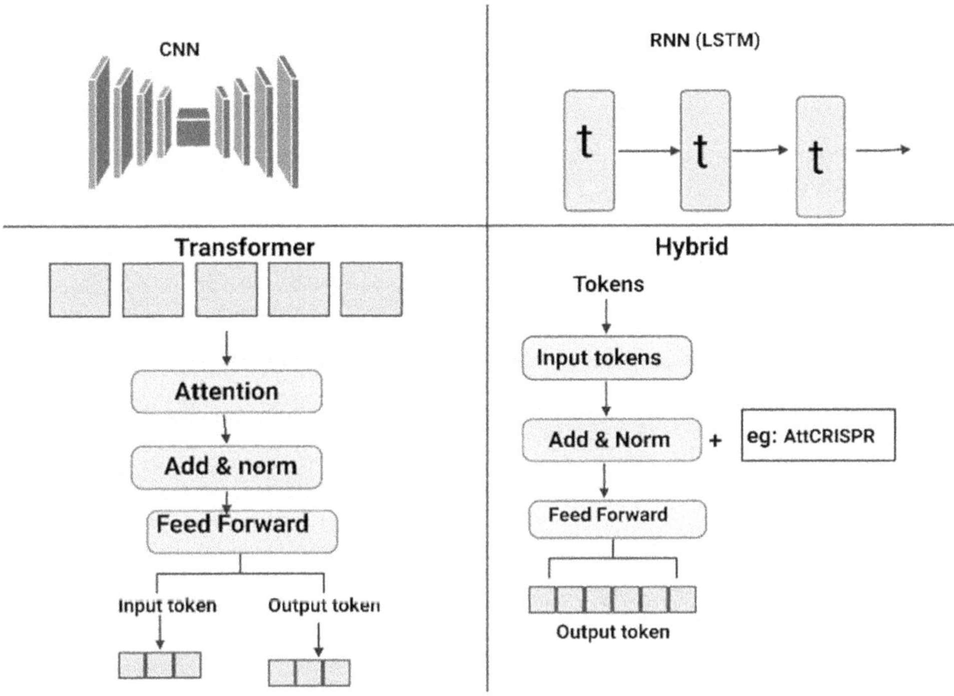

FIGURE 15.2 Comparison of different deep learning architectures including CNN, RNN (LSTM), transformer, and hybrid models used in genomic research.

To contextualize these architectures, Figure 15.2 contrasts the core model classes used for on-target and off-target prediction, including CNNs, RNNs/LSTMs, transformers, and hybrid attention-based designs such as AttCRISPR.

Conceptual comparison of common architectures used in CRISPR modeling: CNNs for local motif detection, recurrent networks (RNNs/LSTMs) for sequential dependencies, transformer blocks with self-attention for long-range context, and hybrid CNN–RNN–attention models (e.g., AttCRISPR) that integrate multiple components for interpretable guide RNA activity prediction.

As illustrated in Figure 15.2, CNNs extract local sequence features, RNNs and LSTMs model positional dependencies, transformers capture long-range context via self-attention, and hybrid designs combine these components to improve accuracy and interpretability for sgRNA scoring.

Deep learning has also been tailored to modified Cas9 enzymes. DeepHF (2019) modeled activity for high-fidelity Cas9 variants such as eSpCas9(1.1) and SpCas9-HF1 (Listgarten et al., 2018; Schneider & Minary, 2024), demonstrating that models can be retrained for specific nucleases or conditions, a concept expanded in Section 15.5 on transfer learning. Table 15.1 summarizes representative deep learning models and their characteristics.

As shown above, deep learning models for on-target prediction generally use a one-hot encoded sequence input of the sgRNA and sometimes the adjacent protospacer or PAM context. The networks (often CNN-based) slide learned filters across the sequence to detect motifs or patterns predictive of high activity (Lin & Wong, 2018). The learned features are then passed through dense layers to output a score or probability of efficient editing. These models have demonstrated that there are inherent sequence biases influencing Cas9 efficacy—for example, GC content, certain dinucleotide patterns, and position-specific nucleotide preferences—which the networks can automatically learn (Cao et al., 2025). Moreover, by incorporating additional features such as DNA accessibility (e.g., DNase-seq signals or ATAC-seq at the target locus),

TABLE 15.1
Representative Deep Learning Models for Predicting CRISPR On-Target Efficiency and Specificity

Model (Year)	Predicted Outcome	Approach	Notable Features and Data
DeepCRISPR (Chuai et al., 2018)	On-target knockout efficiency (and off-target scoring)	CNN ensemble	Learns sequence determinants of activity; integrates epigenetic features to improve prediction. Trained on large-scale knockout screening data.
DeepSpCas9 (Kim et al., 2019)	On-target indel frequency (SpCas9)	CNN (one-hot input)	Trained on ~13k guides from high-throughput assays; reported in Science Advances and achieved high generalization on independent datasets; baseline model for SpCas9 activity prediction.
DeepHF (Wang et al., 2019)	On-target activity for high-fidelity Cas9 variants	CNN+transfer learning	Tailored for eSpCas9(1.1) and SpCas9-HF1 variant guides. Retrained on variant-specific data to account for altered sequence preferences. Improved design for high-specificity nucleases.
AttCRISPR (Abbaszadeh & Shahlai, 2025)	On-target cleavage efficiency	Hybrid CNN–LSTM+Att.	Incorporates an attention layer highlighting important sequence positions for activity (enhanced interpretability). Demonstrated comparable accuracy with added explainability in guide scoring.
CNNoffSpotter (Lin & Wong, 2018)	Off-target cleavage likelihood	CNN and deep feedforward neural network (FNN)	Among first deep learning off-target models; uses a dual-CNN and feedforward network to classify off-target sites vs. non-sites. Trained on GUIDE-seq identified off-targets (CRISPOR dataset), achieving ~97% AUC.

CNN: convolutional neural network; FNN: feedforward neural network; LSTM: long short-term memory; Att.: attention mechanism. Epigenetic features include chromatin accessibility or histone marks at target locus.

Note: The last row (CNNoffSpotter by Lin & Wong) is an off-target model included here for comparison of approaches. Off-target prediction is detailed in Section 15.3.

some models improved performance, indicating that an open-chromatin state tends to correlate with higher editing efficiency (Chuai et al., 2018). The use of epigenomic data was a key innovation of DeepCRISPR, which autonomously identified that certain chromatin marks were correlated with sgRNA success (Kim et al., 2024). This multifactor awareness is why deep learning models are highly robust because they can integrate heterogeneous data modalities beyond the guide sequence itself. In general, deep learning predictors of on-target efficiency have become invaluable tools to guide RNA design. As a result, researchers can choose the best guides in the selection of the best candidates, making tests harder and less likely to have results where they prefer to test out the targeted target. In comparative studies, deep learning models like DeepSpCas9 and DeepCRISPR consistently rank among the top-performing algorithms for predicting guide efficacy (Chen & Wang, 2022; Yuan et al., 2025). As more data (including from new genome editors and cell types) become available, these models can be further refined. Techniques such as transfer learning and meta-learning (using knowledge from one dataset to improve predictions on another) are actively being explored to leverage cross-domain data and address cases where training data is limited for a particular context (Lin & Wong, 2018). In Section 15.5, we will discuss how such approaches, as well as efforts to make predictions more interpretable, are enhancing the utility and trustworthiness of on-target efficacy models in both research and clinical genome-editing applications.

15.3 PREDICTING OFF-TARGET EFFECTS WITH DEEP LEARNING

Off-target cleavage is a major constraint on CRISPR specificity, as Cas9–sgRNA complexes can cut partially matched genomic sites. Earlier rule-based schemes such as the Massachusetts Institute of Technology specificity score and the cutting frequency determination (CFD) score (Chuai et al., 2018), captured only simple mismatch penalties and did not fully account for nonlinear sequence-context interactions. Deep learning models address these limitations by learning directly from experimentally validated GUIDE-seq and CRISPOR datasets.

A foundational CNN model by Lin and Wong (2018) classified sgRNA–DNA pairs with an Area Under the Curve (AUC) of ~0.97, outperforming traditional mismatch scoring (Anzalone et al., 2019; Lin & Wong, 2018). It captured mismatch number, position, and combinatorial effects, confirming the central role of PAM-proximal seed complementarity (Hsu et al., 2013).

More advanced architectures incorporate paired encoders and transformers. CCLMoff uses a pre-trained RNA language model, achieving strong cross-dataset generalization (Cao et al., 2025; Du et al., 2025) and providing attention-based interpretability identifying key seed-region features (Du et al., 2025). Reviews confirm that CNN, RNN, and transformer models surpass mismatch-based methods, especially when integrating chromatin accessibility (Cao et al., 2025; Grünewald et al., 2019).

Genome-wide scoring pipelines enumerate candidate loci and rank them using models such as Elevation (Listgarten et al., 2018), CRISPR-Net (Chuai et al., 2018), and DeepCRISPR, which incorporate sequence, bulges, and GUIDE-seq evidence (Chuai et al., 2018; Listgarten et al., 2018).

Class imbalance remains a key challenge since true off-targets are rare. Curated negatives, oversampling, multitask learning, and transfer learning (pretraining on large GUIDE-seq datasets and fine-tuning on smaller ones) improve robustness (Cao et al., 2025; Charlier et al., 2025; Grünewald et al., 2019).

Interpretability methods such as SHapley Additive exPlanations (SHAP) and integrated gradients identify mismatch patterns that drive predictions, providing mechanistic insight and improving sgRNA safety assessment (Du et al., 2025). Overall, deep learning has markedly advanced off-target prediction, enabling selection of highly specific guides before experimental validation.

15.4 PREDICTING DNA REPAIR OUTCOMES: INDELS, BASE EDITING, AND PRIME EDITING

This is important for understanding and controlling CRISPR edits by accurately modeling the outcome of DNA repair. Beyond determining if and where Cas9 cuts, deep learning models now predict what repair outcome is produced, including indels from NHEJ and product distributions from base and prime editing. These models move genome editing closer to deterministic design by forecasting the frequency of specific editing events.

Building on these ideas, Figure 15.3 summarizes how deep learning models translate training data on repair profiles into predictions that can feed an iterative, AI-driven design loop for genome-editing experiments.

Diagram showing how training data on indel, base editing, and prime editing outcomes are used to train deep learning models, which in turn predict repair outcome distributions. These predictions inform the design and experimental testing of new sgRNAs and editor strategies, forming a closed design–build–test–learn loop for more precise genome editing. As depicted in Figure 15.3, experimentally measured repair outcomes are used to train predictive models, which then guide the selection and testing of new sgRNAs or editor configurations, progressively refining both the models and the underlying repair hypotheses.

15.4.1 Predicting Indel Distributions from NHEJ Repair

CRISPR-induced double-strand breaks generate diverse insertions and deletions, once believed to be largely stochastic. Shen et al. (2018) instead showed that indel patterns are reproducible and strongly shaped by local sequence context (Shen et al., 2018). Their inDelphi model, trained on

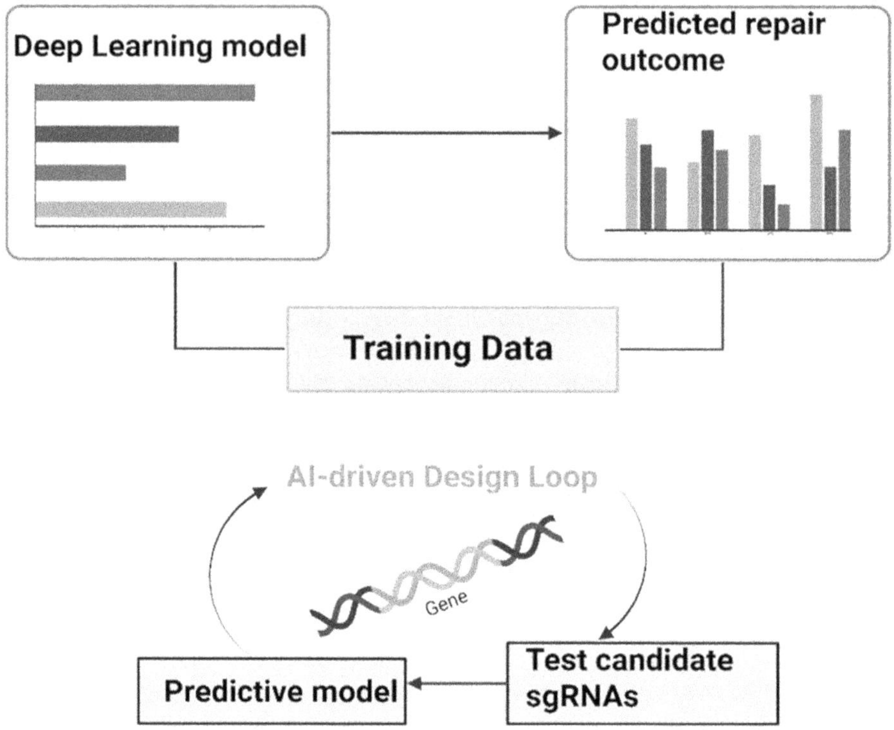

FIGURE 15.3 The AI-driven design loop for optimizing sgRNAs through iterative training data collection and predictive modeling.

>1,500 experimentally derived indel profiles, accurately predicted ~95% of total indel frequencies and ~80% of individual genotypes (Cloney, 2019), revealing that microhomology usage and nearby nucleotides significantly bias NHEJ repair.

Subsequent models extended this capability. SPROUT introduced an attention-based sequence-to-outcome approach with improved context sensitivity (Liu et al., 2022). DeepIndel applied a BERT-style transformer to ~60 bp surrounding the cut site and achieved strong agreement with measured outcomes (Pearson's $r > 0.94$ for multiple classes) (Zhang et al., 2024). SHAP analyses highlighted the importance of nucleotides immediately flanking the break (–4 to +4) and microhomology regions, reflecting known mechanistic biases. DeepIndel also exceeded CROTON by 8%–15% across several benchmarks and generalized well across cell types (Li et al., 2021; Zhang et al., 2024).

These tools allow identifying targets sites with frameshift-rich knockout profiles, the avoidance of in-frame deletions, and the interpretation of the atypical repair signatures. They also support "guided end-joining" practices that make silent sequence changes a proxy for repair toward desired outcomes. Deep learning is the quantitative basis for predicting and potentially acting on CRISPR repair specificity.

15.4.2 Base Editing Outcome Prediction

Base editors (BEs) enable single-nucleotide conversions without double-strand breaks, but their efficiencies and byproduct profiles vary with sequence context. Deep learning has therefore become central to predicting both editing rates and product distributions for cytosine and adenine BEs.

A major advance came from multitask models trained across 63 BEs, enabling prediction of overall efficiency and product fractions at each target site (Sun et al., 2025; Yuan et al., 2024).

DeepBE learned generalizable sequence–activity rules across editors, including reduced ABE performance when multiple adenines compete in the editing window or when local motifs disfavor deaminase activity (Marquart et al., 2020; Pallaseni et al., 2022). It also supports in silico comparison of editor variants, helping to select the most suitable enzyme for each target (Sun et al., 2025).

Earlier tools such as DeepABE and DeepCBE (Park & Kim, 2023) were trained on thousands of targets for ABE7.10 and BE4/AncBE4max, predicting editing fractions at each editable position while capturing known deaminase biases. These models guide selection of sgRNAs expected to produce high editing efficiency while avoiding unwanted bystander edits.

More recently, BEDICT 2.0 incorporated both in vitro and in vivo datasets—including mouse liver editing—to improve prediction accuracy across biologically diverse contexts (Kissling et al., 2025). Reported correlations ranged from R=0.60–0.94 in cell lines and R=0.62–0.81 in vivo, emphasizing the value of training on multiple cellular environments when designing therapeutic BEs. Table 15.2 summarizes deep learning models for repair outcome prediction, including indel, base editing, and prime editing frameworks.

As shown above, deep learning now predicts not only whether an edit will occur but also the full distribution of repair outcomes. Prime editing presents additional modeling challenges due to its multi-component pegRNA design. PRIDICT, described by Mathis et al. (2023), showed that attention-based

TABLE 15.2
Deep Learning Models for Predicting DNA Repair Outcomes in Genome Editing

Model (Year)	Editing Mode	Predicted Outcome	Methodology	Key Features/Findings
inDelphi (Shen et al., 2018)	CRISPR nuclease (NHEJ)	Indel-type frequencies (insertions, deletions)	Neural network modules	Predicts ~95% of total indel frequencies and ~80% of genotype profiles; reveals sequence-determined repair patterns.
DeepIndel (Zhang et al., 2024)	CRISPR nuclease (NHEJ)	Multiple indel classes (e.g., 1-bp ins/del, frameshifts)	Bidirectional Encoder Representations from Transformers (BERT)-based transformer (multilabel)	Strong performance (AUC ~0.88; r ~0.94); SHAP interpretability highlights important sequence features.
DeepABE/ DeepCBE (Park & Kim, 2023)	Base editing (ABE, CBE)	Editing efficiency and byproduct distribution	CNN/RNN hybrid	Predicts per-base editing percentages; models sequence-context effects; guides selection with fewer bystander edits.
DeepBE (Sun et al., 2025)	Multiple base editors	Editing efficiency and outcome spectra	Multitask deep learning	Trained on 63 editors; supports cross-editor comparison; improves editor selection through pooled multi-editor learning.
PRIDICT (Mathis et al., 2023)	Prime editing (PE)	Prime editing efficiency and purity	Bidirectional RNN+attention	High correlation (Spearman r = 0.85); >10× improvement with top-ranked pegRNAs validated in vivo.
PrimeNet (Liao et al., 2025)	PE	pegRNA activity; intended vs. undesired edits	Deep learning	Incorporates primer binding site (PBS) and reverse transcriptase (RT)-template context; improves pegRNA design accuracy.

RNNs trained on large pegRNA-outcome datasets can accurately forecast prime editing efficiency and product purity, capturing factors such as pegRNA structure, target context, and primer binding site compatibility. High-scoring pegRNAs produced up to 12-fold higher editing in cells and 10-fold improvements in vivo, demonstrating strong predictive value (Mathis et al., 2023).

Deep learning is also being applied to pegRNA design itself, accounting for tunable features such as extension length and sequence mismatches. Emerging generative and optimization-based models, including PrimeNet (Liao et al., 2025), highlight growing interest in AI systems that both predict and design prime editing components.

In general, deep learning has also shifted away from predictive genome editing that will select the optimal editors and pegRNA for optimal placements and optimal outcome distributions. Not fully automated, end-to-end AI-guided prime editing is yet to be completely automated and most of the key components are fast approaching.

15.5 INTEGRATING MULTI-OMICS DATA AND ENHANCING MODEL RELIABILITY

Many models to be used for genome editing are now based on epigenomic, transcriptomic, and 3D genome features and the processes for deep learning, and genome modification aims at capturing more of the structure of the cells in order to understand what happens in a text. These layers regulate Cas9 binding, nuclease activity, and DNA repair choice providing useful information for a context-specific prediction. Early studies such as DeepCRISPR showed that the addition of chromatin accessibility improved on-target prediction and open-chromatin sites had higher editing performance (Chuai et al., 2018). In off-target modeling, the integration of accessibility or replication timing, as discussed in Elevation, improved specificity as well (Cao et al., 2025).

The role of transcription-coupled repair may also affect outcomes, especially at the high expression range, where transcription-coupled repair is a determinant of outcome (Batra et al., 2024). The structure of RNA is important for RNA-targeting editors like Cas13. While 3D genome architecture (e.g., Hi-C contact maps) is still evolving, further refinements of predictions could be made by capturing locus proximity and chromatin folding in 3D geography (Abbaszadeh & Shahlai, 2025).

Transfer learning and federated learning are further strengthening model robustness. Transfer learning allows models trained with large datasets to be trained on new cell types or conditions, and thus can perform off-target prediction at scale (Cao et al., 2025). Federated learning can be used as an alternative to sharing data among clinic CRISPR datasets, and this is particularly important for clinical CRISPR datasets (Horst et al., 2025; Pati et al., 2024).

Explainable artificial intelligence (XAI) is increasingly needed for model trust and regulatory acceptance. Attention maps in AttCRISPR (Abbaszadeh & Shahlai, 2025), SHAP analyses in DeepIndel and CCLMoff (Du et al., 2025), and the focus of scientific knowledge in AttCRISPR and its application in DeepIndel and DeepIndel.

Reviewers have suggested that most models are now nearing the limits of experimental reproducibility itself (Cao et al., 2025), and need to have richer and more multimodal datasets.

Overall, multi-omics integration, transfer and federated learning, and XAI-driven transparency are improving the reliability of genome-editing models. The substantial gains shown in applications such as PRIDICT-optimized pegRNAs (Mathis et al., 2023) demonstrate that AI-guided genome editing is rapidly becoming more practical, robust, and predictive.

15.6 FUTURE DIRECTIONS AND EMERGING TRENDS

Deep learning and genome editing are converging rapidly, opening new directions in model-driven design, automated experimentation, and expanded editor development. Figure 15.4 provides a high-level schematic overview of the major technological directions shaping the next generation of AI-enabled genome editing. These directions include: (1) generative models for Cas

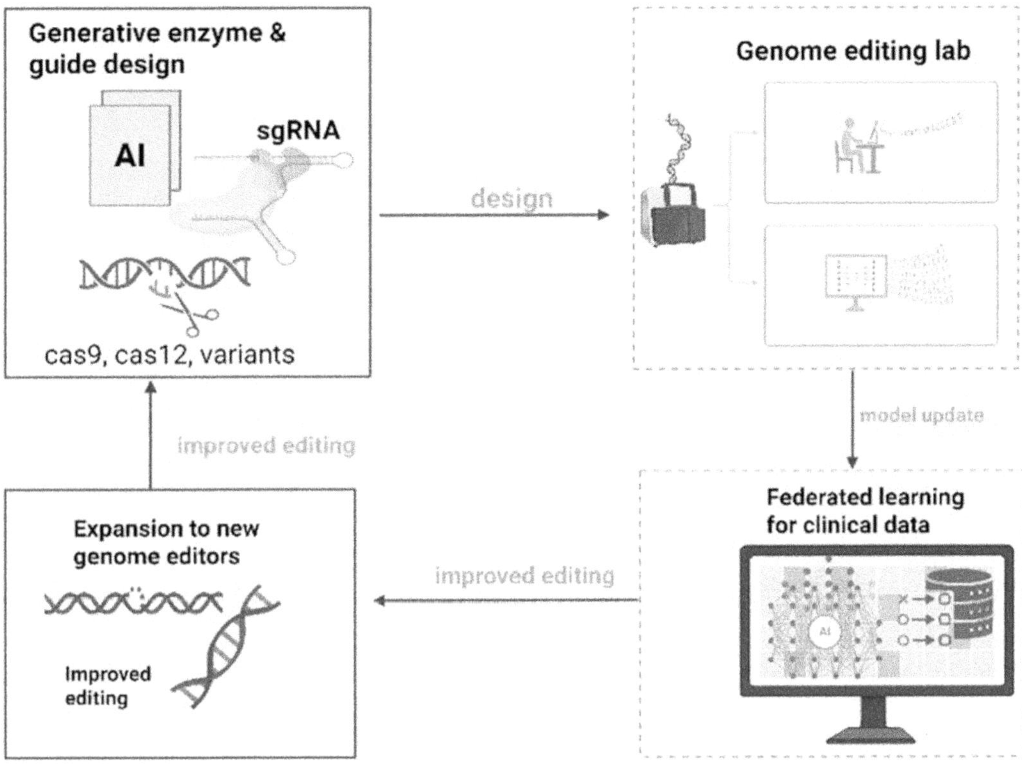

FIGURE 15.4 A collaborative ecosystem for genome editing featuring AI-led enzyme design, lab testing, and federated learning for clinical data updates.

enzyme engineering and sgRNA design; (2) AI-guided optimization of base and prime editors; (3) federated learning for privacy-preserving integration of clinical and genomic data; and (4) self-driving experimental pipelines that combine active learning, automation, and iterative model refinement. Collectively, these developments indicate a transition from standalone predictive models toward closed-loop, AI-guided genome-editing systems. The following subsections expand each component in detail, beginning with generative models for designing new CRISPR enzymes and guides.

Figure 15.4 emerging directions at the intersection of deep learning and genome editing.

15.6.1 Generative Design of CRISPR Enzymes and Guides

Beyond predicting activity, deep learning is now being used to generate new CRISPR enzymes and guides. Generative protein language models have enabled large-scale exploration of Cas protein sequence space. For example, ProGen2-based designs expanded Cas9-like sequence diversity by approximately 4.8-fold, with many variants predicted to fold correctly (Ruffolo et al., 2025; Zeng et al., 2022). These AI-derived sequences offer candidates with altered PAM preferences, reduced size, or potentially improved fidelity (Ruffolo et al., 2025). Similarly, the CasGen project uses transformer-based generative modeling to propose Cas9/Cas12 variants that preserve functional motifs while introducing novel properties (Nammi et al., 2025). Generative models can also design optimized sgRNAs under constraints such as minimized off-target activity-particularly useful for repetitive or viral genomes where manually identifying unique targets is difficult.

15.6.2 AI-Guided Protein and Base Editor Engineering

AI is increasingly applied to engineer deaminases for BEs and reverse transcriptases for prime editors. Neural networks trained on mutational scans of enzymes such as TadA can predict beneficial mutations and altered context preferences. Similarly, ML models are helping to optimize pegRNA architecture by predicting how primer binding site length or secondary structure affects prime editing efficiency (Alipanahi et al., 2024). These tools point toward a future where AI suggests not only the guide RNA but also the most suitable enzyme variant for each target.

15.6.3 Federated and Privacy-Preserving Learning for Clinical Data

As gene editing progresses into clinical trials, federated learning provides a framework for training models on distributed patient-derived editing data without sharing raw genomic information (Horst et al., 2025; Pati et al., 2024). Each site trains the model locally and shares only weight updates, enabling a "shared global model" informed by multiple trials. This approach could reveal patient-specific predictors—such as SNPs or disease-related chromatin states—that affect editing efficiency.

15.6.4 Automated Experimentation and Active Learning

Closed-loop "self-driving labs" combine AI with robotics to accelerate genome-editing research. In this design–build–test–learn workflow, models propose informative edits, automated systems perform experiments, and results update the model. Active learning ensures each round of experiments resolves model uncertainties efficiently. This paradigm is already emerging in synthetic biology and could be used to iteratively evolve editors or pegRNA designs.

15.6.5 Broadening to Other Genome Editors

Deep learning approaches are readily adaptable to other systems beyond Cas9, including Cas12a, Cas13 RNA editing, and older platforms like TALENs and ZFNs (Jinek et al., 2012). With sufficient training data, model architectures such as CNNs and transformers can generalize to new editor types. Future "meta-models" may compare the predicted performance of multiple editing strategies—for example, HDR versus base editing—on a given target mutation.

15.6.6 Ethical and Safety Considerations

As AI-designed enzymes and guides become more capable, dual-use concerns grow. Generative models could theoretically produce harmful constructs, necessitating governance and oversight. Conversely, AI improves biosecurity by identifying risky guides or predicting off-targets (policy discussions highlighted by several groups). Balancing innovation with safety will become increasingly important as AI-driven genome-editing tools advance.

15.7 CONCLUSION

Deep learning has become central to contemporary genome-editing research, enabling increasingly accurate prediction of on-target efficiency, off-target risks, and repair outcome profiles for CRISPR systems. By leveraging large experimental datasets, models such as CNNs, RNNs, and transformers capture sequence and context patterns that earlier rule-based tools could not, allowing researchers to choose effective and safer sgRNAs with greater confidence.

Incorporating multi-omics data—such as chromatin accessibility and epigenomic signals—along with explainability tools like attention maps and SHAP has improved model robustness and biological insight. Transfer learning is helping to adapt models to new editors and cell types where data is limited. The field is now moving from prediction to AI-guided design. Generative models are beginning to propose new guides, pegRNAs, and even Cas variants, while automated experimentation and active learning promise closed-loop systems that iteratively design and test edits. These advances accelerate tool development and help overcome challenges such as PAM constraints and context-dependent editing efficiency.

Continued collaboration between computational and experimental teams is essential, as model predictions must be validated and refined with real data. Together, these efforts are transforming genome editing into a more predictable and engineering-driven discipline. As deep learning integrates more biological context and becomes embedded in automated pipelines, genome editing will increasingly resemble a software-guided process—designed in silico and executed with growing precision in vivo. This shift lays the foundation for safer, more reliable research applications and next-generation therapeutic gene editing.

REFERENCES

Abbaszadeh, A., & Shahlai, A. (2025). *Artificial Intelligence for CRISPR Guide RNA Design: Explainable Models and Off-Target Safety*. https://arxiv.org/abs/2508.20130

Adablanu, S., Barman, U., & Das, D. (2025). Advancing deep learning for automated stroke detection: A review. *Brain Hemorrhages*. https://doi.org/10.1016/j.hest.2025.07.002

Alipanahi, R., Safari, L., & Khanteymoori, A. (2024). DTMP-prime: A deep transformer-based model for predicting prime editing efficiency and PegRNA activity. *Molecular Therapy - Nucleic Acids*, 35(4), 102370. https://doi.org/10.1016/j.omtn.2024.102370

Anzalone, A. V., Randolph, P. B., Davis, J. R., Sousa, A. A., Koblan, L. W., Levy, J. M., Chen, P. J., Wilson, C., Newby, G. A., Raguram, A., & Liu, D. R. (2019). Search-and-replace genome editing without double-strand breaks or donor DNA. *Nature*, 576(7785), 149–157. https://doi.org/10.1038/s41586-019-1711-4

Batra, S. S., Cabrera, A., Spence, J. P., Goell, J., Anand, S. S., Hilton, I. B., & Song, Y. S. (2024). Predicting the effect of CRISPR-Cas9-based epigenome editing. *eLife*. https://doi.org/10.7554/eLife.92991.2

Cao, M., Brennan, A., Lee, C. M., Park, S., & Bao, G. (2025). Deep learning based models for CRISPR/Cas off-target prediction. *Small Methods*, *9*(7). https://doi.org/10.1002/smtd.202500122

Charlier, J., Sherkatghanad, Z., & Makarenkov, V. (2025). Similarity-based transfer learning with deep learning networks for accurate CRISPR-Cas9 off-target prediction. *PLOS Computational Biology*, 21(10), e1013606. https://doi.org/10.1371/journal.pcbi.1013606

Chen, Y., & Wang, X. (2022). Evaluation of efficiency prediction algorithms and development of ensemble model for CRISPR/Cas9 gRNA selection. *Bioinformatics*, 38(23), 5175–5181. https://doi.org/10.1093/bioinformatics/btac681

Chuai, G., Ma, H., Yan, J., Chen, M., Hong, N., Xue, D., Zhou, C., Zhu, C., Chen, K., Duan, B., Gu, F., Qu, S., Huang, D., Wei, J., & Liu, Q. (2018). DeepCRISPR: Optimized CRISPR guide RNA design by deep learning. *Genome Biology*, 19(1), 80. https://doi.org/10.1186/s13059-018-1459-4

Cloney, R. (2019). The oracle of inDelphi predicts Cas9 repair outcomes. *Nature Reviews Genetics*, 20(1), 4–5. https://doi.org/10.1038/s41576-018-0077-z

Doench, J. G., Fusi, N., Sullender, M., Hegde, M., Vaimberg, E. W., Donovan, K. F., Smith, I., Tothova, Z., Wilen, C., Orchard, R., Virgin, H. W., Listgarten, J., & Root, D. E. (2016). Optimized sgRNA design to maximize activity and minimize off-target effects of CRISPR-Cas9. *Nature Biotechnology*, 34(2), 184–191. https://doi.org/10.1038/nbt.3437

Du, W., Zhao, L., Diao, K., Zheng, Y., Yang, Q., Zhu, Z., Zhu, X., & Tang, D. (2025). A versatile CRISPR/Cas9 system off-target prediction tool using language model. *Communications Biology*, 8(1), 882. https://doi.org/10.1038/s42003-025-08275-6

Grünewald, J., Zhou, R., Garcia, S. P., Iyer, S., Lareau, C. A., Aryee, M. J., & Joung, J. K. (2019). Transcriptome-wide off-target RNA editing induced by CRISPR-guided DNA base editors. *Nature*, 569(7756), 433–437. https://doi.org/10.1038/s41586-019-1161-z

Horst, A., Loustalot, P., Yoganathan, S., Li, T., Xu, J., Tong, W., Schneider, D., Löffler-Perez, N., Di Renzo, E., & Renaudin, M. (2025). Federated learning: a privacy-preserving approach to data-centric regulatory cooperation. *Frontiers in Drug Safety and Regulation*, *5*. https://doi.org/10.3389/fdsfr.2025.1579922

Hsieh, M.-L., Kiel, N., Jenkins, L. M. M., Ng, W.-L., Knipling, L., Waters, C. M., & Hinton, D. M. (2022). The Vibrio cholerae master regulator for the activation of biofilm biogenesis genes, VpsR, senses both cyclic di-GMP and phosphate. *Nucleic Acids Research*, 50(8), 4484–4499. https://doi.org/10.1093/nar/gkac253

Hsu, P. D., Scott, D. A., Weinstein, J. A., Ran, F. A., Konermann, S., Agarwala, V., Li, Y., Fine, E. J., Wu, X., Shalem, O., Cradick, T. J., Marraffini, L. A., Bao, G., & Zhang, F. (2013). DNA targeting specificity of RNA-guided Cas9 nucleases. *Nature Biotechnology*, 31(9), 827–832. https://doi.org/10.1038/nbt.2647

Human embryo research policy update. (2018). *Nature Biotechnology*, 36(6), 477–477. https://doi.org/10.1038/nbt.4171

Jinek, M., Chylinski, K., Fonfara, I., Hauer, M., Doudna, J. A., & Charpentier, E. (2012). A programmable dual-RNA–guided DNA endonuclease in adaptive bacterial immunity. *Science*, 337(6096), 816–821. https://doi.org/10.1126/science.1225829

Kim, H. K., Kim, Y., Lee, S., Min, S., Bae, J. Y., Choi, J. W., Park, J., Jung, D., Yoon, S., & Kim, H. H. (2019). SpCas9 activity prediction by DeepSpCas9, a deep learning–based model with high generalization performance. *Science Advances*, *5*(11). https://doi.org/10.1126/sciadv.aax9249

Kim, N., Choi, S., Kim, S., Song, M., Seo, J. H., Min, S., Park, J., Cho, S.-R., & Kim, H. H. (2024). Deep learning models to predict the editing efficiencies and outcomes of diverse base editors. *Nature Biotechnology*, 42(3), 484–497. https://doi.org/10.1038/s41587-023-01792-x

Kissling, L., Mollaysa, A., Janjuha, S., Mathis, N., Marquart, K. F., Weber, Y., Moon, W. J., Lin, P. J. C., Fan, S. H. Y., Muramatsu, H., Vadovics, M., Allam, A., Pardi, N., Tam, Y. K., Krauthammer, M., & Schwank, G. (2025). Predicting adenine base editing efficiencies in different cellular contexts by deep learning. *Genome Biology*, 26(1), 115. https://doi.org/10.1186/s13059-025-03586-7

Lee, M. (2023). Deep learning in CRISPR-Cas systems: A review of recent studies. *Frontiers in Bioengineering and Biotechnology*, 11. https://doi.org/10.3389/fbioe.2023.1226182

Li, V. R., Zhang, Z., & Troyanskaya, O. G. (2021). CROTON: An automated and variant-aware deep learning framework for predicting CRISPR/Cas9 editing outcomes. *Bioinformatics*, 37(Supplement_1), i342–i348. https://doi.org/10.1093/bioinformatics/btab268

Liao, X., Liu, Q., & Chuai, G. (2025). PrimeNet: rational design of prime editing pegRNAs by deep learning. *Briefings in Bioinformatics*, *26*(3). https://doi.org/10.1093/bib/bbaf293

Lin, J., & Wong, K.-C. (2018). Off-target predictions in CRISPR-Cas9 gene editing using deep learning. *Bioinformatics*, 34(17), i656–i663. https://doi.org/10.1093/bioinformatics/bty554

Listgarten, J., Weinstein, M., Kleinstiver, B. P., Sousa, A. A., Joung, J. K., Crawford, J., Gao, K., Hoang, L., Elibol, M., Doench, J. G., & Fusi, N. (2018). Prediction of off-target activities for the end-to-end design of CRISPR guide RNAs. *Nature Biomedical Engineering*, 2(1), 38–47. https://doi.org/10.1038/s41551-017-0178-6

Liu, X., Wang, S., & Ai, D. (2022). Predicting CRISPR/Cas9 repair outcomes by attention-based deep learning framework. *Cells*, *11*(11), 1847. https://doi.org/10.3390/cells11111847

Lu, S., Dai, M., Hu, X., Yi, H., & Zhang, Y. (2021). A new survival model based on ion channel genes for prognostic prediction in hepatocellular carcinoma. *Genomics*, 113(1), 171–182. https://doi.org/10.1016/j.ygeno.2020.12.028

Marquart, K. F., Allam, A., Janjuha, S., Sintsova, A., Villiger, L., Frey, N., Krauthammer, M., & Schwank, G. (2020). Predicting base editing outcomes with an attention-based deep learning algorithm trained on high-throughput target library screens. *bioRxiv (Cold Spring Harbor Laboratory)*. https://doi.org/10.1101/2020.07.05.186544

Mathis, N., Allam, A., Kissling, L., Marquart, K. F., Schmidheini, L., Solari, C., Balázs, Z., Krauthammer, M., & Schwank, G. (2023). Predicting prime editing efficiency and product purity by deep learning. *Nature Biotechnology*, 41(8), 1151–1159. https://doi.org/10.1038/s41587-022-01613-7

Mello, V. H., Wald, J., Marlovits, T. C., & Sartori, P. (2025). Elastic analysis bridges structure and dynamics of an AAA+ molecular motor. *PLOS Computational Biology*, 21(10), e1013596. https://doi.org/10.1371/journal.pcbi.1013596

Nammi, B., Jayasinghe-Arachchige, V. M., Madugula, S. S., Artiles, M., Radler, C. N., Pham, T., Liu, J., & Wang, S. (2025). CasGen: A regularized generative model for CRISPR Cas protein design with classification and margin-based optimization. *bioRxiv (Cold Spring Harbor Laboratory)*. https://doi.org/10.1101/2025.02.28.640911

Pallaseni, A., Peets, E. M., Koeppel, J., Weller, J., Vanderstichele, T., Ho, U. L., Crepaldi, L., van Leeuwen, J., Allen, F., & Parts, L. (2022). Predicting base editing outcomes using position-specific sequence determinants. *Nucleic Acids Research*, 50(6), 3551–3564. https://doi.org/10.1093/nar/gkac161

Park, J., & Kim, H. K. (2023). *Prediction of Base Editing Efficiencies and Outcomes Using DeepABE and DeepCBE* (pp. 23–32). Methods in Molecular Biology, Humana Press. https://doi.org/10.1007/978-1-0716-2879-9_3

Pati, S., Kumar, S., Varma, A., Edwards, B., Lu, C., Qu, L., Wang, J. J., Lakshminarayanan, A., Wang, S., Sheller, M. J., Chang, K., Singh, P., Rubin, D. L., Kalpathy-Cramer, J., & Bakas, S. (2024). Privacy preservation for federated learning in health care. *Patterns*, 5(7), 100974. https://doi.org/10.1016/j.patter.2024.100974

Ruffolo, J. A., Nayfach, S., Gallagher, J., Bhatnagar, A., Beazer, J., Hussain, R., Russ, J., Yip, J., Hill, E., Pacesa, M., Meeske, A. J., Cameron, P., & Madani, A. (2025). Design of highly functional genome editors by modelling CRISPR–Cas sequences. *Nature*, 645(8080), 518–525. https://doi.org/10.1038/s41586-025-09298-z

Schneider, L., & Minary, P. (2024). Be-dataHIVE: A base editing database. *BMC Bioinformatics*, 25(1), 330. https://doi.org/10.1186/s12859-024-05898-0

Shen, M. W., Arbab, M., Hsu, J. Y., Worstell, D., Culbertson, S. J., Krabbe, O., Cassa, C. A., Liu, D. R., Gifford, D. K., & Sherwood, R. I. (2018). Predictable and precise template-free CRISPR editing of pathogenic variants. *Nature*, 563(7733), 646–651. https://doi.org/10.1038/s41586-018-0686-x

Sun, Y., Qu, K., Corsi, G. I., Anthon, C., Pan, X., Xiang, X., Jensen, L. J., Lin, L., Luo, Y., & Gorodkin, J. (2025). Deep learning models simultaneously trained on multiple datasets improve base-editing activity prediction. *Nature Communications*, 16(1), 9821. https://doi.org/10.1038/s41467-025-65200-5

Wang, D., Zhang, C., Wang, B., Li, B., Wang, Q., Liu, D., Wang, H., Zhou, Y., Shi, L., Lan, F., & Wang, Y. (2019). Optimized CRISPR guide RNA design for two high-fidelity Cas9 variants by deep learning. *Nature Communications*, 10(1), 4284. https://doi.org/10.1038/s41467-019-12281-8

Yuan, H., Song, C., Xu, H., Sun, Y., Anthon, C., Bolund, L., Lin, L., Benabdellah, K., Lee, C., Hou, Y., Gorodkin, J., & Luo, Y. (2025). An overview and comparative analysis of CRISPR-SpCas9 gRNA activity prediction tools. *The CRISPR Journal*, 8(2), 89–104. https://doi.org/10.1089/crispr.2024.0058

Yuan, T., Wu, L., Li, S., Zheng, J., Li, N., Xiao, X., Zhang, H., Fei, T., Xie, L., Zuo, Z., Li, D., Huang, P., Feng, H., Cao, Y., Yan, N., Wei, X., Shi, L., Sun, Y., Wei, W., … Zuo, E. (2024). Deep learning models incorporating endogenous factors beyond DNA sequences improve the prediction accuracy of base editing outcomes. *Cell Discovery*, *10*(1), 20. https://doi.org/10.1038/s41421-023-00624-1

Zeng, X., Wang, F., Luo, Y., Kang, S., Tang, J., Lightstone, F. C., Fang, E. F., Cornell, W., Nussinov, R., & Cheng, F. (2022). Deep generative molecular design reshapes drug discovery. *Cell Reports Medicine*, 3(12), 100794. https://doi.org/10.1016/j.xcrm.2022.100794

Zhang, G., Xie, H., & Dai, X. (2024). DeepIndel: an interpretable deep learning approach for predicting CRISPR/Cas9-mediated editing outcomes. *International Journal of Molecular Sciences*, 25(20), 10928. https://doi.org/10.3390/ijms252010928

16 Genome Editing in Zebrafish
Revealing Functional Genomics, Regenerative Potential, and Therapeutic Discovery

Rutupurna Das, Sanjoy Majumder, and Gagan Kumar Panigrahi

16.1 INTRODUCTION

Over the last several decades, the *Danio rerio* known as zebrafish, has developed into an incredibly versatile and potent vertebrate model suitable for both biological and biomedical studies. It has since been discovered to establish itself as a fundamental tool in developmental research, in the areas of regulatory genomics, regenerative biology, toxicology, and therapeutic discovery. Having achieved preliminarily popularity, zebrafish have solidified their niche in functional genomic studies, regenerative biology, tox screens, and therapeutic discovery. Their peculiar biological features are manifested in the scope of their usage. A number of features make the zebrafish a unique experimental model that spans the chasm between classical invertebrate systems, e.g., *Drosophila melanogaster* and *Caenorhabditis elegans*, and vertebrates, especially mammals. The congruence of their genes between zebrafish and human beings is one such attribute that is quite strong. Among the strongest arguments that support why zebrafish should be widely used in research is its exceptional genetic closeness to human beings. About 70% of all human genes have at least an ortholog in zebrafish, and about 84% of the genes identified as linked to human disease have functional counterparts in zebrafish.

This conservation of the genes of zebrafish allows manipulation of the genes to interrogate the basic mechanisms underlying common human diseases, such as cardiovascular diseases, cancers, neurodegenerative disorders, and metabolic syndromes. Because of the presence of a completely sequenced and well-annotated genome and other rich genetic and genomic resources, such as the Zebrafish Information Network, gene-targeted studies and functional validation are made possible (Carrington et al., 2022). Such resources can be used by researchers to conduct accurate genetic perturbations and test the phenotypic outcomes in a timely fashion. Another remarkable asset is the clarity of zebrafish embryos, which can be directly observed to undergo organogenesis, cellular differentiation, and morphogenetic movements in real time under simple light or fluorescent microscopy (Figure 16.1).

Due to the external nature of the development process and clear appearance of embryos at the initial levels of development, they are ideal subjects of in vivo imaging of dynamic biological processes like angiogenesis, neurogenesis, and wound healing (Fricke et al., 2020). The zebrafish develops quickly as most of the key organ systems are developed in 24–48 hours after fertilization. As a result, the developmental process and the phenotypic effects of genetic manipulations can be explored much faster than ever before, and hence, hypothesis testing and data generation are expedited. Zebrafish also has good applications in high-throughput screening. Their low size, great fecundity, laying of hundreds of eggs during mating, and simplicity of culture allow microtiter plate arraying embryos to undergo large-scale chemical or genetic screening. This scaling makes

 DOI: 10.1201/9781003735113-16

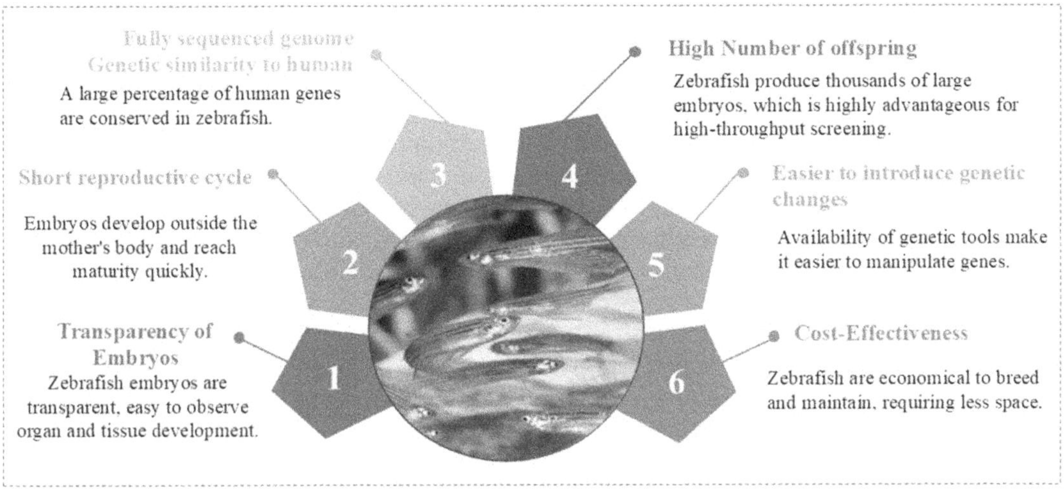

FIGURE 16.1 Key biological advantages of using zebrafish (*Danio rerio*) as a model organism for genetic and high-throughput screening.

zebrafish an effective drug discovery and toxicological studies. Combining transgenic reporter lines with automated imaging systems also increases their application in screening phenotypes and in quantitative analysis.

Genetic reporters utilized in combination with high-throughput image acquisition allow researchers to perform systematic phenotype analysis and obtain quantitative measures of large compound libraries or genetic perturbations (Huang et al., 2024). Functional genomics is a unifying field that seeks to explain how genes and their molecular products are implicated in biological processes, phenotypic phenomena, and pathogenesis. However, unlike classical genetics, which aim mostly to determine the impact that the changes on each gene singly may have, functional genomics focuses on extensive studies of the expression of genes, intergenic effects, and regulatory networks in the full genomic context (Juan et al., 2025).

This area is dependent on high-throughput methodologies, such as transcriptomics, proteomics, metabolomics, and genome-editing technologies (among others) to uncover dynamic genotype–phenotype interactions. Functional genomics offers a powerful platform on which the functions of genes are dissected in zebrafish in developmental, physiological, and pathological processes. The amenability of the organism to forward and reverse genetic stresses, spanning chemical mutagenesis to targeted gene silence, has allowed defining key regulators of tissue patterning, organ development, and cellular homeostasis (Kalaichelvan et al., 2025). Massive mutagenesis screens have historically revealed many developmental genes, whereas new genome-editing technologies like transcription activator-like effector nucleases (TALENs), zinc-finger nucleases (ZFNs), and CRISPR/Cas have transformed functional studies with more precise studies than ever before. Zebrafish functional genomics has played a significant role in disease modeling, especially in human genetic diseases. Disease phenotypes are recapitulated in zebrafish embryos or adults. By creating conditional point mutations or knock-ins introducing patient variants, scientists are able to test their molecular basis in vivo. Finally, the of zebrafish as a functional genomics model system not only helps to uncover more basic biological processes but also acts as a translational interface with the mammalian system and human health.

16.2 GENOME EDITING AND REGENERATIVE MEDICINE

With the development of the genome-editing technologies, modern biology and medicine have changed dramatically. These tools have offered a unique chance to tightly control genes and deconstruct how they influence development, physiology, and regeneration in zebrafish studies. The most popular of them is the CRISPR/Cas9 system, which has become the simplest, efficient, and versatile method to use. Using guide RNAs with reverse complementary sequences to the target sequences, researchers can cause site-specific double-strand breaks (DSBs), resulting in insertions, deletions, or specific edits through homology-directed repair (HDR; Liu et al., 2025).

Zebrafish genome editing has made possible the generation of loss-of-function and reporter lines and disease models which recapitulate human genetic disease. As an example, conditional knockouts of tumor-suppressor gene (e.g., *tp53*) or oncogenes (e.g., *kras*) have been useful in the study of cancer initiation and progression. Similarly, knock-ins based on CRISPR have been utilized to insert fluorescent reporters into loci of interest in order to dynamically visualize the patterns of gene expression and cellular reactions during regeneration. Among the potential uses of zebrafish genome editing, one can identify the sphere of regenerative medicine (Medishetti et al., 2022). Zebrafish have remarkable regenerative capacity when compared to mammals, having the capacity to regenerate large areas of the heart following damage, repair damaged retinal neurons, regenerate amputated fin rays, and repair spinal lesions (Peng et al., 2024).

Through genome-editing technologies, scientists have started to determine important genes and molecular pathways including fgf, Wnt, and notch signaling that coordinate these regenerative reactions. The comparative study of these pathways of zebrafish and mammals has provided important information on the reasons behind the lack of regenerative capacity in higher vertebrates and how it can be reinstated therapeutically (Pourshaban--Shahrestani et al., 2025). In addition to this, zebrafish provide a cost-effective, ethically preferential platform to conduct preclinical screening of regenerative drugs and gene therapies. The small molecules that were identified using a genetic screen in zebrafish have been demonstrated to stimulate the proliferation of cardiomyocytes, stimulate neural regeneration, or to regulate stem cell differentiation, highlighting its translational potential and forming the foundation of new therapeutic interventions based on human regenerative pathways (Parthasarathy et al., 2025). On a larger therapeutic scale, genome editing in zebrafish is used to validate the function of candidate genes previously identified through next-generation sequencing or human genome-wide association studies. Disease-susceptibility genes of interest can be easily edited in zebrafish to determine their roles in the organism, thus removing the gap between genetic association and mechanistic insights. This integrative method is an influential way of linking human genomic data with biological process and therapeutic discovery.

16.3 FUNCTIONAL GENOMICS APPROACHES

Functional genomics goes beyond sequence analysis to determine biological functions for genes and their interactions and precisely how the activity of genes leads to a phenotype (Rajan et al., 2024). Such a goal in zebrafish is accomplished by several experimental approaches: gene knockouts, knock-ins, transgenesis, and large-scale mutagenesis. These strategies enable targeted as well as genome-wide analyses of gene function to elucidate developmental pathways and disease mechanisms.

16.3.1 Gene Knockouts and Knock-ins

Zebrafish gene knockout technologies have dramatically improved with the development of programmable nucleases, including TALENs and, more recently, CRISPR/Cas9. These reagents enable efficient editing of the zebrafish genome through the induction of targeted DSBs that, following repair, either disrupt gene function or introduce specific mutations. The CRISPR/Cas9 system, in particular, has transformed genetic work in zebrafish because of its ease of use, low cost, and high-throughput potential (Sharma et al., 2021). It facilitates the generation of loss-of-function mutants for nearly any gene, thus greatly accelerating reverse genetic approaches. Knock-in experiments, defined here as the insertion of exogenous DNA sequences at specified genomic loci, are increasingly feasible in zebrafish. By applying HDR or new variants of CRISPR-based technologies, such as prime editing and base editing, researchers can introduce point mutations, fluorescent reporters, or epitope tags into endogenous genes (Wei et al., 2025). These engineered lines allow the real-time tracking of gene expression, protein localization, or lineage tracing during embryonic and postembryonic development. Such precise genetic alterations have provided unprecedented insight into gene regulation and cell type–specific functions.

16.3.2 High-Throughput Mutagenesis and Functional Screening

High-throughput mutagenesis strategies have been another key driver of functional genomics in zebrafish, enabling the systematic dissection of genetic mechanisms underlying development and disease. Early large-scale forward genetic screens using chemical mutagens, such as ENU, yielded thousands of mutants with diverse phenotypes that helped uncover a number of essential genes for organ formation, patterning, and signaling. Pioneering work by Nusslein-Volhard and colleagues in the 1990s established zebrafish as a genetic model organism, similar to *Drosophila* or *Caenorhabditis elegans*, but at vertebrate complexity. Modern sequencing and genome-editing tools now enable targeted reverse genetic screens in which collections of mutants for particular gene families or pathways can be rapidly generated.

CRISPR-based multiplex editing enables the simultaneous targeting of multiple genes, making zebrafish a very attractive model for the elucidation of genetic redundancy and epistatic interactions (Shin et al., 2023). Moreover, several approaches that combine genetic screens with transcriptomic and proteomic profiling allow a holistic view to be taken of how gene perturbations affect cellular networks and organismal phenotypes. Zebrafish embryos are also perfect for chemical genetics: high-throughput screening of small molecules can be used to modulate gene function or rescue mutant phenotypes (Patton et al., 2021). This approach marries basic genomics with drug discovery, identifying compounds targeting particular molecular pathways. The transparency of embryos and their compatibility with automated imaging systems render zebrafish one of the most efficient platforms for phenotype-based screening.

16.4 EVOLUTION OF GENOME-EDITING TECHNOLOGIES

With the advent of genome-editing technologies, it has been possible to radically remodel functional genomics in zebrafish, allowing the accurate manipulation of genes to determine their functions in development, physiology, and pathology. Zebrafish have become a model vertebrate to use and perfect to apply these tools because of their transparent embryos, outward development, and genetic acceptability. To date, CRISPR/Cas9, TALENs, and ZFNs represent the most used genome-editing technologies, whereas more recent technologies like base editing and prime editing are also becoming increasingly popular (Nipu et al., 2025). Both systems create specific DNA double strand or base breakages that are thereafter mended by cellular own repair mechanisms to induce the intended genetic change. Figure 16.2 represents functional genomics strategies in zebrafish to dissect gene function in regeneration, cardiac and neural development.

16.4.1 CRISPR/Cas9

The most notable and most commonly used genome-editing system in zebrafish research is CRISPR/Cas9 system which is based on bacterial adaptive immune system. Cas9 nuclease creates a site-specific double-strand cut in genomic DNA directed to an assembled synthetic single-guide RNA, the sequence of which is complementary to a target sequence adjacent to a protospacer adjacent motif (PAM, usually NGG in *Streptococcus pyogenes* Cas9). Repair of the induced DSB can occur through non-homologous end joining (NHEJ), which can commonly lead to small insertions or deletions that cause gene disruption, or through HDR, which allows the insertion of accurately replenished sequence when a repair template is present. Common procedures related to zebrafish include

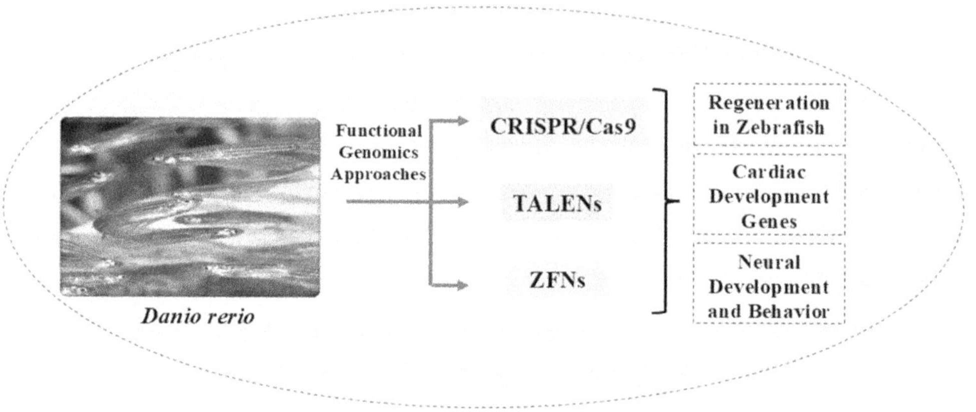

FIGURE 16.2 Functional genomics approaches in zebrafish focusing on CRISPR/Cas9, TALENs, and ZFNs to study regeneration and development.

microinjecting one-cell embryos of zebrafish with Cas9 mRNA or protein along with sgRNA. The resultant reagents diffuse across the forming embryo, causing elevated abundance of genome-editing occasions that most often result in mosaic F0 animals (often referred to as crispants).

These crispants may be directly phenotyped or crossed to generate stable mutant lines in the second generation. CRISPR/Cas9 works especially well with zebrafish due to their external embryonic development, fast development, and mosaic tolerance at early generation (Yin et al., 2025). The mutation rates in injected embryos with sgRNAs designed to cut well followed in the mutagenic environment are generally found to be very high (greater than 70–90%). Various parameters such as sgRNA sequence, accessibility of the targets, chromatin, and quality of the injection affect editing efficiency. CHOPCHOP, CRISPR scan, and CRISPR are bioinformatic tools that can be used to design sgRNAs in the most optimized way possible to the zebrafish genome (Figure 16.2). However, because CRISPR/Cas9 is highly efficient, it also may induce off-target effects, in which nontarget genomic loci are cleaved with partial sequence homology.

This off-target behavior is likely to complicate the phenotypic study and thus should be strictly addressed. The ways to reduce off-target events include the use of variants of Cas9 with high fidelity, truncated sgRNAs, and in vitro or in silico validation of potential off-target loci, and subsequent confirmation by sequencing (Zheng et al., 2022). Recent advances in CRISPR technologies, e.g., Cas12a and Cas13 systems, offer enhanced flexibility, specificity, and extended PAM. CRISPR/Cas9 system has been effectively used to produce loss-of-function mutants of a wide range of zebrafish genes representing a wide range of biological conditions. Examples can be the knockout of mitfa to create pigmentless fish useful in imaging; knockout of tp53 that has been useful in modeling tumorigenesis; and mutants in a gene, dystrophin (dmd), that can be used to recapitulate phenotypes of Duchenne muscular dystrophy (DMD). It has also been implemented on functional screening, where a complex combination of separate sgRNAs is injected to perturb many genes simultaneously, allowing gene networks and pathway interactions to be interrogated quickly.

16.4.2 Transcription Activator-Like Effector Nucleases

Before the introduction of CRISPR/Cas9, TALENs were the most common technology of targeted genome editing in zebrafish. TALENs are tailored nucleases that consist of customizable DNA-binding domains based on *Xanthomonas* transcription activator-like effectors complex with the FokI endonuclease catalytic region. The recognition of a single nucleotide by each repeat is based on a set of repeat variable diresidues which enables the design of TALEN pairs that specific repeat pairs recognize a specific set of flanking DNA sequences around the site of intended cleavage. Upon binding to each other, the FokI domains dimerize, forming a site-specific DSB (Zhu et al., 2021).

TALENs possess a highly flexible, modular architecture that enables them to target virtually any genomic locus. Several tools (e.g., TALEN Targeter, TAL Effector-Nucleotide Targeter) can help construct TALENs efficiently with the best possible spacer length and composition of bases to be highly active and have minimal off-target effects. At around 2011, the first zebrafish adaptations of TALENs came out and became immediately useful as a system to generate targeted mutations. Similar to CRISPR/Cas9, TALEN mRNA is injected into embryos at the one-cell stage, causing DSBs which are repaired through NHEJ to create small insertions or deletions at the target locus. Initial seminal work was able to knock down several genes, such as the golden (slc24a5) gene and tyrosinase, which induced pigmented phenotypes, and confirmed the effectiveness of this method. The TALENs are highly target-specific and has low toxicity to off-target cleavage due to the length of their recognition sequences (30–40 base pairs) (Figure 16.2). This makes them especially convenient to use in editing genomic regions where high fidelity is needed or in regions of the genome where appropriate PAM sequences to use with CRISPR are not available. TALEN assembly, however, is less efficient and time-consuming than CRISPR/Cas9, making it impractical to use on a large scale to perform mutagenesis.

16.4.3 Zinc-Finger Nucleases

The former group of intentional genome editors are ZFNs, which were the predecessors of other advances like TALENs and CRISPR. ZFNs are made up of DNA-binding domain that is made up of multiple tandem zinc-finger motifs that are each bound to three base pairs and are attached to the FokI nuclease domain. Dimerization of the FokI domains and the formation of a DSB occur upon the binding of two ZFN monomers to the opposite faces of the strands of DNA at adjacent sites, and the DSB is repaired by the NHEJ or HDR. ZFNs were initially used in zebrafish in the late 2000s to produce heritable mutations, and later started being used to target genome modification in the species (Choi et al., 2021). They include knockouts of target gene using ZFN of gold and no tail (ntl), one of the first examples of targeted mutagenesis of vertebrate embryos. However, the technology had a number of disadvantages such as context-dependent binding, complicated assembly, and increased rate of off-target activity, thus limiting its use even more when more efficient systems were developed. Still, ZFNs are relevant to provide evidence-of-concept of targeted genome editing in zebrafish and other vertebrates.

16.5 INTEGRATION OF TRANSCRIPTOMIC AND PROTEOMIC STUDIES WITH GENOME EDITING

Genome editing is on par with causative genetic changes, but when combined with transcriptomic and proteomic studies, it connects the gap between genotype and phenotype. When paired with RNA sequencing and proteomics, coupling CRISPR/Cas9-mediated perturbations allows mapping the downstream molecular consequences and discovering mechanisms of compensation or secondary regulation. CRISPR-induced mutants in zebrafish can be subjected to single-cell RNA sequencing, which measures cell type–specific transcriptional phenotypes (de Vrieze et al., 2021). This enables the explanation of the networks of genes that coordinate the development or regeneration of organs. As an example, transcriptomic analysis of gata1 mutant zebrafish embryos has shown dysregulation in the heme biosynthesis and erythropoietic pathways, thereby explaining the mechanism through which the loss of this transcription factor disrupts lineage commitment. Proteomics provides information into post-translational changes and protein–protein interactions disturbed by genetic changes. The use of quantitative mass spectrometry on zebrafish mutants, e.g., tp53 knockouts, has shown alterations in stress response and apoptosis-related proteins, contributing to the knowledge on tumor-suppressor networks.

16.6 CASE STUDIES

16.6.1 Cardiac Development Genes

The zebrafish has been of great use in clarifying the genetic controlling factors of vertebrate developing heart. Using CRISPR-mediated mutagenesis, mutation of tbx5, gata4, and hand2 illustrates the functions of these genes in regulating chamber morphogenesis, looping, and contractile activity. Phenotypes of these mutants indicate the appearance of similar patterns to congenital human heart defects such as septal defects and arrhythmias (Cheung et al., 2025). Functional genomics has also found that microRNAs and long non-coding RNAs regulate cardiac transcriptional networks. The combination of transcriptomic data of CRISPR perturbations in cardiomyocytes has outlined downstream signaling controlled by nkx2.5, providing mechanistic understanding of cardiac differentiation and regeneration.

16.6.2 Neural Development and Behavior

Fundamental roles of the synaptic genes, mef2ca, syn2a, and grin2b, regarding locomotor behavior and learning were clarified, using the CRISPR/Cas9 gene knockout, and this shows conserved neuronal pathways. Genetic interference with genes associated with neuropsychiatric disorders, such as discl and scn1lab, restored epileptic and schizophrenia-like phenotypes, allowing real-time visualization of circuits and pharmacological interrogation (Chahardehi et al., 2023). Transcriptomic analysis of these mutants demonstrates dysregulation of neurogenetic pathways in neurotransmission and synaptic plasticity, which supports the use of the zebrafish as a neurogenetic translational model.

16.6.3 Regenerative Potential: Zebrafish

Zebrafish has established itself as a forefront of the regenerative biology of vertebrates because of its ability to regenerate a variety of complex tissues and organs during the course of its existence. Zebrafish regenerate cardiac, hepatic, fin, retinal, and spinal cord tissues with high anatomical and functional fidelity unlike mammals, which have limited regenerative potential. As a result, zebrafish forms an indispensable system in the context of understanding the cellular and molecular processes, which form the basis of tissue renewal and repair. Recent technological gains in genome engineering, in particular, CRISPR/Cas9, make it easier to specifically interrogate genes and signaling pathways involved in regenerative processes.

16.6.4 Heart Regeneration

The cardiac tissue regeneration is an example of zebrafish regenerative capacity following an injury. After surgical removal of about 20% of the ventricle, zebrafish regenerate the removed myocardium in two months. This regeneration is mostly based on dedifferentiation and replication of the already existing cardiomyocytes, but not the recruitment of the stem cells. The re-entered cardiomyocytes re-express embryonic markers, migrate to the lesion area, and remodel functional myocardium (Chakraborty et al., 2025). It requires reactivation of important transcription factors – gata4, hand2, and mef2c – in order to induce redifferentiation of cardiac cells. The further study of these processes promises to develop regenerative therapies of human heart failure where the cardiomyocyte growth is still detrimentally limited.

16.6.5 Liver Regeneration

Both hepatocytes and biliary epithelial cells coordinate liver regeneration in zebrafish. After a partial hepatectomy or chemical-induced injury, hepatocytes increase in number to replenish liver bulk.

In the cases of extensive damage with massive depletion of populations of hepatocytes, biliary epithelial cells transdifferentiate into hepatocytes. Sox9b and hhex are tight regulators of this cellular plasticity. The zebrafish model is, therefore, a special platform to study in vivo cellular reprogramming and lineage conversion, which can supply information that can be used in regenerative therapies of hepatic diseases.

16.6.6 Fin Regeneration

One of the most widely examined vertebrate regenerative models is fin regeneration. The amputation induces wound closure, and then blastema (proliferative mass of the progenitor cells, i.e., dedifferentiated osteoblasts, fibroblasts, and epidermal cells) appears. The blastema mediates outgrowth and patterning reconstituting the original fin architecture, bone, connective tissue, nerves, and vasculature. Fibroblast growth factor (FGF), Wnt, and Hedgehog signaling pathways are necessary to maintain and differentiate blastemal cells. The system offers a workable background to analyzing the patterning, morphogenesis, and positional memory in vertebrate tissues.

16.6.7 Zebrafish Disease Modeling with Genome-Edited Animals

Zebrafish can be utilized as a disease model because it is capable of replicating all complex pathology under a living vertebrate situation. Genome-editing technology enables the accurate insertion, deletion, or substitution of sequences of genes, thus enabling scientists to model human disease mutations. The ability has been used to cover a range of disease areas such as cardiovascular, neurodegenerative, muscular, and metabolic diseases.

16.6.8 Cardiovascular Disorders

Morphologically and physiologically, the zebrafish heart is analogous to the human heart, and they share the signaling pathways that are conserved to control the contractility and the rhythm of the cardiac muscle. Zebrafish that are genome-edited have previously been utilized to model both acquired and congenital heart diseases. Knockout (KO) or mutation of the cardiac troponin T (tnt2) gene causes a silent heart phenotype with loss of contractility, and is a model of dilated cardiomyopathy. In the same manner, CRISPR-based knockout of kcnh2, the zebrafish ortholog of the human hERG potassium channel, results in sustained cardiac repolarization in the same way the electrophysiological characteristics of Long QT syndrome are recapitulated. Besides these structural disorder models, more recent zebrafish models of myocardial infarction and heart failure have been developed, allowing the study of regenerative responses and the screening of cardioprotective drugs (Cai et al., 2021). As an example, transcription factors gata4 and hand2 mutants have been shown to fail at regenerating after an injury, revealing possible therapeutic genes of use to promote cardiac healing in humans. Fluorescent labeling of living animals allows real-time imaging of cardiac morphology, blood flow, and the abnormalities in the rhythm.

16.6.9 Neurodegenerative Disorders

Zebrafish can be used as an effective model to research on neurodegenerative diseases, including Parkinson's disease, Alzheimer's disease, and amyotrophic lateral sclerosis. Knockout of lrk2, pink1, and parkin by CRISPR/Cas9 triggers the demise of dopaminergic neurons and locomotor deficits that are akin to the phenotypes of Parkinson's disease. Similarly, mutations in the APP and tau genes cause plaque formation and cell death, which recapitulate the pathogenic processes underlying Alzheimer's disease. Moreover, the neuronal degeneration, malformation of the axon transport, and synaptic abnormality can be observed directly via the optical transparency of the zebrafish using fluorescent reporters (Balzamino et al., 2025). Such models are priceless in the screening of neuroprotective compounds that cause neuronal integrity or mito dynamics. The sod1 and tardbp mutants are amyotrophic lateral sclerosis (ALS) replicants through which they cause degeneration of the motor neurons and atrophy of the muscles, thus facilitating high-throughput screening of therapeutics that postpone the progression of the disease.

16.6.10 Muscular Disorders

Zebrafish has been successfully used as a model of muscular pathologies, such as DMD. The species sapje mutant has a mutation in the dystrophin gene, and this causes degeneration of the muscle fibers, poor swimming, and poor survival which are characteristic of DMD. Models based on CRISPR can now introduce targeted human variant-corresponding mutations, which are a platform to test exon-skipping approaches, gene-therapy vectors, and small molecules that stabilize muscle architecture. Zebrafish models of neuromuscular diseases, such as congenital myopathies and myasthenic syndromes, allow researchers to directly observe muscle repair and innervation behaviors, thereby contributing to drug discovery.

16.7 CONCLUSION

Genome editing has transformed studies based on zebrafish, which has placed the small vertebrate at the center as a flexible instrument in functional genomics, regenerative biology, and therapeutic discovery. CRISPR/Cas9, TALENs, and base editors have allowed DNA editing of the genome of zebrafish to be carried out with high speed and accuracy, therefore supporting loss continuously and gain continuously of function studies, high-throughput mutagenesis screens, and the modeling of complex human diseases. Zebrafish drug screens have already helped in the discovery of candidate compounds in epilepsy, cardiomyopathies, metabolic disorders, and cancer and therefore, their use is translational. New technologies in genome editing like prime editing and epigenome editing will also be introduced in the next generation, which can expand the functional genomics toolkit in zebrafish. Combining multi-omics methodologies, including single-cell transcriptomics, proteomics, metabolomics, and epigenomics, should provide a comprehensive picture of how cells react to genetic manipulations. These data layers may be combined with regenerative phenotypes or disease in zebrafish to establish important gene networks, signaling centers, and molecular checkpoints. Integrative studies of this kind will become educational not only in fundamental biological knowledge but also in rational design of regenerative therapies and targeted drugs. Particularly interesting, edited zebrafish models show specific potential in regenerative medicine, particularly through understanding the molecular secretions of tissue repair, and through providing a preclinical model system of compounds or genetic interventions that can stimulate regeneration. Also, zebrafish offer a scalable in vivo platform to be used in precision therapeutics and synthetic biology, including engineering tissue-specific gene circuits, testing gene-therapy vectors, and modeling human–organoid interactions. The above abilities demonstrate the possible role of zebrafish serving as an interface between experimental biology and clinical translation due to its rapid, inexpensive, and ethically viable alternative to mammalian studies. Overall, genome-edited zebrafish have become a mainstay in contemporary biomedical studies. With the ongoing improvements in genome editing alongside the incorporation of multi-omics and the ability to assess the functionality of genes in vivo, zebrafish are poised to continue to lead in the clarification of gene functions, the acceleration of regenerative approaches, and supporting the discovery of therapeutic approaches of the next generation. Combining these technologies makes zebrafish not just a model organism, but also a predictive and translational platform that can close the divide between laboratory findings and clinical practice.

REFERENCES

Balzamino, B. O., Severino, M., Cafiero, C., Coassin, M., Di Zazzo, A., & Micera, A. (2025). The zebrafish as a model for ocular translational research: From retinal repair to regeneration. *Cells*, 14(17), 1405. https://doi.org/10.3390/cells14171405

Cai, W. J., Li, J., Li, L., Chen, X., Wei, J. R., Yin, Z., … Liang, X. F. (2021). Knockout of t1r1 gene in zebrafish (*Danio rerio*) by CRISPR/Cas9 reveals its roles in regulating feeding behavior. *Aquaculture*, 545, 737189. https://doi.org/10.1016/j.aquaculture.2021.737189

Carrington, B., Bishop, K., & Sood, R. (2022). A comprehensive review of indel detection methods for identification of zebrafish knockout mutants generated by genome-editing nucleases. *Genes*, 13(5), 857.https://doi.org/10.3390/genes13050857

Chahardehi, A. M., Hosseini, Y., Mahdavi, S. M., & Naseh, I. (2023). Zebrafish, a biological model for pharmaceutical research for the management of anxiety. *Molecular Biology Reports*, 50(4), 3863–3872.https://doi.org/10.1007/s11033-023-08263-1

Chakraborty, P., Pal, S., Galstyan, D. S., Petersen, E. V., Amstislavskaya, T. G., Kolesnikova, T. O., … Kalueff, A. V. (2025). Zebrafish models for studying central nervous system effects of Ayurvedic medicinal plants. *Journal of Ayurveda and Integrative Medicine*, 16(4), 101169. https://doi.org/10.1016/j.jaim.2025.101169

Cheung, M. Y., Jiang, C., Hassan, I. U., Wang, H., Guo, D., Dio, D. W., … Zhao, H. (2025). Knockout of thyroid hormone receptor alpha a (thraa) enhances cardiac regeneration in zebrafish through metabolic and hypoxic regulation. *Cell Communication and Signaling*, 23(1), 1–21.https://doi.org/10.1186/s12964-025-02350-5

Choi, T. Y., Choi, T. I., Lee, Y. R., Choe, S. K., & Kim, C. H. (2021). Zebrafish as an animal model for biomedical research. *Experimental & Molecular Medicine*, 53(3), 310–317.https://doi.org/10.1038/s12276-021-00571-5

Fricke, T., Smalakyte, D., Lapinski, M., Pateria, A., Weige, C., Pastor, M., … Bochtler, M. (2020). Targeted RNA knockdown by a type III CRISPR-Cas complex in zebrafish. *The CRISPR Journal*, 3(4), 299–313. https://doi.org/10.1089/crispr.2020.0032

Huang, W., Xiong, T., Zhao, Y., Heng, J., Han, G., Wang, P., … Zhang, Q. C. (2024). Computational prediction and experimental validation identify functionally conserved lncRNAs from zebrafish to human. *Nature Genetics*, 56(1), 124–135. https://doi.org/10.1038/s41588-023-01620-7

Juan, T., Molina, T., Xie, L., Papadopoulou, S., Cardoso, B., Jha, S. G., & Stainier, D. Y. (2025). A recombinase-activated ribozyme to knock down endogenous gene expression in zebrafish. *Plos Genetics*, 21(2), e1011594. https://doi.org/10.1371/journal.pgen.1011594

Kalaichelvan, A., Nadarajapillai, K., Sellaththurai, S. R., Arachchi, U. P. E., Kim, M. J., Jung, S., & Lee, J. (2025). CRISPR/Cas9-induced knockout of tumor necrosis factor-alpha-type I augments viral infection in zebrafish. *Fish & Shellfish Immunology*, 157, 110092. https://doi.org/10.1016/j.fsi.2024.110092

Liu, H., Sun, N., Liu, Z., Li, J., & Zhang, X. (2025). Knockout of bcas3 gene causes neurodevelopment defects in zebrafish. *Biological Research*, 58(1), 34.https://doi.org/10.1186/s40659-025-00615-4

Medishetti, R., Balamurugan, K., Yadavalli, K., Rani, R., Sevilimedu, A., Challa, A. K., … Chatti, K. (2022). CRISPR-Cas9-induced gene knockout in zebrafish. *STAR Protocols*, 3(4), 101779. https://doi.org/10.1016/j.xpro.2022.101779

Nipu, N., Wei, L., Thomas, J., & Mennigen, J. A. (2025). Methylene blue is a widely used antifungal agent that confounds behavioral toxicity assays in larval zebrafish. *Environmental Science & Technology Letters*, 12(2), 130–136. https://doi.org/10.1021/acs.estlett.4c00955

Parthasarathy, S., Aly, S. H., Tharumasivam, S. V., Siva, D., Krishnasamy, G., Chavaan, A., & El-Shazly, M. (2025). Zebrafish in the spotlight: expanding Frontiers in toxicology and drug discovery. *Toxicology Research*, 14(4), tfaf095. https://doi.org/10.1093/toxres/tfaf095

Patton, E. E., Zon, L. I., & Langenau, D. M. (2021). Zebrafish disease models in drug discovery: From preclinical modelling to clinical trials. *Nature Reviews Drug Discovery*, 20(8), 611–628. https://doi.org/10.1038/s41573-021-00210-8

Peng, W., Zhang, Y., Song, B., Yang, P., & Liu, L. (2024). Developmental delay and male-biased sex ratio in esr2b knockout zebrafish. *Genes*, 15(5), 636. https://doi.org/10.3390/genes15050636

Pourshaban-Shahrestani, A., Rezazadeh, A., & Hassan, J. (2025). Zebrafish as a model for assessing biocide toxicity: A comprehensive review. *Toxicology Reports*, 101980. https://doi.org/10.1016/j.toxrep.2025.101980

Rajan, A. R. D., Huang, Y., Stundl, J., Chu, K., Irodi, A., Yang, Z., … Bronner, M. E. (2024). Generation of a zebrafish neurofibromatosis model via inducible knockout of nf2a/b. *Disease Models & Mechanisms*, 17(12), dmm050862. https://doi.org/10.1242/dmm.050862

Sharma, P., Sharma, B. S., & Verma, R. J. (2021). CRISPR-based genome editing of zebrafish. *Progress in Molecular Biology and Translational Science*, 180, 69–84. https://doi.org/10.1016/bs.pmbts.2021.01.005

Shin, M., Yin, H. M., Shih, Y. H., Nozaki, T., Portman, D., Toles, B., … Lawson, N. D. (2023). Generation and application of endogenously floxed alleles for cell-specific knockout in zebrafish. *Developmental Cell*, 58(22), 2614–2626. https://doi.org/10.1016/j.devcel.2023.07.022

de Vrieze, E., de Bruijn, S. E., Reurink, J., Broekman, S., van de Riet, V., Aben, M., … van Wijk, E. (2021). Efficient generation of knock-in zebrafish models for inherited disorders using CRISPR-Cas9 ribonucleoprotein complexes. *International Journal of Molecular Sciences*, 22(17), 9429.https://doi.org/10.3390/ijms22179429

Wei, Z., Zha, C., & Xiong, M. (2025). Gene editor: Innovative use of zebrafish models in inflammatory bowel disease research. *Cell Biology Research*, 6(2), 22–26. https://doi.org/10.18063/cbr.v6i2.907

Yin, C., Wen, J., Wang, S., Xie, T., Chen, Z., Wang, Z., … Fang, M. (2025). CRISPR/Cas9-mediated ehd3 knockout enhances high unsaturated fatty acid content in zebrafish muscle. *Food Bioscience*, 107164. https://doi.org/10.1016/j.fbio.2025.107164

Zheng, J., Long, F., Cao, X., Xiong, B., & Li, Y. (2022). Knockout of Katnal2 leads to autism-like behaviors and developmental delay in zebrafish. *International Journal of Molecular Sciences*, 23(15), 8389. https://doi.org/10.3390/ijms23158389

Zhu, J., Xu, H., Song, H., Li, X., Wang, N., Zhao, J., … Xia, H. (2021). CRISPR/Cas9-mediated grna gene knockout leads to neurodevelopmental defects and motor behavior changes in zebrafish. *Journal of Neurochemistry*, 157(3), 520–531. https://doi.org/10.1111/jnc.15307

17 Genome Editing for Remodeling of Root System Architecture (RSA) to Alleviate Drought Stress in Crops

Gyanendra Kumar Patra, Upasana Mahopatra,
Appanna V. Koppad, and Gowsiya Shaik

17.1 INTRODUCTION

17.1.1 Definition of Drought and Its Effect on Agriculture

Drought is generally understood as a sustained period of water deficit that has an adverse impact on plant growth and productivity (Blum, 2017). It is a complex phenomenon that is influenced by environmental factors, such as the variable precipitation, the water storage capacity of the soil, the evapotranspiration rate and high temperature stress. Unlike water stress, which may last for only a few days, drought can go on for weeks or months, leading to cumulative damage to agricultural systems. From a physiological point of view, drought leads to a decrease in soil water availability and to a decrease of turgor pressure, to stomata closure and to the inhibition of photosynthesis (Farooq et al., 2009). Water deficit affects cellular homeostasis as it destroys membrane integrity, protein stability and enzyme activity. Prolonged drought promoted an oxidative stress state by increasing the accumulation of reactive oxygen species (ROS) and inducing lipid, protein and nucleic acid oxidation (Mittler, 2017). Agronomically, drought disrupts all stages of plant growth, from germination to grain filling. Seedling emergence can be delayed, root development limited, leaf growth stunted and reproductive development ruined. Consequently, drought poses severe constraints on the crop yield potential as well as the quality.

17.1.2 Statistics of Crop Yield Losses Due to Drought Stress

Global crop production is subject to significant losses because of drought. Estimates indicate that drought is responsible for yield reduction of up to 50% in maize, 30%–40% in wheat and 20%–30% of rice under severe stress condition (Lobell et al., 2014). In legumes, like chickpea and soybean, terminal drought is a major cause of pod abortion and decreased seed weight resulting in economic losses to farmers (Farooq et al., 2009). Oilseed crops such as sunflower and rapeseed have also been found to have a decrease in oil content and alteration in fatty acid (FA) composition under water stress (Barnabás et al., 2008). Statistical analyses show that about 40% of global cropland is impacted by drought per year, and such losses are expected to be amplified under a climate change scenario which is likely to result in more frequent and intense occurrences of water-deficit-related events (Lesk et al., 2016). Particularly in agricultural areas that are rainfed, where irrigation facilities are limited, drought becomes a major factor that determines the instability of yield.

DOI: 10.1201/9781003735113-17

17.1.3 Economic Impact of Crop Failures due to Drought

Drought has serious economic implications for both smallholder farmers and national economies. Reduced yields result in lower marketable surplus, higher prices of commodities and financial stress on producers. For example, the 2010–2012 droughts in the US mid-west caused billions of dollars losses in maize and soybean production (Smith et al., 2013). Similarly, subsistence farmers in Africa and South Asia experience the problem of food insecurity caused by failure of the monsoon (rainfall); maize, wheat, rice and pulses serve as staple foods.

In addition to direct yield losses, drought causes secondary economic effects, including the increase in irrigation and fertilizer costs and disruption to supply chains. At the national level, recurrent drought events have a negative impact on the contributions of GDP made by the agricultural sector, highlighting the urgent need for drought-resilient crop varieties (Cattivelli et al., 2008).

17.2 PLANTS' RESPONSE TO DROUGHT STRESS

Plants have evolved a wide range of adaptive strategies to allow them to survive and be productive under water deficit conditions. These responses are working on many levels: morphological, physiological, biochemical and molecular.

17.2.1 Physiological Changes of Plants during Drought

- **Stomatal Regulation:** Stomatal closure under abscisic acid (ABA) is the first physiological reaction to water deficit. It decreases the rate of transpiration and loss of water but reduces CO_2 assimilation, the production of photosynthate (Cutler et al., 2010).
- **Leaf and Root Adjustments:** Drought causes leaf area reduction and leaf thickness, which reduces water loss per unit area. Root systems also tend to stretch deeper into soil layers to tap moisture (Lynch, 2013). These morphological changes improve survival at the expense of growth rates under chronic stress.
- **Osmotic Adjustment:** Plants concentrate compatible solutes like proline, glycine betaine and soluble sugars to maintain cell turgor and protect the enzymatic activities (Szabados & Savoure, 2010). Osmotic adjustment is also responsible for continued root elongation in drying soils.
- **ROS Detoxification:** Water deficit increases the levels of ROS. Plants increase the activation of antioxidant enzymes, such as superoxide dismutase (SOD), catalase (CAT) and peroxidases, to counter oxidative damage (Gill & Tuteja, 2010).

17.2.2 Molecular Mechanisms That Are Activated in Response to Water Deficit

Drought triggers both ABA-dependent and ABA-independent signaling pathways:

- **ABA-Dependent Pathway**: ABA biosynthesis is induced by water stress through the action of 9-cis-epoxycarotenoid dioxygenase (NCED). ABA is a ligand for PYR/PYL/RCAR receptors, a stress hormone that inhibits PP2C phosphatases, activates SnRK2 kinases and leads to stress-responsive gene expression through ABF/AREB transcription factors (TFs, Shinozaki & Yamaguchi-Shinozaki, 2007).
- **ABA-Independent Pathway**: Dehydration-responsive element binding (DREB) proteins control the expression of stress-induced genes in an ABA-independent manner. NAC and bZIP TFs are also involved in downstream reactions such as synthesis of osmoprotectants and accumulation of LEA proteins and antioxidant defense (Hu et al., 2006; Osakabe et al., 2014).
- **Epigenetic Regulation**: DNA methylation, histone modification and microRNAs (miRNAs; miR164, miR393) dynamically regulate drought-responsive gene networks to provide stress memory and priming for subsequent drought stress (Sunkar et al., 2012).

17.2.3 Importance of Understanding Plant Drought Responses toward Crop Improvement

Identifying the mechanisms of drought response is crucial in order to rationally design crop improvement strategies, for example, targeting of genome editing; identification of key TFs and signaling nodes; root morphological adaptations, which provide information for breeding for better water acquisition; and the understanding of osmotic adjustment and ROS detoxification, which facilitates metabolic engineering strategies. By combining physiological, molecular and genetic information, researchers will be able to speed up the development of drought-tolerant crops that can maintain yields during water-limited conditions.

17.3 CONTRIBUTION OF ROOT SYSTEM ARCHITECTURE TO DROUGHT STRESS

17.3.1 Definition and Components of Root System Architecture

Root system architecture (RSA) describes the spatial arrangement and pattern of root structure and is responsible for the ability of plants to acquire water and nutrients. Components of RSA include the following:

- **Primary Root Length**: Defines how deep the root can penetrate the soil.
- **Lateral Root Density and Angle**: Control horizontal soil investigation.
- **Root Hairs**: Enhance the surface area for water and nutrient absorption.
- **Crown and Adventitious Roots**: Add to the total root biomass and hardiness.

17.3.2 Adaptive Alterations in RSA to Drought Condition

Plants adapt RSA to water deficit through plastic responses:

- Deep root growth of primary roots results in more access to deeper moisture.
- Decreased development of lateral roots near to the surface saves resources.
- Increased root hair growth increases absorption efficiency.
- Altered root angle and gravitropism lead to growth in water-rich areas (Uga et al., 2013).

These adaptations comprise a steep deep cheap root ideotype that is optimized for drought water foraging.

17.3.3 Importance of RSA Plasticity in the Drought Tolerance

RSA plasticity also enables crops to respond to heterogeneous soil moisture profiles, and increases Water Use Efficiency (WUE) (Comas et al., 2013). High-RSA plasticity genotypes grow and yield under water deficit, while low-RSA plasticity genotypes suffer yield collapse. Therefore, RSA is a key target for crop improvement, both by conventional breeding and transgenics and genome editing.

17.4 NECESSITY FOR CROP IMPROVEMENT TO BE DROUGHT-TOLERANT

17.4.1 Existing Constraints on Drought-Tolerant Crop Varieties

Despite decades of research, many crops are not sufficiently drought-tolerant, owing to the following:

- Complex genetic structure of drought tolerance.
- Little genetic variation for root traits in elite germplasm.
- Environmental dependence of stress response (Cattivelli et al., 2008).

Conventional breeding and molecular approaches have helped to some extent to address these problems, but the pace of rapid climate change calls for more rapid and accurate approaches.

17.4.2 Benefits of Work on Creating Drought-Stressed Crops

- Agronomic stability in water-limited systems to guarantee food security.
- Freshwater resources conserved by using less irrigation.
- Improved resilience of farmers in rainfed agriculture.
- Improved nutrition by stressing seed development.

17.4.3 Need for Addressing Drought Tolerance in the Context of Climate Change

Climate models anticipate a rise in the frequency, intensity and duration of droughts, especially in tropical and subtropical regions (Lesk et al., 2016). Development of crop varieties more tolerant to drought is therefore a strategic requirement for sustainable agriculture, which can cushion yield fluctuations and guarantee food security to the world in the context of a changing climate.

17.5 APPROACHES TO IMPROVE CROP PRODUCTIVITY

17.5.1 Respectable Proceedings of Traditional Breeding Techniques

Conventional plant breeding involves a process of selective matings of good-looking varieties to produce offspring with enhanced qualities. The general strategies are as follows:

- Thus, hybridization involves crossing genetically diverse parents to blend advantageous traits together.
- **Selection**: Selecting offspring with the potential best drought tolerance characteristics (root depth, leaf water retention and so on)
- **Backcrossing**: Introgressing specific traits into elite cultivars with superior agronomic performance and higher potential.
- **Mutation Breeding**: Treating seeds or tissues with chemical mutagens or radiation to bring about diversity.
- Conventional breeding methods have yielded a large number of improved cultivars with increased yield, disease resistance and abiotic stress tolerance (Campos et al., 2004). Conventional selection likely has focused on phenotypical traits such as root depth, canopy temperature and osmotic adjustment for drought tolerance.

17.5.2 Limitations to Conventional Drought-Resistant Breeding:

Successful crossbreeding was hampered by several challenges:

- **Complexity of Drought Tolerance**: Drought tolerance is a complex polygenic trait determined by root architecture, stomatal control, osmolyte buildup and stress-responsive transcript networks. All selection based on phenotype is often inefficient and environment-dependent (Cattivelli et al., 2008).
- **Time Limits**: To stabilize a new variety, it can take anywhere from 8 to 12 years for the test.
- **Environmental Variability**: Drought tolerance cannot be replicated between different climates or soil types.
- **Limited Genetic Diversity**: Elite cultivars tend to have narrow genetic origins, which limits the magnitude of available pool for stress-tolerance alleles.

17.5.3 Lack of Time and Resources for Conventional Breeding Programs

Field trials are very expensive in terms of labor, land and water resources especially when evaluating drought response. Controlled environments (rainout shelters, hydroponics) make up some of the difference in stress simulation, but the stress environments are often not consistent between controlled nursery/research (CNR) plots and their translation to field environments. Nevertheless, while traditional breeding will continue to play a major role, it is also vital that faster and more accurate methods be available to achieve the goals set out for climate-resilient agriculture.

17.6 MARKER-ASSISTED SELECTION (MAS)

17.6.1 Principles of the Marker-Assisted Selection

Molecular markers associated with traits of interest are used in MAS to make breeding decisions. In the identification of alleles associated with stress resistance potential, breeders no longer need to rely on phenotypic selection to identify stress-resilient offspring at the DNA level.

Key concepts include the following:

- **Quantitative Trait Loci (QTL) Mapping**: Mapping QTL for traits related to drought tolerance (e.g., root depth, osmotic adjustment).
- **Marker Validation**: Ensuring good predictive merit of molecular markers for phenotypic performance under different environments.
- **Backcrossing Using MAS**: Fast introgression of good alleles into elite cultivars.

MAS increases the speed of selection for complex traits such as drought tolerance by decreasing dependence on the use of environmental screening.

17.6.1.1 Examples of Uses of Breeding for Drought Tolerance

MAS has been successfully applied to increase drought tolerance in major crops:

- **Rice**: Deeper roots through introgression of DRO1 enhanced water acquisition during terminal drought (Uga et al., 2013).
- Wheat MAS for TaDREB alleles of wheat improved root biomass and osmotic adjustment (Wang et al., 2020).
- **Maize**: QTLs for root angle and crown root number have been used for selection of drought-tolerant hybrids (Tuberosa et al., 2002).

17.6.2 Benefits and Constraints of MAS

17.6.2.1 Advantages

- Accurate selection for stress-tolerant alleles.
- Cut effects on field generations needed.
- QTLs can be combined for different polygenic traits.

17.6.2.2 Limitations

- Insufficient number of validated markers far reaching.
- It becomes essential to remember/record, the effects of QTL which may differ through environments (interactions between genotypes and environments; G×E).
- MAS can never breed a new allele as it needs existing genetic entities.

17.7 GENETIC ENGINEERING

17.7.1 Recent Advances in Transgenic Approaches to Drought Tolerance

Genetic engineering provides the tool to insert directly into a plant's genetic makeup the specific genes responsible for drought resilience without the limitations of the conventional crossing method. Strategies include the following:

- Overexpression of TFs that function in response to stress (DREB, NAC, bZIP) to activate downstream protective genes.
- Deployment of osmoprotectant-producing genes (proline, glycine betaine, etc.).
- Engineering elements of ABA signaling system for improved stomatal control and WUE.

Transgenic technologies have been able to provide marked enhancements in stress tolerance in model and crop species (Kasuga et al., 1999; Hu et al., 2006).

17.7.1.1 Examples of Successful Drought-Tolerant Genetically Changed Crops

- **Rice**: To date, overexpression of OsDREB1A conferred increased dehydration and osmotic stress tolerance (Dubouzet et al., 2003).
- **Maize**: Transgenic lines overexpressing ZmNF-YB2 had more root biomass and grain yield in the presence of water deficit (Nelson et al., 2007).
- **Tomato**: The overexpression of SlNAC family TFs enhanced the drought stress tolerance and water-use efficiency (Hu et al., 2006).

These successes demonstrate that a selected gene manipulation can be used to achieve physiological and morphological improvement that is impossible by conventional breeding.

17.7.2 Regulatory and Social Acceptance Issues

Although technically successful, genetic modified crops are facing regulations and public doubts:

- **Regulatory Background**: Transgenic crops must undergo strict safety assessments for environmental interactions, gene flow and human consumption.
- **Public Perception**: Fearing "GMOs" adoption is less than desired, especially in Europe and within parts of Asia.
- **Cost and Time**: Regulatory compliance is expensive and delay-driven for the commercialization.

As such, alternatives to transgenic genome editing based on CRISPR are gaining popularity because edited crops can be obtained without the introduction of foreign DNA.

17.7.3 Use of Genetic Modification to Create Better Organisms

17.7.3.1 Brief History of the Tools Used for Genome Editing (ZFNs, TALENs)

Genome engineering began with engineered nucleases, which can bring about these DNA breaks at targets of choice:

- **Zinc Finger Nucleases (ZFNs)**: Combining DNA-binding domains with zinc fingerdomains and FokI nuclease to allow targeted cleavage.
- **Transcription Activator Like Effector Nucleases (TALENs)**: T-DNA arrangements that consist of heterologous DNA-binding domains of *Xanthomonas* TALE effector protein conjugated to nucleases.

These tools enabled gene knockouts, insertions and replacements, but required complex protein engineering of each target sequence (Voytas, 2013).

17.8 INTRODUCTION TO CRISPR/CAS9 AND IT'S IMPORTANCE IN CROP IMPROVEMENT FOR DROUGHT RESISTANCE

CRISPR/Cas ushered a new age into genome editing because of its simplicity and versatility. Key components are as follows:

- **Cas9 Nuclease**: Produces double-strand breaks (DSBs) at the target site.
- **Single-Guide RNA (sgRNA)**: Leads to the binding of Cas9 to complementary DNA sequences around a protospacer adjacent motif (PAM).
- **Repair Mechanisms**: Non-homologous end joining (NHEJ) can cause indels (gene knockout), whereas homology-directed repair (HDR) can cause the precise replacement of DNA.

Other types, such as Cas12a, base editors and prime editors, broaden targeting scope and enable precise editing with no DSBs.

17.8.1 Advantages of CRISPR vs. Other Genome Editing Methods

- **Ease of Design**: sgRNAs can be synthesized with ease and can target several genes at the same time.
- **High Efficiency**: Ability to produce homozygous lines in a single generation with high frequencies of editing.
- **Precision**: Target specific, less off-target effects with optimized sgRNA design.
- **Applicable Sites**: miRNAs, coding and regulatory regions, epigenetic editing.
- **No Foreign DNA**: Edits come with the option of separating out the editing machinery and creating transgenic-free plants.

17.8.2 CRISPR for Crop Improvement

17.8.2.1 Principles and Applications of CRISPR/Cas9 Technology

17.8.2.1.1 Process of CRISPR/Cas9 Gene Editing

The CRISPR/Cas9 gene editing technology is a modified adaptation of a prokaryotic adaptive immune system. The system consists of three basic steps:

- **Target Recognition**: PG1 20 base-pairs complementary to the genomic target are embedded in the sgRNA. Cas9 is able to recognize this sequence both within a PAM, usually NGG in the case of *Streptococcus pyogenes* Cas9 (Jinek et al., 2012).
- **DNA Cleavage**: Cas9 catalyzes the cleavage of the target site causing a DSB.
- **DNA Repair**: The DSB is repaired by either of the following two mechanisms of the plant cell.
  - **Non-Homologous End Joining (NHEJ)**: Error-prone repair which results in small insertions/deletions, causing loss-of-function genes.
  - **Homology-Directed Repair (HDR)**: High-fidelity repair using the donor template and permitting for gene replacement or insertion (Puchta & Fauser, 2014).

Such a tool allows highly precise manipulation of genes involved in determining root architecture, root stress responses, metabolic pathways and so on.

17.8.3 Areas of CRISPR/Cas9 System

CRISPR/Cas9 genome editing involves a process that needs three essential elements:

- **Cas9 Nuclease**: A DNA-cutting enzyme of programmable efficacy.
- **sgRNA**: Takes Cas9 to the target genomic site.
- **Delivery System**: Makes sure the components are efficiently delivered into plant cells.

Optional parts are Cas versions (Cas12a/Cpf1, Cas9 nickase), base editors (for point mutations) and prime editors (for precise insertions, deletions and substitutions) (Anzalone et al., 2019).

17.8.4 Methods of Establishing CRISPR Elements in Plants

The effective delivery of the CRISPR machinery is critical for appropriate editing. Common approaches include the following:

- ***Agrobacterium*-Mediated Transformation**: Commonly used in dicots; utilizes editing determinants which integrate into the plant genome.
- **Particle Bombardment (Biolistics)**: Directly injects DNA or ribonucleoprotein (RNP) complexes into plant cells; appropriate only for monocot plants.
- **Protoplast Transfection**: Transient transfection of isolated cells; enables rapid screening of the efficiency of sgRNA use.
- **RNP Delivery**: Delivers assembled Cas9–sgRNA complexes; reduces the off-target effects and results in non-transgenic edited plants (Liang et al., 2018).

17.8.5 Applications for Using CRISPR in Crop Improvement

17.8.5.1 Transgenic Manipulation for Maximizing Traits

CRISPR provides targeted loss-of-function mutagenesis, and is therefore useful for the following:

- **Negative Regulators of the Stress Response**: Mutants that eliminate genes that repress populations of drought-responsive genes (e.g., some ABA catabolic genes) can make plants more tolerant to drought.
- **Root Architectural Modulators**: The genes that are able to restrict root elongation/lateral branching are selectively removed for efficient RSA for water acquisition (Uga et al., 2013).
- **Susceptibility Genes in Pathogens**: Gene knockout makes plants less susceptible to diseases while indirectly enhancing plant vigor under abiotic stress.

For example, overexpression of the ABA receptor binding in the protein A from rice (OsERA1) enhanced ABA sensitivity and improved the drought tolerance without yield penalty (Lim et al., 2015).

17.9 GENE INSERTION AND REPLACEMENT STRATEGIES

Through HDR, the CRISPR enables the insertion of:

- Isolation of stress-tolerant alleles from other genotypes or other species (e.g., drought-tolerant TFs or osmoprotectant genes).
- Regulatory elements for fine-tuning gene expression in response to water deficit, such as drought-inducible promoters.

Additionally, although the efficiency of HDR is low in plants, recently reported methods based on the use of geminiviral replicons and Cas variants are increasing targeted recombination abundance (Baltes et al., 2014).

17.9.1 Base Editing and Prime Editing Applications in Crops

Base Editing: Makes single nucleotide modification without DSBs. Applications include the following:

- Generating gain-of-function alleles using mutagenesis of the stress response genes.
- Reducing deleterious polymorphism types in elite cultivars.

Prime Editing: Cas9 nickase transposase fused to a reverse transcriptase will introduce insertions, deletions or substitutions to the target. It allows

- multiplex editing of regulatory motifs regulating root development and drought stress tolerance
- precisely optimistic modifier engineering of promoters to raise expression of RSA-related genes.

These technologies reduce the off-target effects and allow the production of non-transgenic crops which can be accepted by regulation.

17.9.2 Benefits of CRISPR for Crop Breeding

17.9.2.1 Accuracy and Effectiveness of Genetic Manipulations

CRISPR can be used to target particular loci, without causing significant damage in the rest of the genome. This precision is especially important for polygenic traits, such as RSA, which is regulated by a number of genes and regulatory encoders which interact to exert regulation on drought response.

17.9.2.2 Time- and Cost-Effective Compared to Traditional Methods

The ability to develop homozygous edited lines using CRISPR in one to two generations, far more rapidly than traditional breeding which takes many generations combined with rigorous field testing, permits cultivar development to proceed at a much faster rate than conventional breeding (Bortesi & Fischer, 2015).

17.9.2.3 Potential for Developing Non-transgenic Engineered Crops

By transient transfection or RNPs of CRISPR components, it is possible to segregate out the transgenic DNA and create plants that are not distinguishable from the conventionally bred ones. This approach can

- reduce regulatory burden,
- increase public acceptance, and
- prevent indirect transgene-related effects.

17.9.2.4 Summary

CRISPR/Cas9 and its variants offer unprecedented possibilities for precision crop improvement, especially for traits, such as RSA, that are of central value for drought adaptation. The mechanistic understanding, direct manipulation of genes of interest, and adaptable delivery systems place CRISPR as a technology of the future in the toolbox of developing agriculturally relevant drought-tolerant crops.

17.9.3 Targets of the CRISPR for Drought Tolerance

17.9.3.1 Genes That Are Related to Drought Response

17.9.3.1.1 ABA Biosynthetic and Signal Transduction Genes

ABA is a central hormone that controls the plant response to drought. Genes involved in ABA biosynthesis, perception and signaling are prime targets for crystallization with the help of a pumice gene.

17.9.3.1.1.1 Biosynthesis Genes NCED is the catalyst for the rate-limiting step in the synthesis of ABA.

- CRISPR-mediated overexpression or activation of NCED leads to enhanced accumulation of ABA, which promotes stomata closure, osmotic adjustment and root growth under water deficit (Iuchi et al., 2001).

17.9.3.1.1.2 Signaling Components PYR/PYL/RCAR receptors are the receptors for ABA, which trigger downstream processes.

- PP2C phosphatases are negative regulators. Therefore, the targeted knockout of PP2C genes specifically can promote ABA effects without an apparent growth penalty (Yoshida et al., 2019).

By modifying these genes, in the form of so-called CRISPR methods, it is possible to calibrate the expression of the ABA signal, increasing the water-use efficiency and drought resistance.

17.9.3.2 Genes for Biosynthesis of Osmolytes

Under conditions of drought, compatible solutes are accumulated by plants as a means to support cell turgor. Key targets include the following:

- *P5CS* (Δ^1-pyrroline-5-carboxylate synthetase) for proline biosynthesis.
- *BADH* (betaine aldehyde dehydrogenase) for glycine betaine synthesis.
- *TPS* (trehalose-6-phosphate synthase) for trehalose accumulation.

CRISPR-induced upregulation or promoter modification of these genes increases the osmotic adjustments and protection of cells and increases drought tolerance life with no major trade-offs on growth (Szabados & Savoure, 2010).

17.9.3.3 Antioxidant System Genes

Drought-induced ROS accumulation leads to membrane, protein and DNA damage. Genes that are targets for the use of the tool called CRISPR include the following:

- **Superoxide Dismutase (SOD)**: Reacts with and converts superoxide radicals into hydrogen peroxide.
- **CAT and Ascorbate Peroxidase**: Neutralize hydrogen peroxide.
- **Glutathione Reductase**: Enzyme to determine the redox balance.

Increasing the level of cellular redox homeostasis by modifying regulatory sequences to upregulate the production of antioxidant genes or knocking out negative regulators can enhance drought tolerance (Gill and Tuteja, 2010).

17.9.4 Transcription Factors

TFs- against stress: the transcription factors are the master regulators of networks of stress response genes.

17.9.4.1 DREB/CBF Family

- DREBs bind dehydration-responsive elements of promoters of stress genes.
- Overexpression or promoter activation using the TF actuator, CRIPSR, enhances osmolyte production, root growth and stress responses that are not regulated by ABA (Lata & Prasad, 2011).

17.9.4.2 NAC Family

- NAC TFs control root development, stomata conductance and cell wall remodeling.
- CRISPR modulation may be able to promote deeper rooting and water uptake (Hu et al., 2006).

17.9.4.3 bZIP Family

- bZIP TFs regulate ABA-mediated gene expression and oxidative stress response.
- Modification of ABA sensitivity and stress-inducible transcription by editing of the bZIP genes enhances drought adaptation (Jakoby et al., 2002).

17.9.5 MicroRNAs

17.9.5.1 miRNAs in Crops Responding to Drought

miRNAs are small non-coding RNAs for the post-transcriptional regulation of targeted genes. Some of the key drought-responsive miRNAs are as follows:

- **miR164**: Confers regulation on NAC TFs regulating root development.
- **miR393**: Negative regulation of auxin signaling to modulate lateral root development.
- **miR398**: A regulator of SOD genes, having effects on ROS detoxification.

17.9.5.2 MicroRNA Targets Associated with Drought Tolerance

- miR164 → NAC TFs → root elongation
- miR393 → TIR1/AFB auxin receptors → lateral root density
- miR398 → CSD1/CSD2 → oxidative stress tolerance

CRISPR can be used to knock out or to mutate miRNA binding sites, good optimization of the gene expression for RSA and adaptation to stress without disturbing global plant development.

17.9.5.3 Strategies for Manipulating miRNA Expression by Using CRISPR

- Knockout of miRNA genes: Stops suppressing stress-related targets to enhance stress tolerance and adaptive responses under drought conditions.
- Mutation of miRNA binding sites in target genes: Confers resistance to miRNA-mediated repression.
- Promoter editing of miRNAs: When induced or stimulated by tissue in drought.

17.9.6 Epigenome Modifications

Epigenetic control of stress remembering and genes expression plasticity makes it a good candidate for being targeted by a CRISPR.

17.9.6.1 DNA Methylation Targets

So there are DNA methyltransferases (MET1, DRM2) or demethylases (ROS1) that could be modulated in order to affect the promoter methylation of drought-related genes, to change the transcriptional response, and that could be edited by DNA chromatin editors (Zhang et al., 2018).

17.9.6.2 Histone Modification Targets

- Histone acetyltransferases and histone deacetylases control chromatin accessibility.
- CRISPR-mediated manipulation of these enzymes can aid in the increased transcription of stress-protective genes or repression of negative regulators.

17.9.6.3 Chromatin Modification Factors

- Chromatin remodelers like SWI/SNF affect root-specific gene expression.
- Targeting remodelers can be used to optimize RSA under water-limited conditions.

17.9.6.4 Summary

CRISPR can be used to specifically target genes, TFs, miRNAs, and epigenetic regulators that coordinate drought response. Through adjustment of the following elements, plants can grow:

- Optimized RSA for water acquisition
- Increased tolerance to osmotic and oxidative stress
- Precisely regulated signaling by hormones.

17.9.7 RSA Modulation by CRISPR

17.9.7.1 Genes That Determine Root Development

RSA is an integral part of water and nutrient acquisition and modulation can seriously improve drought tolerance. CRISPR has been used to edit genes that regulate primary root growth, lateral root growth and root hair growth.

Elongation genes located at the primary root level.

DRO1 (Deeper Rooting 1): Controls gravitropic response of roots. Editing DRO1 in rice by use of CRISPR to produce more profound primary roots, and thus potentially allow plants to tap into subsoil moisture during drought (Uga et al., 2013).

EXPA (*Expansin*) Genes: Incidentally enhance cell wall loosening and root growing. An increase in the rate of root elongation in response to water deficit through upregulation mediated by the trans-acting elements of the CRISPR pathway has been found (Cosgrove, 2005).

17.9.7.2 Lateral Root Development and Branching Genes

- **LBD (Lateral Organ Boundaries Domain) Family**: Regulates lateral root induction. Knockout of negative regulators increases lateral root density, increasing soil exploration (Zhang et al., 2020).
- **Auxin Response Factor (ARF) Genes**: Regulate auxin signaling in the determination of lateral root development. Recognition of the characterized ARFs for drought resistance–specific CRISPR editing strategies is optimal for modulating branching patterns (Goh et al., 2012).

17.9.7.3 Root Hair Development Genes

- **Root Hair Defective Six-Like Protein (RSL1) Genes**: Control the initiation and growth of root hair. Edited lines indicate the increased surface area of the roots for better uptake of water and nutrients in case of restricted availability of moisture (Yi et al., 2010).
- **CPC/GL2 Genes**: Determine the density and length of root hairs. Adaptive RSA can be fine-tuned by CRISPR knockout or promoter modification of root hair development.

17.9.7.4 Hormonal Regulation of RSA

Hormones have an important role as regulators of root growth and plasticity. In addition, CRISPR can be used to excise genes associated with hormone and optimize RSA under drought.

17.9.7.5 Auxin-Related Genes

- The root elongation and branching is determined by auxin gradients.
- Target genes include PIN-FORMED (PIN) transporters, AUX/IAA repressors and ARFs.
- Example: The root angle and lateral root distribution in root of rice were changed by the editing of OsPIN1a gene using the technique of the clustered regularly homologous site-specific integration (CRT) of transgenic technology (Iida & Terada, 2005).

17.9.7.6 Cytokinin-Related Genes

- Cytokinins have the opposite effects on root elongation and shoot growth.
- Crown root primordia (CRP) knockout of CKX (cytokinin oxidase/dehydrogenase) genes promotes root growth but not severely affecting shoots (Werner et al., 2003).
- This adaptation makes it more drought-resistant through improved deep and extensive roots.

17.9.7.7 Genes Related to Ethylene and ABA

- Ethylene controls root hair induction and lateral root development.
- ABA exhibits cross-talk with auxin and cytokinin pathways in root elongation during water deficit.
- CRISPR-mediated modification of ethylene signaling components (e.g., EIN2) and ABA receptors are able to balance stress response with root development, promoting RSA plasticity (Stepanova et al., 2005).

17.9.7.8 Nutrients Sensing and Uptake Genes

Expression of root function under drought also requires efficient nutrient acquisition which contributes to growth under water-limited conditions.

17.9.7.9 Nitrate Transporter Genes

- Nitrate uptake and root forging are regulated by members of NRT1 and NRT2 families.
- CRT editing improves expression or changes transporter kinetics for improved nitrogen acquisition during drought that leads to sustained root growth (Krouk et al., 2010).

17.9.7.10 Gene Coding for Transporters for Phosphate Transport

- Transporters of the PHT1 family control the uptake of phosphate.
- Editing the regulatory elements has the potential to boost the root surface phosphate uptake, which indirectly contributes to water uptake through root proliferation (Chiou & Lin, 2011).

17.9.7.11 Potassium Channel Genes

- K^+ channels are responsible for osmotic regulation and root turgor.
- CRISPR-mediated changes enhance water holding capacity and drought tolerance through stabilizing root hydraulic conductivity (Ahmad et al., 2016).

17.9.8 Methods for RSA Optimization

17.9.8.1 Growing Deeper Roots for Better Ability to Acquire Water

- For example, the genes DRO1 and expansins can increase primary root growth into deeper layers of soils.
- Combining the root deepening with lateral branching increase leads to dual water and mineral foraging especially on heterogeneous soils.

17.9.8.2 Nutrient Uptake with Increased Lateral Root Density

- Editing of LBD and ARF family genes increases lateral root number and density, increasing the root–soil contact area.
- Deeper roots with the enhanced lateral root density aid surface-level water and nutrient absorption that complements deep primary roots.

17.9.8.3 Adjusting Angle at Roots for Maximum Soil Penetration

- Root growth angle is the key for balancing between shallow exploitation of soil and deep access to water.
- A recent study reported that modification of the PIN and DRO genes by CRISPR can bring the PIN and DRO gene regulation to align with soil water content and crop performances, so that the shallow root architecture can be more appropriate for drier conditions and the deep root architecture for wetter conditions (Uga et al., 2013).

17.9.8.4 Integration of Multiple RSA Traits

- The combination of edits in primary root elongation, branching and root hair density gives maximal adaptive RSA.
- Multiplex CRISPR enables targeting of multiple loci simultaneously producing optimized root system cultivars for drought tolerances (Abdallah et al., 2025).

17.9.8.5 Summary

CRISPR-mediated RSA modulation allows root architecture to be precisely regulated:

- **Developing by Root**: Genetic control of growth pattern and branching.
- Hormonal regulation takes into account environmental signals to fine-tune RSA.
- Nutrient sensing genes assure the best resource acquisition during stress.
- This should make it possible to select drought-tolerant crops with improved water-use efficiency and nutrient use.

17.9.9 Opportunities and Insights

17.9.9.1 Technical Challenges toward CRISPR-Based RSA Modification

While precise and versatile, there are technical challenges still needed to overcome for efficient RSA engineering:

17.9.9.2 Off-Target Effects and Measures to Reduce Them

- CRISPR/Cas9 has some possibility to bind to nontarget genomic regions, which can lead to off-target mutations.
- Strategies to minimize off-target effects include the following:
  - Using bioinformatics tools to design highly specific sgRNAs.
  - Using Cas variants with higher fidelity (i.e., eSpCas9 and SpCas9-HF1).
  - Transient transfection with RNP complex.
  - Targeting of root architecture genes is particularly important because unintended editing may interfere with other developmental traits.

17.9.9.3 Delivery of CRISPR Components into Plant Cells

- Most crops, especially monocot crops, are recalcitrant to transformation.
- These methods (*Agrobacterium*, biolistics, protoplasts, RNPs) have species-specific constraints.
- A key to high efficiency of genome edits is the optimization of tissue culture protocols and in planta delivery techniques.

17.9.9.4 Regeneration of Genetically Modified Plants

- Regeneration and editing of whole plants from transformed cells are a prerequisite for genetic modification of non-model species.
- **Roots and Shoots**: Development in plants is coordinated in part by the genetic makeup, so abdomen regeneration failure can cause the formation of developmental abnormalities in the roots and shoots.
- However, root traits are pleiotropic, meaning that more robust tissue culture techniques and selection systems will be necessary to obtain viable edited plants with favorable root traits.

17.9.10 Legislative and Ethical Issues

17.9.10.1 Regulatory Landscape around the World

- Regulatory requirements for genome-edited crops are different worldwide:
  - **United States**: Foreign DNA-free GMO regulations may not catch some GE crops.
  - **European Union**: All crops that contain a transgene (without regard to the nature of the transgene) are subject to the GMO laws.
  - **Asia**: Mixed regulatory strategies—Some countries are moving toward less strict regulatory systems for non-transgenic modification (Schaart et al., 2016).

17.9.10.2 Social Awareness and Acceptance

- Adoption can be affected by societal concerns over safety, environmental impact and modifications that are considered as unnatural changes.
- Public acceptance requires education about the advantages and demonstration of non-transgenic CRISPR strategies (a major threat to the public acceptance of all other new technologies) and their outputs (improved crop production, drought tolerance).

17.9.10.3 Legal Problems Regarding Intellectual Property

- The CRISPR technology is patented and in order to use it commercially, one must license the technology.
- The practical reality of breeding programs, particularly in developing countries, is the struggle to gain access and freedom-to-operate.

17.9.10.4 Recommendations for RSA Enhancement and Development

- Improved Drought Resistance Using Multiple Structural Design of Rice Agent
  Multiplex CRISPR makes it possible to edit the following:
  - Root primary elongation genes (DRO1, EXPAs)
  - Lateral root and branching genes (ARFs, LBDs)
  - Root hair growth genes (RSL, CPC/GL2).
- Such integrations can be used to enhance water-use and nutrient-use efficiency, generating resilient crops for drought conditions.

17.9.10.5 High-Throughput Phenotyping for RSA Assessment

- Root characteristics are hard to measure in field conditions.
- Editorial RSA phenotypic traits can be assessed quickly with the help of new phenotyping tools (rhizotrons, X-Ray Computed Tomography (X-HT) and Magnetic Resonance Imaging (MRI)) linking genotype to phenotype.

17.9.10.6 Combined Potential of Base and Prime Editing for High-Precision RSA Modification

- Editing of key regulatory sites is flashy by base and prime editors, which are devoid of pleiotropic effects.

- Future studies can focus on promoter regions, miRNA binding sites and regulatory motifs for optimal RSA for context-specific settings.

17.9.10.7 Wider Impact of RSA Amendment

17.9.10.7.1 Effects on Sustainable Agriculture and Food Security

- RSA-enhanced crops are more water-use efficient which helps in growing the crop in marginal or drought-prone regions.
- Improved nutrient utilization, thus lowering the need for chemical fertilizers, which in turn boosts low-input agriculture.

17.9.10.8 Unused Potential to Enhance Other Abiotic Stress Tolerances

- Many of the RSA traits are cross-tolerant to salinity, nutrient deficiency and soil compaction.
- Therefore, we hypothesize that editing of root architecture genes would result in enhanced multi-stress tolerance, which would increase crop stability.

17.9.10.9 Trade-off of Above-Ground and Below-Ground Characteristics

- From the above, it is clear that the optimization of RSA has to be performed with trade-offs with shoot growth and reproductive development.
- CRISPR facilitates additive breeding for above-ground and below-ground traits simultaneously such as yield, tolerance of stress and efficient use of resources.

17.9.10.10 Summary

Furthermore, CRISPR-mediated RSA modification has unparalleled potential to create drought-tolerant crops, yet technical, regulatory and ethical issues still remain. "These new developments in high-fidelity genome editing, multiplexing and phenotyping integration will pave a path for sustainable and precise crop improvement." With proper prioritization of traits and involvement of stakeholders, these approaches have great potential to complement global food security against climate change.

17.10 CONCLUSION

Genome editing offers a powerful and precise strategy for reshaping RSA to enhance crop resilience under drought stress. By targeting genes that regulate root depth, branching patterns, hydraulic conductivity and nutrient uptake efficiency, researchers can develop crop varieties better adapted to water-limited environments. Such modifications enable improved soil moisture exploration, optimized water-use efficiency and sustained productivity under climatic variability. As functional genomics knowledge expands and gene editing tools become more efficient and accessible, integrating genome-based RSA engineering into crop improvement programs holds great promise for strengthening global food security in the face of increasing drought frequency and intensity.

REFERENCES

Abdallah, N. A., Elsharawy, H., Abulela, H. A., Thilmony, R., Abdelhadi, A. A., & Elarabi, N. I. (2025). Multiplex CRISPR/Cas9-mediated genome editing to address drought tolerance in wheat. *GM Crops & Food*, 16(1), 1–17. https://doi.org/10.1080/21645698.2022.2120313

Ahmad, I., Mian, A., & Maathuis, F. J. (2016). Overexpression of the rice AKT1 potassium channel affects potassium nutrition and rice drought tolerance. *Journal of Experimental Botany*, 67(9), 2689–2698. https://doi.org/10.1093/jxb/erw103

Anzalone, A. V., Randolph, P. B., Davis, J. R., Sousa, A. A., Koblan, L. W., Levy, J. M., Chen, P. J., Wilson, C., Newby, G. A., Raguram, A., & Liu, D. R. (2019). Search-and-replace genome editing without double-strand breaks or donor DNA. *Nature*, 576(7785), 149–157. https://doi.org/10.1038/s41586-019-1711-4

Baltes, N. J., Gil-Humanes, J., Cermak, T., Atkins, P. A., & Voytas, D. F. (2014). DNA replicons for plant genome engineering. *The Plant Cell*, 26(1), 151–163. https://doi.org/10.1105/tpc.113.119792

Barnabás, B., Jäger, K., & Fehér, A. (2008). The effect of drought and heat stress on reproductive processes in cereals. *Plant, Cell & Environment*, 31(1), 11–38. https://doi.org/10.1111/j.1365-3040.2007.01727.x

Blum, A. (2017). Osmotic adjustment is a prime drought stress adaptive engine in support of plant production. *Plant, Cell & Environment*, 40(1), 4–10. https://doi.org/10.1111/pce.12800

Bortesi, L., & Fischer, R. (2015). The CRISPR/Cas9 system for plant genome editing and beyond. *Biotechnology Advances*, 33(1), 41–52. https://doi.org/10.1016/j.biotechadv.2014.12.006

Campos, H., Cooper, M., Habben, J. E., Edmeades, G. O., & Schussler, J. R. (2004). Improving drought tolerance in maize: A view from industry. *Field Crops Research*, 90(1), 19–34. https://doi.org/10.1016/j.fcr.2004.07.003

Cattivelli, L., Rizza, F., Badeck, F. W., Mazzucotelli, E., Mastrangelo, A. M., Francia, E., … Stanca, A. M. (2008). Drought tolerance improvement in crop plants: An integrated view from breeding to genomics. *Field Crops Research*, 105(1–2), 1–14. https://doi.org/10.1016/j.fcr.2007.07.004

Chiou, T. J., & Lin, S. I. (2011). Signaling network in sensing phosphate availability in plants. *Annual Review of Plant Biology*, 62, 185–206. https://doi.org/10.1146/annurev-arplant-042110-103849

Comas, L. H., Becker, S. R., Cruz, V. M. V., Byrne, P. F., & Dierig, D. A. (2013). Root traits contributing to plant productivity under drought. *Frontiers in Plant Science*, 4, 442. https://doi.org/10.3389/fpls.2013.00442

Cosgrove, D. J. (2005). Growth of the plant cell wall. *Nature Reviews Molecular Cell Biology*, 6(11), 850–861. https://doi.org/10.1038/nrm1746

Cutler, S. R., Rodriguez, P. L., Finkelstein, R. R., & Abrams, S. R. (2010). Abscisic acid: Emergence of a core signaling network. *Annual Review of Plant Biology*, 61, 651–679. https://doi.org/10.1146/annurev-arplant-042809-112122

Dubouzet, J. G., Sakuma, Y., Ito, Y., Kasuga, M., Dubouzet, E. G., Miura, S., … Yamaguchi-Shinozaki, K. (2003). OsDREB genes in rice, Oryza sativa L., encode transcription activators that function in drought-, high-salt- and cold-responsive gene expression. *The Plant Journal*, 33(4), 751–763. https://doi.org/10.1046/j.1365-313X.2003.01661.x

Farooq, M., Wahid, A., Kobayashi, N., Fujita, D., & Basra, S. M. A. (2009). Plant drought stress: Effects, mechanisms and management. *Agronomy for Sustainable Development*, 29(1), 185–212. https://doi.org/10.1051/agro:2008021

Gill, S. S., & Tuteja, N. (2010). Reactive oxygen species and antioxidant machinery in abiotic stress tolerance in crop plants. *Plant Physiology and Biochemistry*, 48(12), 909–930. https://doi.org/10.1016/j.plaphy.2010.08.016

Goh, T., Kasahara, H., Mimura, T., Kamiya, Y., & Fukaki, H. (2012). Multiple AUX/IAA–ARF modules regulate lateral root formation: the role of Arabidopsis SHY2/IAA3-mediated auxin signalling. *Philosophical Transactions of the Royal Society B: Biological Sciences,* 367(1595), 1461–1468. https://doi.org/10.1098/rstb.2011.0232

Hu, H., Dai, M., Yao, J., Xiao, B., Li, X., Zhang, Q., & Xiong, L. (2006). Overexpressing a NAM, ATAF, and CUC (NAC) transcription factor enhances drought resistance and salt tolerance in rice. *Proceedings of the National Academy of Sciences*, 103(35), 12987–12992. https://doi.org/10.1073/pnas.0604882103

Iida, S., & Terada, R. (2005). Modification of endogenous natural genes by gene targeting in rice and other higher plants. *Plant Molecular Biology,* 59(1), 205–219. https://doi.org/10.1007/s11103-005-2162-x

Iuchi, S., Kobayashi, M., Taji, T., Naramoto, M., Seki, M., Kato, T., … Shinozaki, K. (2001). Regulation of drought tolerance by gene manipulation of 9-cis-epoxycarotenoid dioxygenase, a key enzyme in abscisic acid biosynthesis in Arabidopsis. *The Plant Journal*, 27(4), 325–333. https://doi.org/10.1046/j.1365-313x.2001.01096.x

Jakoby, M., Weisshaar, B., Dröge-Laser, W., Vicente-Carbajosa, J., Tiedemann, J., Kroj, T., & Parcy, F. (2002). bZIP transcription factors in Arabidopsis. *Trends in Plant Science*, 7(3), 106–111. https://doi.org/10.1016/S1360-1385(01)02223-3

Jinek, M., Chylinski, K., Fonfara, I., Hauer, M., Doudna, J. A., & Charpentier, E. (2012). A programmable dual-RNA–guided DNA endonuclease in adaptive bacterial immunity. *Science*, 337(6096), 816–821. https://doi.org/10.1126/science.1225829

Kasuga, M., Liu, Q., Miura, S., Yamaguchi-Shinozaki, K., & Shinozaki, K. (1999). Improving plant drought, salt, and freezing tolerance by gene transfer of a single stress-inducible transcription factor. *Nature Biotechnology*, 17(3), 287–291. https://doi.org/10.1038/7036

Krouk, G., Lacombe, B., Bielach, A., Perrine-Walker, F., Malinska, K., Mounier, E., … Gojon, A. (2010). Nitrate-regulated auxin transport by NRT1.1 defines a mechanism for nutrient sensing in plants. *Developmental Cell*, 18(6), 927–937. https://doi.org/10.1016/j.devcel.2010.05.008

Lata, C., & Prasad, M. (2011). Role of DREBs in regulation of abiotic stress responses in plants. *Journal of Experimental Botany*, 62(14), 4731–4748. https://doi.org/10.1093/jxb/err210

Lesk, C., Rowhani, P., & Ramankutty, N. (2016). Influence of extreme weather disasters on global crop production. *Nature*, 529(7584), 84–87. https://doi.org/10.1038/nature16467

Liang, Z., Chen, K., Li, T., Zhang, Y., Wang, Y., Zhao, Q., ... Gao, C. (2018). Efficient DNA-free genome editing of bread wheat using CRISPR/Cas9 ribonucleoprotein complexes. *Nature Communications*, 8(1), 14261. https://doi.org/10.1038/ncomms14261

Lim, C. W., Baek, W., Jung, J., Kim, J. H., & Lee, S. C. (2015). Function of ABA in stomatal defense against biotic and drought stresses. *International Journal of Molecular Sciences*, 16(7), 15251–15270.

Lobell, D. B., Schlenker, W., & Costa-Roberts, J. (2014). Climate trends and global crop production since 1980. *Science*, 333(6042), 616–620. https://doi.org/10.1126/science.1204531

Lynch, J. P. (2013). Steep, cheap and deep: An ideotype to optimize water and N acquisition by maize root systems. *Annals of Botany*, 112(2), 347–357. https://doi.org/10.1093/aob/mcs293

Mittler, R. (2017). ROS are good. *Trends in Plant Science*, 22(1), 11–19. https://doi.org/10.1016/j.tplants.2016.08.002

Nelson, D. E., Repetti, P. P., Adams, T. R., Creelman, R. A., Wu, J., Warner, D. C., ... Bensen, R. J. (2007). Plant nuclear factor Y (NF-Y) B subunits confer drought tolerance and lead to improved corn yields on water-limited acres. *Proceedings of the National Academy of Sciences*, 104(42), 16450–16455. https://doi.org/10.1073/pnas.0707193104

Osakabe, Y., Osakabe, K., Shinozaki, K., & Tran, L. S. P. (2014). Response of plants to water stress. *Frontiers in Plant Science*, 5, 86. https://doi.org/10.3389/fpls.2014.00086

Puchta, H., & Fauser, F. (2014). Synthetic nucleases for genome engineering in plants: Prospects for a bright future. *The Plant Journal*, 78(5), 727–741. https://doi.org/10.1111/tpj.12338

Schaart, J. G., van de Wiel, C. C. M., Lotz, L. A. P., & Smulders, M. J. M. (2016). Opportunities for products of new plant breeding techniques. *Trends in Plant Science*, 21(5), 438–449. https://doi.org/10.1016/j.tplants.2015.11.006

Shinozaki, K., & Yamaguchi-Shinozaki, K. (2007). Gene networks involved in drought stress response and tolerance. *Journal of Experimental Botany*, 58(2), 221–227. https://doi.org/10.1093/jxb/erl164

Smith, A. B., Katz, R. W., & Bates, G. T. (2013). U.S. billion-dollar weather and climate disasters: Data sources, trends, accuracy and biases. *Natural Hazards*, 67(2), 387–410. https://doi.org/10.1007/s11069-013-0566-5

Stepanova, A. N., Hoyt, J. M., Hamilton, A. A., & Alonso, J. M. (2005). A link between ethylene and auxin uncovered by the characterization of two root-specific ethylene-insensitive mutants in Arabidopsis. *The Plant Cell*, 17(8), 2230–2242. https://doi.org/10.1105/tpc.105.033365

Sunkar, R., Li, Y. F., & Jagadeeswaran, G. (2012). Functions of microRNAs in plant stress responses. *Trends in Plant Science*, 17(4), 196–203. https://doi.org/10.1016/j.tplants.2012.01.010

Szabados, L., & Savouré, A. (2010). Proline: A multifunctional amino acid. *Trends in Plant Science*, 15(2), 89–97. https://doi.org/10.1016/j.tplants.2009.11.009

Tuberosa, R., Salvi, S., Sanguineti, M. C., Landi, P., Maccaferri, M., & Conti, S. (2002). Mapping QTLs regulating morpho-physiological traits and yield: Case studies, shortcomings and perspectives in drought-stressed maize. *Annals of Botany*, 89(7), 941–963.

Uga, Y., Sugimoto, K., Ogawa, S., Rane, J., Ishitani, M., Hara, N., ... Yano, M. (2013). Control of root system architecture by DEEPER ROOTING 1 increases rice yield under drought conditions. *Nature Genetics*, 45(9), 1097–1102. https://doi.org/10.1038/ng.2725

Voytas, D. F. (2013). Plant genome engineering with sequence-specific nucleases. *Annual Review of Plant Biology*, 64, 327–350. https://doi.org/10.1146/annurev-arplant-042811-105552

Wang, W., Vinocur, B., & Altman, A. (2020). Plant responses to drought, salinity and extreme temperatures: Towards genetic engineering for stress tolerance. *Planta*, 218(1), 1–14. https://doi.org/10.1007/s00425-003-1105-5

Werner, T., Motyka, V., Laucou, V., Smets, R., Van Onckelen, H., & Schmülling, T. (2003). Cytokinin-deficient transgenic Arabidopsis plants show multiple developmental alterations indicating opposite functions of cytokinins in the regulation of shoot and root meristem activity. *The Plant Cell*, 15(11), 2532–2550. https://doi.org/10.1105/tpc.014928

Yi, K., Menand, B., Bell, E., & Dolan, L. (2010). A basic helix-loop-helix transcription factor controls cell growth and size in root hairs. *Nature Genetics*, 42(3), 264–267. https://doi.org/10.1038/ng.529

Yoshida, T., Christmann, A., Yamaguchi-Shinozaki, K., Grill, E., & Fernie, A. R. (2019). Revisiting the basal role of ABA–roles outside of stress. *Trends in Plant Science*, 24(7), 625–635. https://doi.org/10.1016/j.tplants.2019.04.008

Zhang, H., Lang, Z., & Zhu, J. K. (2018). Dynamics and function of DNA methylation in plants. *Nature Reviews Molecular Cell Biology*, 19(8), 489–506. https://doi.org/10.1038/s41580-018-0016-z

Zhang, Y., Li, Z., Ma, B., Hou, Q., & Wan, X. (2020). Phylogeny and functions of LOB domain proteins in plants. *International Journal of Molecular Sciences*, 21(7), 2278. https://doi.org/10.3390/ijms21072278

18 Ethical, Regulatory and Biosafety Considerations in Genome Editing

Ratikanta Mohapatra, Sougat Satapathy, Vishal, and Yogesh Dashrath Naik

18.1 INTRODUCTION

By 2050, the global population is expected to reach 10 billion and thus bring unprecedented pressure on the agricultural systems to meet the growing food and economic stability demands (Tilman et al., 2011). The resulting increase in demand will necessitate an increase in crop production. However, even now the agricultural systems experience a significant impact of climate change, soil erosion, water deficit and an even greater prevalence of pests and diseases (Reynolds et al., 2015). Such issues are further exacerbated by the limited supply genetic diversity of crop plants that is a key component in crop improvement. Traditional breeding strategies, such as crossbreeding, mutation breeding and transgenic breeding, have significantly contributed to agricultural improvement. However, these approaches are often time-consuming, labor-intensive, and sometimes controversial (Ahmad, 2023). The weaknesses of the above-mentioned limitations support the need for more accurate and efficient approach.

Genome editing, particularly clustered regularly interspaced short palindromic repeats (CRISPR) systems, has emerged as a promising solution. It allows scientific researchers to introduce specific and precise modifications in the genome. The method has potential in the treatment of human diseases, such as hematologic and muscular diseases, cardiovascular, renal, genetic, stem cell and ocular diseases (Morshedzadeh et al., 2024). Moreover, it can enhance crop resilience, disease resistance and yields more precisely and efficiently than traditional approaches (Wang et al., 2025). Regulations pertaining to its usage in agriculture have recently been relaxed due to its ability to make genomic alterations in plants without necessitating the permanent insertion of DNA from other species. Elimination of transgene/foreign DNA can be achieved through three main methods. These include segregation-based elimination, transient expression of DNA vectors and DNA-free elimination through RNA or Cas9-gRNA ribonucleoproteins (Gu et al., 2021). However, its use in agriculture raises concerns of off-target effects, delivery mechanisms and potential environmental or human exposure to engineered or nuclease protein (Venkataraman et al., 2025). Public confusion also persists regarding the difference between genome-edited crops and "GMO", especially when no foreign DNA present in an edited crop. This lack of clarity sparked discussions about long-term effects, safety and ethical issues.

Regulatory challenges further complicate the adoption of genome-edited crops, as countries classify them in different ways (Buchholzer and Frommer, 2023). For example, genome-edited plants are sometimes classified as genetically modified organisms (GMOs) when they involve the introduction of DNA or genes from a different species or genus. In contrast, some regulatory frameworks and perspectives consider them equivalent to natural mutations, similar to those achieved through conventional selective breeding (Fernández Ríos et al., 2025). Some countries have adopted more "flexible regulatory policies". For example, in Japan, SDN1 (site-directed nucleases1) genome-edited crops are exempt from GMO regulations but still require regulatory confirmation

DOI: 10.1201/9781003735113-18

before commercialization. Genome editing using SDNs is categorized into three types: SDN1 (which makes precise small genetic changes through spontaneous repair), SDN2 (where specific DNA sequences are incorporated) and SDN3 (which involves inserting larger foreign genes) (Buchholzer and Frommer, 2023). In India, SDN1 and SDN2 crops are exempt from GMO regulations, enabling quicker approval for agricultural use. In the Philippines, genome-edited products that do not introduce new genetic combinations are certified through a simplified process. Meanwhile, China has introduced a streamlined approval process for genome-edited crops, promoting their wider use in agriculture (Bugnosen, 2023). This chapter aims to explore this less discussed aspect of genome editing, i.e., its ethical, regulatory and biosafety concerns. These topics have been approached in an easily understandable manner for the readers in order to establish the importance of knowing about these topics as well as updating them regarding the developments in the regulatory landscape governing the growth and application of this technology. The objectives are not only to inform the audience about the ethical and biosafety regulations involved in genetic engineering but also to sensitize them about the same and provide future directions for better policy and governance with respect to the genetic manipulation and its practical utility.

18.2 BIOSAFETY CONCERNS AND EMERGING RISKS

Biosafety refers to the mitigation of the unintended risk to the safety of human life and environment stemming from the genesis of life forms which are either genetically altered or synthetic in nature, via relevant measures and precautions (Li et al., 2021). This concern is an unwanted yet unavoidable consequence of the wide variety of genome editing platforms. Despite the various possible beneficial applications, gene-editing platforms have an inherent issue which severely jeopardizes their safe use. Off-target effects are generally defined as accidental changes in the genome in regions other than the one intended during gene editing because of sequence-level similarities and nonspecific interactions (Sturme et al., 2022). Such problems have been commonly found in case of crops with large and complex genomes where the probability of the occurrence of similar sequences increases (Lassoued et al., 2019). The major safety concern arising from off-targets is unintentional mutation which might further lead to destabilization of genomes, errors in DNA repair or a loss of functional genes. In the context of human health, unintentional mutation is the not the only issue. There is clinical evidence of even unexpected immune response elicited by certain components of the CRISPR gene-editing machinery along with cases of toxicity caused by vectors used in gene therapy (El-Mounadi, 2020; Li et al., 2025). Undoubtedly, the risk of unwanted changes at the genetic and epigenetic level, not only due to off-targets but also due to transgenic construct integration into the genome or lapses in the protocol followed for micropropagation especially during regeneration of transgenic plants (Eckerstorfer et al., 2019). However, in a broader ecological context, there also exists the imminent danger of horizontal gene flow between species in a shared environment or due to the escape of a transgenic organism from the lab because of human error (Li et al., 2025). This possibility poses a grave threat to the existing balance in biodiversity due to anthropogenic interference. Certain engineered traits such as enhanced tolerance to herbicides or biotic and abiotic stresses, if transferred horizontally to unwanted plants, can enhance their reproductive fitness from an evolutionary point of view and confer selection advantages which might even lead to the rise of new invasive species (Eckerstorfer et al., 2019). The risk is particularly high due to the use of an interesting technology called "Gene Drive" which bypasses the mechanisms of inheritance according to Mendelian laws and favours the passing down of specific traits down the generations, thus posing a threat to the natural balance of genotypes, phenotypes and environment (Lunshof and Birnbaum, 2017).

The threat is limited not only to the harmful effects on biodiversity but also to some unexpected consequences in terms of the food we consume. This particular biosafety risk arises due to the changes in genetic makeup causing corresponding alterations in the natural composition of crop-based foods leading to the creation of novel allergens and toxic compounds (Movahedi et al., 2023). The technology of gene editing, like all other transformative scientific advancements throughout human history, is not safe from misuse and malicious applications. The rise of synthetic biology has opened up a multitude of avenues in biological research. However, with the increasing accessibility of the knowledge and tools of genomic engineering to the general public and the spread of the DIY (Do It Yourself) culture, the threat of this knowledge being put to use for malevolent causes has also increased alarmingly. De-extinction of ancient pathogens or synthesis of novel biological weapons of war is no more a matter of science fiction. The danger of bioterrorism is frighteningly real and needs serious attention (Ahteensuu, 2017). The risks to biosafety are limited not only to the engineered genes in organisms but also to the data and information about these experiments stored in databases and as online resources. Cyber-biosecurity deals with these risks at the junction of biological science, digital infrastructure and bio-informatic data. The easy availability of such data to people with ill-intent can seriously threaten the safety of human health and food systems. These issues need immediate attention especially the organisms with multiplex edits which increase the difficulty of assessment of biosafety risks (Gulyaeva, 2025; Li et al., 2021).

18.3 GLOBAL REGULATORY FRAMEWORKS FOR GENOME EDITING

Globally, regulatory approaches for biotechnological products generally follow three systems: process-based, product-based or case-by-case assessment (Sprink and Wilhelm 2024; Figure 18.1). In case-by-case assessment, each product is evaluated individually. The evaluation considers the

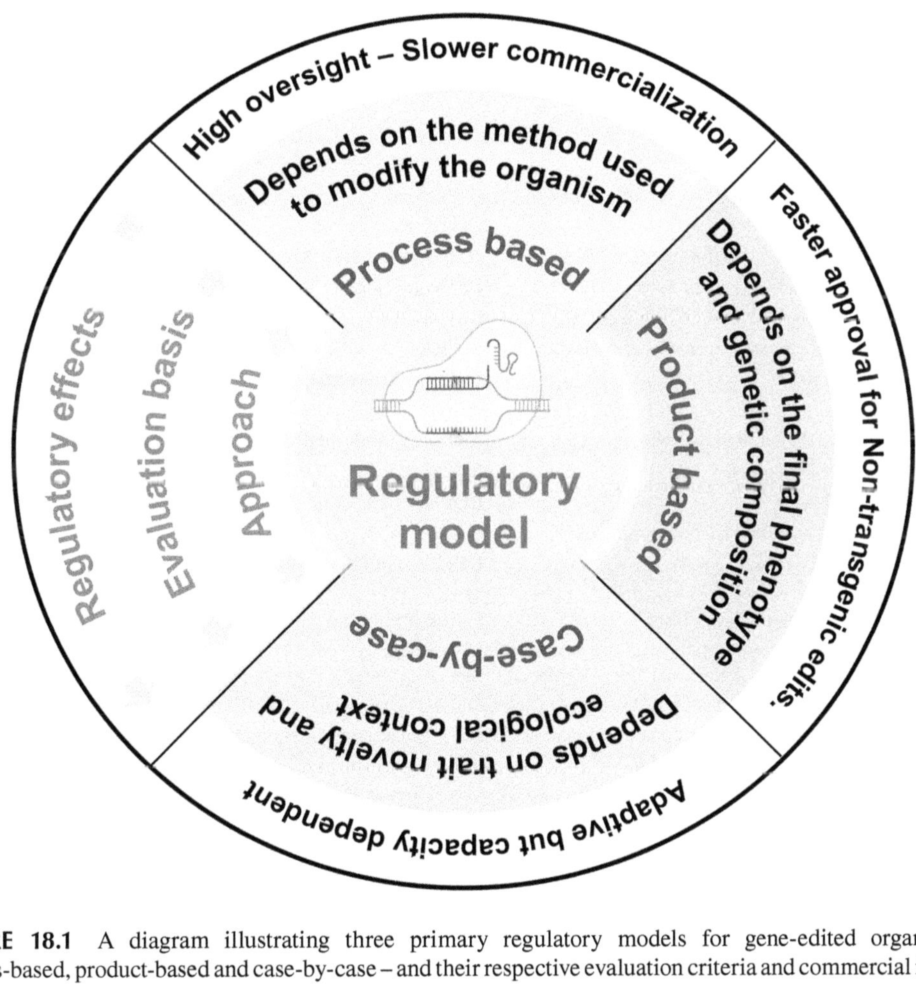

FIGURE 18.1 A diagram illustrating three primary regulatory models for gene-edited organisms – process-based, product-based and case-by-case – and their respective evaluation criteria and commercial impacts.

novelty of traits, potential environmental impacts and the context of use. Process-based regulation emphasizes the technique used to modify the organism, such as "recombinant DNA technology" or CRISPR-mediated editing and aligns with the Cartagena Protocol on Biosafety (Vengadesen et al., 2025). The regulation of genome-edited plants differs markedly across countries due to variations in national GMO policies, creating challenges for international harmonization (Table 18.1). These regulatory frameworks can be classified based on key criteria: whether genome-edited products are considered "GMOs" and the level of regulatory oversight applied (Tachikawa and Matsuo, 2024). While the choice between "process" and "product"-based frameworks provides a conceptual basis for regulation, the practical implementation of these systems varies widely across countries (Table 18.1; Vengadesen et al., 2025), due to differences in institutional regulatory structures, policy priorities and risk assessment strategies.

North America has generally adopted flexible and product-based approaches. In the United States, GMOs and related products are regulated by the "Department of Agriculture" (USDA), the "Food and Drug Administration" (FDA) and the "Environmental Protection Agency" (EPA). A product may need approval from one or more of these agencies depending on its traits. The EPA regulates GMOs with pesticidal properties (to ensure they do not pose unreasonable risks to environment), while foods derived from GMOs are typically reviewed by both the USDA and FDA (Globus and Qimron, 2018). Canada evaluates products based on "novel traits", regardless of breeding method, meaning some genome-edited organisms may still require assessment. South American countries,

TABLE 18.1
Regulatory Status and Oversight of Genome-Edited Products across Countries

Country	Gene-Edited Treated as GMO or Non-GMO	Important Agencies	Product-Based or Process-Based Regulation
Argentina	Non-GMO (If no foreign DNA)	CONABIA	Product-based
Australia	SDN1 = non-GMO	OGTR and FSANZ	Process-based
Brazil	Non-GMO (If no foreign DNA)	CTNBio and ANVISA	Case-by-case assessment
Canada	Non-GMO (If no foreign DNA)	CFIA and Health Canada	Product-based
Chile	Non-GMO (If no foreign DNA)	SAG	Case-by-case assessment
China	Many edits treated as non-GMO	MARA	Process-based
European Union	GMO	EFSA and European Commission	Process-based
India	SDN1 and SDN2 = non-GMO	MOEFCC, DBT and GEAC	Process-based
Japan	SDN1 = non-GMO	MAFF and MHLW	Case-by-case assessment
New Zealand	GMO	EPA and FSANZ	Process-based
Philippines	Non-GMO (If no foreign DNA)	DA and NCBP	Case-by-case assessment
United Kingdom	SDN1 and SDN2 = non-GMO	DEFRA and FSA	Product-based
United States	Generally non-GMO	USDA, FDA, EPA	Product-based

Note: ANVISA –Agência Nacional de Vigilância Sanitária; CFIA – Canadian Food Inspection Agency; CONABIA – Comisión Nacional Asesora de Biotecnología Agropecuaria; CTNBio – Comissão Técnica Nacional de Biossegurança; DA – Department of Agriculture; DBT – Department of Biotechnology; DEFRA – Department for Environment, Food and Rural Affairs; EPA – Environmental Protection Agency; EFSA – European Food Safety Authority; FSA – Food Standards Agency; FSANZ – Food Standards Australia New Zealand; FDA – Food and Drug Administration; GEAC – Genetic Engineering Appraisal Committee; MAFF – Ministry of Agriculture, Forestry and Fisheries; MARA – Ministry of Agriculture and Rural Affairs; MHLW – Ministry of Health, Labour and Welfare; MOEFCC – Ministry of Environment, Forest and Climate Change; NCBP – National Committee on Biosafety; OGTR – Office of the Gene Technology Regulator; SAG – Servicio Agrícola y Ganadero; USDA – United States Department of Agriculture.

Argentina, Brazil, Chile and Colombia, have embraced genome editing by adopting clear procedures that determine whether a product contains foreign genomic material.

If no new genetic combinations remain in the organism, it is typically not regulated as a "GMO". These policies may help accelerate or promote domestic development and commercialization. As the European Union (EU) regards most genome-edited plants as GMOs under the 2018 CJEU regulation, there is an ongoing proposal in 2023 to grant exemptions on specific targeted genome edits, making them exempt from full GMO classification. Following its exit from the EU, the UK shifted to a more lenient approach with the Precision Breeding Act of 2023 that allows easier regulation for plants and animals that have undergone successful genome editing. Australia does not regulate SDN1 edits but regulates edits to templates (and New Zealand has extremely strict GMO regulations and makes subject of genome-edited organisms subject to full control). Food Standards Australia New Zealand (FSANZ), based in Australia and New Zealand, evaluates and grants exemptions on some of the genome-edited foods depending on the risk. In Asia, there are increased regulatory reforms since 2019. Japan waives the environmental and food safety rules on SDN1 products but demands notification and disclosure. China has announced principles permitting reduced testing of low-risk genome-edited crops, which indicates a strategic move towards commercialization. India did not impose GMO regulations on SDN1 and SDN2 products without foreign DNA and put in place elaborate guidelines regarding risk assessment. Another system that was implemented in the Philippines is the notification-based system, under which genome-edited plants that do not contain a foreign genetic material are not subjected to complete regulation as GMOs. As a bloc, the Asian region is shifting towards lean and innovation-friendly systems, although they have maintained control over areas that may present risks. Regulatory status and oversight of genome-edited products across countries are indexed in Table18.1.

18.3.1 Ethical Considerations in Genome Editing

The evolutionary potential of CRISPR technology has caused an intense debate about the technology legal, moral and social effects around the globe. Human germline editing is one of the most controversial concepts, which suggests altering genes of embryos, in order to cause genetic alternation to be inherited. In the long run, this kind of progress may lead to unexpected side effects and the development of a population of designer babies, even though it may one day cure hereditary diseases. Global bioethics along with governmental organizations, is therefore working on policies and frameworks to ensure the responsible use of gene-editing technologies. The major ethical priorities are making an informed consent, keeping biosafety and risk assessment, establishing transparency in research and ensuring fair access to emerging technologies. Specialized agency like World Health Organization (WHO) and the U.S. National Academics are taking the forefront to ensure that they develop global ethical standards and apply especially in clinical settings. Informed consent and community involvement are the foundation of ethical decision-making in biotechnology sector. This requires the process of obtaining the approval of every stakeholder including farmers, consumers and local communities, as well as including their opinion in genetic engineering programmes (Nawaz et al., 2024). The participation strategies and the clear decision-making methods should be discussed during designing and implementing the genetically modified (GM) crops (Resnik, 2018; Eastwood et al., 2019). Ethical evaluation of potential environmental hazards is also crucial, especially with regard to biodiversity, ecosystem services and unexpected impacts (Abushal et al., 2021). A combination of both theoretical and practical models should be used to evaluate and prevent these risks (Mampan et al., 2011). Limits in adoption and benefit sharing must be addressed in order to provide fair access to biotechnological innovations. This involves looking into programmes that promote social justice and equality in relation to the spread of GM technology (Devos et al., 2008). Likewise, research integrity and transparency are a two-principle frame doubted by the public. To maintain ethical research standards, there should be open data sharing, unbiased publication of findings and conflicts of interest management (Harfouche et al., 2021).

More ethical investigation is required about the long-term health effects of eating biofortified or genetically engineered food, which requires strict and ongoing safety assessments (Nawaz et al., 2023). The ethics of bioengineering are also considered completely, and the sociocultural effects of bioengineering such as how it affects traditional food systems, livelihoods and cultural practices are also examined (Shahab et al., 2023). Human germline editing: As increasingly complex gene-editing systems, like CRISPR-Cas9, become more accurate, the debate surrounding human germline editing has gained more attention. Although it has the potential to cure genetic diseases, numerous scholars claim that it is not safe to be used as a reproductive drug as it may have some unacceptable genetic defects that are inheritable. The major scientists of the International Summit on Human Gene Editing have noted that more research should be undertaken to prove its safety before clinical use is taken into consideration. And some people even argue that this safety might never be possible (Schmerker, 2024). The other major ethical concern is the issue of consent, since embryos and their offspring are incapable of providing informed consent for genome editing (Savulescu et al., 2015). It has been compared to parental choices in assisted reproduction such as IVF (in-vitro fertilisation), where future consent is also unfeasible (Schmerker, 2024). In addition, it is hard to separate therapeutic interventions (treating genetic diseases) and enhancement (treating people by altering their height or intelligence). With germline editing as a commercial offering, it might be hard to draw a line between the case of medical intervention and the case of an enhancement process, which will only contribute to the development of more social disparities. In line with these worries, the WHO Director-General in 2019 stated that any attempt to seek clinical germline genome editing at the point would be irresponsible (World Health Organization, 2019). Ecological Impacts of GMOs Implementation of CRISPR-Cas9 in biotechnology in plants and animals has created several advantages, such as pest-resistant crops, better food security and pest resistance in livestock (Movahedi et al., 2023). Indicatively, a series of drought-resistant and nutrient-fortified crops have helped people sustain in places with poor climates. Similarly, the GM crop Rainbow Papaya in Hawaii has controlled the spread of the ringspot virus (Kaushal et al., 2022). Genetic alterations may result in unwanted protein interference, off-target mutations and the emergence of new undesirable traits that endanger natural ecosystems (CBAN, 2023).

18.4 INTEGRATION OF ETHICS, REGULATION AND BIOSAFETY

The integration of ethics, regulation and biosafety in plant genome editing remains one of the primary conceptual and practical challenges shaping the future of agricultural biotechnology. Genome editing systems, such as CRISPR/Cas, have enabled precise and efficient modification of plant genomes, including the introduction of single-nucleotide changes that can resemble naturally occurring variation (Vishal et al., 2024). Nonetheless, these developments have not made the social and political issues of biotechnology any easier. They have instead only deepened the discussions about the adequate level of human intervention in the genome of plants, the legitimacy and categorization of new breeding methods within the framework of the current legal framework and the conditions under which environmental and human health safety can be guaranteed. Consequently, ethics, regulation and biosafety are interdependent elements of governance, and all of them should be correlated to ensure the confidence of the population and the safe use of genome editing.

Ethical implications of genome editing go beyond the risk evaluation to include cultural, philosophic and symbolic viewpoints of plants and nature. Kuntz (2024)assert that this tends to be the case in the context of public mistrust of genome editing, especially in Europe, which is not determined so much by quantifiable biological risk but rather by more profound social notions of what is natural, human agency and ecological harmony. Plants have cultural significance as agricultural commodity and part of local identity, heritage and landscape. This can lead to the genome editing being viewed as a technical intervention, but as a form of intrusion on the ever-changing relationship between humans and ecosystems. Opposition still exists where ethical and cultural aspects are not well handled even in cases where the scientific evidence shows a high level of precision and low biosafety risk. This demonstrates that scientific innovation cannot work on its own, since, in the absence of trust, and meaningful interaction by the people, safe and beneficial genome-edited crops are hardly likely to be adopted into the real world (Folta, 2025). Thus, the ethical governance should include the idea of the public memory, the cultural identity and ecological values, shifting from the one-way communication towards the participatory dialogue. Equity also remains a problem making the ethical context more difficult. According to Ricroch (2024), genome editing is in a complicated intellectual property setting, comprising multinational corporations, educational institutions and research companies. The cost of a licence and proprietary technology can limit access for researchers in low-income and middle-income countries and also for public breeding programmes in low regional crops of interest. Genome editing has the potential to strengthen the existing inequalities in agricultural development without conscious effort. It also has the effect of bringing minimal advantages to smallholder farmers and marginalized communities. Ethical responsibility therefore goes beyond cultural acceptability to material justice which involves coming up with choices not just on how the genomes are altered, but on who the beneficiaries are and in which circumstances.

The main aim of biosafety evaluation is to determine the environmental and health effects of the genome-edited plants. Genome editing has the ability to produce identical changes to those that are created through the traditional breeding or random mutation. On this account, biosafety governance starts appreciating the need to assess traits and not approaches. According to Ricroch et al. (2024), the environmental risk assessment is to be performed regarding the gene flow, ecological interactions, consequences and reproductive fitness of nontarget organisms. Lack of foreign DNA is not a guarantee of ecological neutrality, and the occurrence of modification is not necessarily an indication of danger. Moreover, Mołodziejko et al. (2023) note that the biosafety ability of different regions differs. Most nations do not have the facilities to engage in long-term surveillance, molecular surveillance and investigations on the effects on the ecosystem. Thus, the efficient global governance of biosafety has to involve investment on regulatory training, laboratory capacity, ecological surveillance mechanism and cross-national data exchange. The responsible biosafety governance balances between innovation and precaution that enhances resilience, uses less chemical inputs and protects the biodiversity. The only way that a coherent governance framework would be established is when the ethics, regulation and biosafety are addressed collectively and not separately. Biosafety assessment helps provide objective data to support regulatory decision-making. Embedding these decisions within a structured regulatory framework, along with ethical considerations, helps ensure that governance processes remain socially acceptable and scientifically robust. The integration requires constant communication between the policymakers, scientists, producers and the population. There should be trust-building, transparency and accountability. Genome editing can benefit agriculture by improving its resilience, making it more sustainable, and enhancing equity in the food system, although only in a world where it is developed and deployed based on shared values in society and environmental sustainability. Genome editing can never promote an equitable and sustainable agriculture in the 21st century without being governed through a scientifically informed, ethically aware and socially responsive approach.

18.5 GLOBAL TRADE BARRIERS FOR GENOME-EDITED PRODUCTS

Dissimilar regulatory practices by different nations are significant problems to the international movement of genome-edited foodstuffs and crops. This difference significantly contributes to the increase of costs, increases approvals and restricts access to the market (Fernández Ríos et al., 2025). A patchwork of regional laws often requires the developers to contend with, and this may necessitate further testing and documentation and compliance procedures. These intricacies delay commercialization and make commercialization to take place more difficult as companies or developers seek to commercialize innovations or developed products more effectively using a sophisticated biotechnological strategy. The latter is especially high risk in the case of small- and medium-sized developers/companies, as they tend to be less resourceful and have more entry barriers (Kalaitzandonakes et al., 2023). The uncertainty in regulation might send negative messages to invest in genome-edited or advanced products of biotechnology. The disparity of pre- and post-market rules in different regions also diminishes the level of transparency, limits the information of the consumers and makes risk management in international trade more difficult. To address these obstacles, analysts and policymakers will always recommend to encourage regulatory harmonization through experience gained in nations where the emphasis is placed on the end-product, and not on the process (Fernández Ríos et al., 2025). The bilateral and multilateral agreements may also help to back up the regulations standards and promote more consistency to assist in creating fewer challenges in trade deals and expanding the use of genome-edited products.

18.6 CONCLUSION

Genome-edited crops have tremendous potential to increase food security in the world as well as ensuring environmental sustainability. Notably, these GM plants are evaluated utilizing structures equivalent to those used to test traditional breed varieties in certain nations. Nevertheless, the regulatory policies differ significantly across countries and keep changing at a fast pace. Flexible processes have been developed in several countries in order to establish a process of authorizing gene-edited products, and other countries have established rigorous and time-consuming protocol requirements that focus on scientific assessment and societal compliance. Regulatory strategies are slowly becoming homogenized: certain regulatory practices have been noticed, particularly in cases where the products of gene-editing technologies are not defined as GMOs. Clear definitions of SDN1, SDN2 and SDN3 are needed to prevent trade disruptions and to implement them. The transparency in the regulatory process, the communication with stakeholders and evidence-based communication also play similarly significant roles in establishing the trust of the people. These and well-developed biosafety provisions, adjustive regulation and surveillance will increase awareness and can possibly assist in improving food security.

REFERENCES

Abushal, L. T., Salama, M., Essa, M. M., Qoronfleh, M. W. (2021). Agricultural biotechnology: Revealing insights about ethical concerns. *Journal of Biosciences*, 46(3), 81. https://doi.org/10.1007/s12038-021-00203-0

Ahmad, M. (2023). Plant breeding advancements with "CRISPR-Cas" genome editing technologies will assist future food security. *Frontiers in Plant Science*, 14, 1133036. https://doi.org/10.3389/fpls.2023.1133036

Ahteensuu, M. (2017). Synthetic biology, genome editing, and the risk of bioterrorism. *Science and Engineering Ethics*, 23(6), 1541–1561. https://doi.org/10.1007/s11948-016-9868-9

Buchholzer, M., & Frommer, W. B. (2023). An increasing number of countries regulate genome editing in crops. *New Phytologist*, 237(1). https://doi.org/10.1111/nph.18333

Bugnosen Z. J. (2023). Global trends of genome editing approaches. *Science Speaks – A Blog by ISAAA*. https://www.isaaa.org/blog/entry/default.asp?BlogDate=2/1/2023

CBAN. (2023). Environmental impacts. *Canadian Biotechnology Action Network*. https://cban.ca/gmos/issues/environmental-impacts/#:~:text=Biodiversity%20Loss% 3A%20The%20use%20of

Devos, Y., Maeseele, P., Reheul, D., Van Speybroeck, L., De Waele, D. (2008). Ethics in the societal debate on genetically modified organisms: A (Re) quest for sense and sensibility. *Journal of Agricultural and Environmental Ethics*, 21(1), 29–61. https://doi.org/10.1007/s10806-007-9057-6

Eastwood, C., Klerkx, L., Ayre, M., Dela Rue, B. (2019). Managing socio-ethical challenges in the development of smart farming: From a fragmented to a comprehensive approach for responsible research and innovation. *Journal of Agricultural and Environmental Ethics*, 32(5), 741–768. https://doi.org/10.1007/s10806-017-9704-5

Eckerstorfer, M. F., Dolezel, M., Heissenberger, A., Miklau, M., Reichenbecher, W., Steinbrecher, R. A., Waßmann, F. (2019). An EU perspective on biosafety considerations for plants developed by genome editing and other new genetic modification techniques (NGMS). *Frontiers in Bioengineering and Biotechnology*, 7. https://doi.org/10.3389/fbioe.2019.00031

El-Mounadi, K., Morales-Floriano, M. L., Garcia-Ruiz, H. (2020). Principles, applications, and biosafety of plant genome editing using CRISPR-CAS9. *Frontiers in Plant Science*, *11*. https://doi.org/10.3389/fpls.2020.00056

Fernández Ríos, D., Quintana, S. A., Gómez Paniagua, P., Arrúa, A. A., Brozón, G. R., Bertoni Hicar, M. S., Castro Alegría, A. and Goberna, M. F. (2025). Regulatory challenges and global trade implications of genome editing in agriculture. *Frontiers in Bioengineering and Biotechnology*, 13, 1609110. https://doi.org/10.3389/fbioe.2025.1609110

Folta, K. M. (2025). Acceptance of crop biotechnology requires a change in communication strategy. *Plant Physiology*, *198*(3). https://doi.org/10.1093/plphys/kiaf167

Globus R, Qimron U. (2018). A technological and regulatory outlook on CRISPR crop editing. *Journal of Cellular Biochemistry*. 119(2), 1291–1298. https://doi.org/10.1002/jcb.26303

Gu, X., Liu, L., Zhang, H. (2021). Transgene-free genome editing in plants. *Frontiers in Genome Editing*, 3, 805317. https://doi.org/10.3389/fgeed.2021.805317

Gulyaeva, E. E. (2025). Problems of biosafety in current international law. *Rev. Direitos Cult*, 2177–1499. https://doi.org/10.31512/rdc.v20i50.2174

Harfouche, A. L., Petousi, V., Meilan, R., Sweet, J., Twardowski, T., Altman, A. (2021). Promoting ethically responsible use of agricultural biotechnology. *Trends in Plant Science*, 26(6), 546–559. https://doi.org/10.1016/j.tplants.2020.12.015

Kalaitzandonakes, N., Willig, C., Zahringer, K. (2023). The economics and policy of genome editing in crop improvement. *The Plant Genome*, 16(2), e20248. https://doi.org/10.1002/tpg2.20248

Kaushal, N., Sood, N., Kaur, M., Singh, T. (2022). Rainbow papaya. In Kour, J., Sharma, V., Khanday, I. (Eds.), *Genetically Modified Crops and Food Security* (pp. 151–161). Routledge. https://doi.org/10.4324/9781003278566-10

Kuntz, M. (2024). What is the problem with Europe in a philosophical point of view?*A Roadmap for Plant Genome Editing*, 533. https://doi.org/10.1007/978-3-031-46150-7_32

Lassoued, R., Macall, D. M., Smyth, S. J., Phillips, P. W., Hesseln, H. (2019). Risk and safety considerations of genome edited crops: Expert opinion. *Current Research in Biotechnology*, 1, 11–21. https://doi.org/10.1016/j.crbiot.2019.08.001

Li, J., Zhao, H., Zheng, L., An, W. (2021). Advances in synthetic biology and biosafety governance. *Frontiers in Bioengineering and Biotechnology*, 9. https://doi.org/10.3389/fbioe.2021.598087

Li, X., Gao, Y., Zhang, Z., Deng, W., Cao, W., Wei, X., Gao, Z., Yao, L., Wang, S., Xie, Y., Li, M. (2025). Biosafety considerations triggered by genome-editing technologies. *Biosafety and Health*. https://doi.org/10.1016/j.bsheal.2025.05.003

Lunshof, J. E., Birnbaum, A. (2017). Adaptive risk management of gene drive experiments. *Applied Biosafety*, 22(3), 97–103. https://doi.org/10.1177/1535676017721488

Mampan, K. P., Hill, J., Saleem, M. (2011). Natural resources management and food security in the context of sustainable development. *Sains Malaysiana*, 40(12), 1331–1340.

Mołodziejko, W., Nowakowska, J., Linkiewicz, A. (2023). The awareness of the polish society on new genomic techniques. In Ricroch, A., Eriksson, D., Miladinović, D., Sweet, J., Van Laere, K., Woźniak-Gientka, E. (Eds.), *A Roadmap for Plant Genome Editing* (pp. 481–491). Cham: Springer Nature Switzerland. https://doi.org/10.1007/978-3-031-46150-7_29

Morshedzadeh, F., Ghanei, M., Lotfi, M., Ghasemi, M., Ahmadi, M., Najari-Hanjani, P., Sharif S., Mozaffari-Jovin, S., Peymani, M., Abbaszadegan. M. R. (2024). An update on the application of CRISPR technology in clinical practice. *Molecular Biotechnology*,66(2):179–197. https://doi.org/10.1007/s12033-023-00724-z

Movahedi, A., Aghaei-Dargiri, S., Li, H., Zhuge, Q., Sun, W. (2023). CRISPR variants for gene editing in plants: Biosafety risks and future directions. *International Journal of Molecular Sciences*, 24(22), 16241. https://doi.org/10.3390/ijms242216241

Nawaz, T., Gu, L., Fahad, S., Saud, S., Harrison, M. T., Zhou, R. (2024). Sustainable protein production through genetic engineering of cyanobacteria and use of atmospheric N_2 gas. *Food and Energy Security*, 13(2), e536. https://doi.org/10.1002/fes3.536

Nawaz, T., Gu, L., Fahad, S., Saud, S., Jiang, Z., Hassan, S., Harrison, M. T., Liu, K., Khan, M. A., Liu, H., El-Kahtany, K., Wu, C., Zhu, M., Zhou, R. (2023). A comprehensive review of the therapeutic potential of cyanobacterial marine bioactives: Unveiling the hidden treasures of the sea. *Food and Energy Security*, 12(5), e495. https://doi.org/10.1002/fes3.495

Resnik, D. B. (2018). Privacy and confidentiality. In D. B. Resnik (Ed.), *The Ethics of Research with Human Subjects* (Vol. 74, pp. 149–163). Cham: Springer Nature Switzerland. https://doi.org/10.1007/978-3-319-68756-8_6

Reynolds, T. W., Waddington, S. R., Anderson, C. L., Chew, A., True, Z., & Cullen, A. (2015). Environmental impacts and constraints associated with the production of major food crops in Sub-Saharan Africa and South Asia. *Food Security*, 7(4), 795–822. https://doi.org/10.1007/s12571-015-0478-1

Ricroch, A. (2024). CRISPR processes patents in green biotechnology: Collaborative licensing models. In Ricroch, A., Eriksson, D., Miladinović, D., Sweet, J., Van Laere, K., Woźniak-Gientka, E. (Eds.), *A Roadmap for Plant Genome Editing* (pp. 453–467). Cham: Springer Nature Switzerland. https://doi.org/10.1007/978-3-031-46150-7

Ricroch, A., Desachy, L.D., Penfornis, M., Akin, M., Kondić-Špika, A., Kuntz, M. and Miladinović, D. (2024). Worldwide study on field trials of biotechnological crops: New promises but old policy hurdles. *Frontiers in Plant Science*, *15*, 1452767. https://doi.org/10.3389/fpls.2024.1452767

Savulescu, J., Pugh, J., Douglas, T., Gyngell, C. (2015). The moral imperative to continue gene editing research on human embryos. *Protein & Cell*, 6(7), 476–479. https://doi.org/10.1007/s13238-015-0184-y

Schmerker, J. (2024). *CRISPR-Cas9: 10 pros and 7 cons*. https://sg.idtdna.com/page/support-and-education/decoded-plus/crispr-cas9-what-are-the-10-pros-and-7-cons/

Shahab, M., Roberto, S. R., Adnan, M., Fahad, S., Koyama, R., Saleem, M. H., Nawaz, T. (2023). Phenolic compounds as a quality determinant of grapes: A critical review. *Journal of Plant Growth Regulation*, 42(9), 5325–5331. https://doi.org/10.1007/s00344-023-10953-w

Sprink, T., Wilhelm, R. (2024). Genome editing in biotech regulations worldwide. A roadmap for plant genome editing. In Agnès Ricroch, Dennis Eriksson, Dragana Miladinović, Jeremy Sweet, Katrijn Van Laere, Ewa Woźniak-Gientka (Eds.), *A Roadmap for Plant Genome Editing* (pp. 437–452). Cham: Springer Nature Switzerland. https://doi.org/10.1007/978-3-031-46150-7

Sturme, M. H. J., Van Der Berg, J. P., Bouwman, L. M. S., De Schrijver, A., De Maagd, R. A., Kleter, G. A., Wilde, E. B. (2022). Occurrence and nature of off-target modifications by CRISPR-CAS genome editing in plants. *ACS Agricultural Science & Technology*, 2(2), 192–201. https://doi.org/10.1021/acsagscitech.1c00270

Tachikawa, M., Matsuo, M. (2024). Global regulatory trends of genome editing technology in agriculture and food. *Breeding Science*, 74(1), 3–10. https://doi.org/10.1270/jsbbs.23046

Tilman, D., Balzer, C., Hill, J., Befort, B. L. (2011). Global food demand and the sustainable intensification of agriculture. *Proceedings of the National Academy of Sciences*, 108(50), 20260–20264. https://doi.org/10.1073/pnas.1116437108

Vengadesen, S., Bin Abdul Aziz, M. F., Jones, M. K. (2025). Rethinking gene-edited crop regulation: Advancing a principle-based framework for modern biotechnology governance. *GM Crops & Food*, 16(1), 852–869. https://doi.org/10.1080/21645698.2025.2576734

Venkataraman, S., Zaruk, D.,Hefferon, K. (2025). Recent developments, challenges and opportunities in genome editing for crop science from a societal perspective. *Frontiers in Genome Editing*, 7, 1568072. https://doi.org/10.3389/fgeed.2025.1568072

Vishal, Mali, S., Verma, V., Zinta, G. (2024). Genome editing in ornamental crops: Progress and challenges. In Bhargava, B., Kumar, P., Verma, V. (Eds,), *Ornamental Horticulture: Latest Cultivation Practices and Breeding Technologies* (pp. 193–209). Singapore: Springer. https://doi.org/10.1007/978-981-97-4028-4_10

Wang, Y., Phelps, A., Godbehere, A., Evans, B., Takizawa, C., Chinen, G., Singh, H., Zhiqing, F., Du, Z. Y. (2025). Revolutionizing agriculture with CRISPR technology: Applications, challenges, and future perspectives. *Biotechnology Journal*, 20(9), e70113. https://doi.org/10.1002/biot.70113

World Health Organization. (2019).https://www.who.int/health-topics/human-genome-editing/#tab=tab_1

Index

Note: **Bold** page numbers refer to tables and *italic* page numbers refer to figures.

For Product Safety Concerns and Information please contact our EU
representative GPSR@taylorandfrancis.com
Taylor & Francis Verlag GmbH, Kaufingerstraße 24, 80331 München, Germany

www.ingramcontent.com/pod-product-compliance
Ingram Content Group UK Ltd.
Pitfield, Milton Keynes, MK11 3LW, UK
UKHW052220190926
475702UK00006B/1130

* 9 7 8 1 0 4 1 2 3 0 4 0 3 *